Horseshoe Crabs
and Velvet Worms

Praise for Richard Fortey's

Horseshoe Crabs
and Velvet Worms

"Fortey has a unique way with the most humble of life forms, an infectious curiosity that can slide into near rapture, coupled with a lack of presumption that many of his peers in the field of evolutionary biology lack entirely." —*London Evening Standard* (UK)

"A wide-ranging survey. . . . Fortey keeps the long discussion lighthearted. . . . Instructive and entertaining." —*Booklist*

"Erudite and engaging." —*Times Literary Supplement* (UK)

"A magnificent book. . . . Fortey's intense, humane passion for everything that lives and has lived is amply proven on every page." —*Literary Review*

"[A] delightful account . . . even those squeamish about worms will find Fortey's enthusiastic excavations charming." —*Publishers Weekly* (starred review)

"Fortey retains his characteristic ability to paint vivid word pictures of times long ago and places far away. . . . Passionate, clear and comprehensive." —*The Telegraph* (UK)

"Fascinating. . . . Taking great joy in his trip back in time, Fortey plays both adventurer and detective. . . . Informative, engrossing and delightful." —*Kirkus Reviews* (starred review)

Richard Fortey

Horseshoe Crabs
and Velvet Worms

Richard Fortey was a senior paleontologist at the Natural History Museum in London. His previous books include *The Hidden Landscape: A Journey into the Geological Past*, which won the Natural World Book Prize in 1993; *Life: A Natural History of Four Billion Years of Life on Earth*; *Trilobite!*, which was shortlisted for the Samuel Johnson Prize and was a *New York Times* Best Book of the Year; *Earth: An Intimate History*; and *Dry Storeroom No. 1: The Secret Life of the Natural History Museum*. He was Collier Professor in the Public Understanding of Science and Technology at the Institute for Advanced Studies at the University of Bristol in 2002. In 2003, he won the Lewis Thomas Prize for Writing About Science from Rockefeller University. He has been a Fellow of the Royal Society since 1997 and was elected as a Fellow in the Royal Society of Literature in 2009.

Horseshoe Crabs and Velvet Worms

THE STORY OF THE ANIMALS AND
PLANTS THAT TIME HAS LEFT BEHIND

Richard Fortey

Vintage Books
A Division of Random House, Inc.
New York

FIRST VINTAGE BOOKS EDITION, DECEMBER 2012

Grateful acknowledgement is made to Curtis Brown, Ltd., for permission
to reprint an excerpt from "The Coelacanth" from *Selected Poems* by Ogden Nash,
copyright © 1972 by Ogden Nash (London: Little, Brown & Co., 1972).

The Library of Congress has cataloged the Knopf edition as follows:
Fortey, Richard A.
Horseshoe crabs and velvet worms: the story of the animals and plants that time has
left behind / by Richard Fortey. — 1st American ed.
p. cm.
"Originally published as *Survivors* in Great Britain by HarperPress, an imprint of
HarperCollins Publishers, London."
Includes index.
1. Arthropoda—Conservation. 2. Invertebrates—Conservation. 3. Limulus
polyphemus—conservation. 4. Worms. 5. Plant conservation. I. Title.
QL434.F67 2012
595—dc23
2011039941

Vintage ISBN: 978-0-307-27553-0

Book design by Virginia Tan
Author photograph © Jo Desmond

www.vintagebooks.com

Printed in the United States of America
10 9 8 7 6 5 4 3

To my sister, with love

CONTENTS

	Acknowledgements	*ix*
	Prologue	*xiii*
	Table of Geological Periods	*xv*
1.	Old Horseshoes	3
2.	The Search for the Velvet Worm	32
3.	Slimy Mounds	61
4.	Life in Hot Water	96
5.	An Inveterate Bunch	124
6.	Greenery	154
7.	Of Fishes and Hellbenders	186
8.	Heat in the Blood	216
9.	Islands, Ice	243
10.	Survivors Against the Odds	273
	Epilogue	297
	Glossary	301
	Illustration Credits	307
	Further Reading	315
	Index	319

ACKNOWLEDGEMENTS

Many people have helped me in my quest to visit survivors from former worlds. My book has benefited immeasurably from their knowledge and enthusiasm. I freely acknowledge their contribution, without allowing them to take responsibility for any errors that may have appeared despite my best intentions.

Those who have helped this book to reach completion are listed below more or less in the order in which their contribution appears in the text: all were equally important in helping the project onwards. Glenn Gauvry's enthusiasm for the horseshoe crab was invaluable during our sojourn around Delaware Bay. He introduced me to Carl Shuster, doyen of *Limulus* studies; I learned many of my crab facts directly from the great man. For unstinting hospitality in Portugal, and a guided tour around the Arouca trilobite site, Dr. Artur Sa is gratefully acknowledged. In New Zealand, our old friends Roger and Robyn Cooper were generous hosts, and Roger introduced me to George Gibbs, who showed me where to find the velvet worm. I could not have managed without his help. In Newfoundland, Dr. Andy Kerr kindly arranged for me to visit the famous fossil site at Mistaken Point, and even arranged the fine weather (most unusual). Thanks are extended to the staff at the Ecological Reserve for their kindness in allowing access to the cliffs. For advice on early microfossils and ancient sea chemistry, Professor Andrew Knoll of Harvard University is an incomparable source of wisdom. I know little about seaweeds, and what little I do know is because of the kindness of Juliet Brodie at the Natural History Museum in London, who directed me to a site at Sidmouth where *Porphyra* grows. Paul and Kay Griew generously invited us to stay with them in their house by the sea while this research was carried out. The staff at the Visitor Center in Yellowstone

National Park at Mammoth Springs kindly passed on information about hot springs and ancient bacteria. In Hong Kong we stayed with my oldest friend. Bob Bunker was at school with me, and now we had the chance to get to know his delightful wife, Sally. Dr. Paul Shin and his colleagues at Hong Kong University generously gave of their time to take us to find *Lingula* in the New Territories; they even supplied us with wellington boots. Rupert McCowan of the Royal Hong Kong Geographical Society contributed to the financing of this part of the expedition. Dr. John Taylor (and Emily Glover) from the Natural History Museum in London kindly directed me towards Australian contacts in the molluscan world, and gave me much information about *Solemya*. Dr. John Hooper of the Queensland Museum could not have been more helpful in taking us to Stradbroke Island in Moreton Bay. He spent much time successfully convincing me of the pivotal importance of sponges. John Hooper also introduced me to Andy Dunstan, who kindly helped me to understand the living *Nautilus* in a way I never had before.

I have known Professor David Bruton and his wife, Anne, since we did fieldwork together several decades ago, but I had never been to their cabin in the mountains until we went in search of the primitive plant *Huperzia*. We had a wonderful time cooking mushrooms and nosing through the forest. My colleague at the Natural History Museum, Dr. Paul Kenrick, gave me much useful advice on plant evolution, as did the former Director of Kew Gardens, Sir Ghillean Prance. I enjoyed his pictures of *Welwitschia*. In the search for ginkgo in China, several of my colleagues at the Nanjing Institute of Geology and Palaeontology could not have been kinder. My old friend Zhou Zhiyi and his brother Zhiyan in particular deserve thanks for setting up my trip within the People's Republic. Paul and Miriam Clifford very kindly allowed us to stay with them in Beijing, and took us to the flea market. Only two years earlier Miriam had taken us around the Brooklyn Botanic Garden in New York. Peter Kind generously took us on a field trip to see where the Australian lungfish lived in rural Queensland, and told me many interesting things about this remarkable survivor. Professor Jean Joss actually allowed me to handle a large specimen fished from her tanks on the roof of Macquarie University. I thank Jean and her husband for giving up a day to explain their research into these lungfish. My sister-in-law Caroline Lawrence was her hospitable self in Sydney, and shared the driving elsewhere. My near namesake and colleague Peter Forey generously shared with me his unrivalled knowledge of coelacanth fishes. Adam Hilliard and his team at the Environment Agency in the UK allowed me to get in their way as they fished up the lamprey from the River Lambourn late in 2010. I am grateful to the British Council for inviting me to lecture in Vilnius, where I had an enjoyable, if unsuccessful hunt for a related species. In New Zealand, Rob Stone led us around Somes Island off Wellington until we found the tuatara, minding its own business.

So we go on to mammals, and back to Australia. I am indebted to Peggy Rismiller for sharing her unrivalled knowledge of the echidna, as we chased it around Kangaroo Island. My old friend Jim Jago, and slightly newer friend Jim Gehling showed me the Cambrian Emu Bay Shale on another part of the same island. Company on this jaunt was provided by the younger Coopers, Alan and Sarah, who also accommodated us in Adelaide; and the wine was delicious. I am grateful to Jørn Hurum for showing me the famous early primate fossil, known in the popular press as Ida, as it hides behind the scenes in the Palaeontological Museum in Oslo, Norway. Professor Lynn Margulis kindly asked me to speak at a World Summit on Evolution on the Galapagos Islands a few years ago, and subsequently I was able to visit the cloud forest in Ecuador—which was a wonderful treat (and the home of the tinamou). Samuel Pinya looks after the interests of the Mallorcan midwife toad, and took my wife and me on a hunt to find some of these delightful animals tucked away remotely in the mountains. It would have been impossible without his help. My encounter with the musk ox goes back to such an early stage in my history that the principal actors are no more, but a posthumous greeting would not be out of place to Brian Harland, who set me on course to become a professional palaeontologist by inviting me on an expedition to Spitsbergen. Like some of the organisms portrayed in this book, luck has played an important part in my life also.

Arabella Pike at HarperCollins has been a great editor for five of my books, and my debt to her is evident. Heather Godwin read the first draft of this book, as she has my previous ones. She has the best critical eye I know, and if we occasionally disagree she has always proved right in the end. She worked very hard on this particular book, and my gratitude to her is immense. Katharine Reeve's very careful reading helped identify any ambiguities or inadequacies that still survived. I sincerely thank her for her meticulous attention to detail and Sophie Goulden for guiding the book so efficiently through the design and editorial process. Professor Derek Siveter from Oxford University applied his eagle eye to the manuscript to spot scientific errors, and earns my unfeigned gratitude for his work. Finally, my wife, Jackie, has made this book possible. She organised nearly all the field trips, a few of which were highly complex. She was the "official photographer" while I busied myself with my notebooks. To cap it all, she was the picture researcher, and added her own editorial input. I doubt whether the book would have happened at all without her help.

PROLOGUE

These anomalous forms may almost be called living fossils;
they have endured to the present day, from having inhabited
a confined area, and from having thus been exposed to less
severe competition.

—Charles Darwin: *The Origin of Species*

Evolution has not obliterated its tracks as more advanced animals
and plants have appeared through geological time. There are, scattered over the globe, organisms and ecologies which still survive from
earlier times. These speak to us of seminal events in the history of life.
They range from humble algal mats to hardy musk oxen that linger
on in the tundra as last vestiges of the Ice Age. The history of life can
be approached through the fossil record; a narrative of forms that
have vanished from the earth. But it can also be understood through
its survivors, the animals and plants that time has left behind. My
intention is to visit these organisms in the field, to take the reader
on a journey to the exotic, or even everyday, places where they live.
There will be landscapes to evoke, boulders to turn over, seas to pad-
dle in. I shall describe the animals and plants in their natural habitats,
and explain why they are important in understanding pivotal points
in evolutionary history. So it will be a journey through time, as well
as around the globe.

I have always thought of myself as a naturalist first, and a palaeontologist second, although I cannot deny that I have spent most of my
life looking at thoroughly dead creatures. This book is something of
a departure for me, with the focus switched to living organisms that
help reveal the tree of life (see endpapers). I will frequently return to
considering fossils to show how my chosen creatures root back into
ancient times. I have also broken my usual rules of narrative. The

logical place to start is at the beginning, which in this case would mean with the oldest and most primitive organisms. Or I could start with the present and work backwards, as in Richard Dawkins' *The Ancestor's Tale.* Instead, I have opted to start somewhere in the middle. This is not perversity on my part. It seemed appropriate to start my exploration in a place, biologically speaking, that is familiar to me. The ancient horseshoe crabs of Delaware Bay were somehow fitting, not least on account of their trilobite connections. Amid all the concern about climate change and extinction, it is encouraging to begin with an organism whose populations can still be counted in their millions. From this starting point somewhere inside the great and spreading tree of life I can climb upwards to higher twigs if I wish, or maybe even delve downwards to find the trunk. Let us begin to explore.

TABLE OF
GEOLOGICAL PERIODS

Ma BP*	ERA	MAJOR DIVISIONS	
(0.01)	QUATER-NARY	HOLOCENE	Ice Ages: northern hemisphere greatly affected, specially adapted mammals
1.64	QUATER-NARY	PLEISTOCENE	
5.2	TERTIARY	PLIOCENE	
23	TERTIARY	MIOCENE	
34	TERTIARY	OLIGOCENE	Alpine evolution
56	TERTIARY	EOCENE	
65	TERTIARY	PALAOCENE	Mammals and birds diversity
145	MESOZOIC	CRETACEOUS	Great Extinction "End of the dinosaurs" Chalk deposited widely
199	MESOZOIC	JURASSIC	Modern oceans widen
251	MESOZOIC	TRIASSIC	
299	UPPER PALAEOZOIC	PERMIAN	Great Extinction Pangaea Supercontinent
359	UPPER PALAEOZOIC	CARBONIFEROUS	"Ice age" Coal swamps
416	UPPER PALAEOZOIC	DEVONIAN	Fish and amphibians
443	LOWER PALAEOZOIC	SILURIAN	Caledonian mountains at zenith, colonisation of the land
488	LOWER PALAEOZOIC	ORDOVICIAN	The ancient ocean Iapetus at its widest
542	LOWER PALAEOZOIC	CAMBRIAN	Trilobites and other marine animals appear
2,500	PRECAMBRIAN	PROTEROZOIC	"Snowball Earth." Multicellular life Oxygenated atmosphere develops
3,500	PRECAMBRIAN	ARCHAEAN	Life — traces in the rocks
4,550	PRECAMBRIAN	(HADEAN)	

Ma BP* Millions of years before present

Horseshoe Crabs
and Velvet Worms

1

Old Horseshoes

Turn seawards off Route 1, Delaware, and a century rolls away. A small road soon leaves the commercial strip and the fast food joints behind, as it moves onto flat fields and marshland that still supports scattered, pretty villages lined with white-painted picket fences. This is how much of America's east coast used to be. The Little Creek Inn is a grander building altogether: a large, foursquare, wooden Victorian farmstead on three floors with shutters at the windows and a fine portico, and inside all polished wood and turned banisters. In the spacious drawing room of the Inn, anticipation is buzzing. The hosts, Bob and Carol Thomas, are serving iced tea to a crew of enthusiasts all dedicated to travelling back much further in time than a mere century or so. This is the chance to come face to face with life as it was millions of years ago. Glenn Gauvry, the local expert, is waving around models of an ancient animal. A small TV crew is there to straighten out their facts before they get down to filming. Two young women biologists have travelled from Canada to see for themselves an event that only happens when the conditions are just right in late May. I am there with my notebook and a fluttering heart. All of us are impatient for darkness to fall.

Deep in the night along the shores of Delaware Bay, the horseshoe crabs are stirring. The tide is now high and there is no moon. Darkness rules, but even in the feeble starlight the overwhelming flatness of the countryside can be made out, except along the rim of the bay where old sand dunes have built up a levee providing foundations for a scattering of wooden beach houses, which loom against the night sky. A path passes between them onto a sandy beach that stretches away into the darkness in a long gentle arc. The shoreline seems to heave with gentle movements.

First, I notice some very odd sounds. There is a general hollow clattering, a tapping and grinding sound, somewhat like that made by knocking coconut shells together (once used on the radio to imitate horses' hooves) but altogether less rhythmic, and with a kind of underlying push. Then, as my eyes get used to the darkness, low shelly mounds the size of inverted colanders can be seen slowly pushing and jostling all along the shore and perhaps six metres up onto the sands. Their bumping and clambering together is the source of those tap-tapping percussive sounds. The flash of an infrared torch reveals more details. The head-shield of the horseshoe crab is domed upwards and carries a few weak spines; at its back end a hinge marks a jointed boundary with a second large plate, spiny at the edge, which can flap downwards; and beyond that again projects a stout triangular spike as long as the head, which can waggle up and down. Here at Kitt's Hummock more crabs are gathered on the mud flats seaward of the sands waiting their turn: strange, green-black, slowly animated lumps. Further offshore again in the shallow seawater, tail spikes project briefly above the gentle waves like raised radio antennae and are gone, showing where still more horseshoe crabs vie with one another to get their place on the sand. There are evidently thousands upon thousands of these large animals gathered together in some sort of compulsive collusion.

One horseshoe crab lies upturned on the sand. Its tail spike waggles feebly, quite unable to perform the task of turning the body back over again. Five pairs of legs twitch ineffectually in a vain attempt to achieve the same end. I find it impossible to resist the temptation to right the poor animal. It is easy to grasp it by the edges of the head-shield. Once righted again, those spindly legs allow the crab to trundle slowly away. Its behaviour seems at once strangely determined, but also apparently random, like the slow progress of a confused old lady on a walker.

Now I see that many of the largest crabs are digging in the sand, their limbs working away beneath the carapace. Some have become almost completely buried, and, although I can detect a kind of deep scrabbling from these animals, they do not seem to be worried by their self-inflicted interment. Other slightly smaller crabs crowd on top of the buried animals. The scrabblers are the females of the species burying their eggs in the sand, while the smaller ones on top are males, competing to fertilise the eggs with their sperm (milt). I realise that there is some kind of order to the apparent mayhem on the beach. A proportion of the horseshoe crabs are paired off, with the lighter male desperately hanging off the tail end of the female, having got a purchase by using his special claspers. However, this right of occupation does not deter other males from having a go at mounting the same burdened female. There is enough of a gap behind the head-shield for some of their interloping milt to have a chance. Much of the clinking noise is a consequence of tussles for dominance. So this gathering of crabs is really an orgy, and an orgy that runs for dozens of miles along the strand, all thickly bordered with scrabbling, lustful animals. As for the poor exhausted females, gravid and overprovided with mates, the moist sand stops their gills drying out, and they may eventually struggle back to the sea when the laying is done—although many do not. Bits and pieces of their carcasses litter the shore.

I have a better chance to scrutinise the horseshoe crabs closely during the day, although most of them have returned to the sea by sunrise. Coastal Delaware is a land of marshes, with gentle wetlands dominated by the reed *Phragmites,* and the cries of wading birds always in the wind. The landscape reminds me of the East Anglian coast in England. Creeks wind their ways inland from the sea, and terminate in small picturesque harbours like Leipsic, where a few fishing boats are tied up to stout piers, with white-painted clapperboard houses landward of the stage. Sambo's is a restaurant with a view of the creek, and well known for its edible crabs, which are consumed on simple tables covered with newspapers. Eating in Sambo's is an audible experience, with everybody bashing lunch out from shells. It is a place of crunching and squishing and little conversation. Some of the shucked piles are prodigious. There is nothing on the menu about horseshoe crabs. The nearby villages of no more than one or two streets include neat little houses dating back to the 1880s, which is ancient by American standards. Delaware car number plates bear the legend "The first state," acknowledging the fact that it was the first to ratify the Constitution. Like several other early American states, but unlike the majority, it is tiny. Nowadays, vehicles on the main roads shoot past; but a mile or two from the freeways little has changed from post-colonial days. I warmed to it immediately.

After the crustaceous lunch, a visit to Port Mahon shows a few stragglers still on the beach at midday, providing a chance to get close. A large female's carapace is about 45 cm across. In the sunlight I can readily see that the creature has nearly semicircular eyes set to either side in the midst of the head-shield, topped by sharp spines, rather like the perky eyebrows I associate with clerics of a certain age. Under high magnification it would be apparent that the eyes are composed of many tiny lenses—they are what are known as compound eyes, similar to those of houseflies or bees. The whole animal is a dull pinkish- to greenish-grey colour, the kind of colour I used to get as a kid when I mixed all my powder paints together. The front of the head-shield is subtly bowed upwards about the middle. The tail (or telson) has a triangular cross section, and it makes a stoutly elegant termination to the animal. The middle part of the body is defined on its top surface as a kind of convex median lobe over about half its length: this is where the muscles that power the legs are lodged. The leading edge of the head-shield is thickened into a prominent marginal rim that is prolonged backwards into short spines; this part of the body needs to be strong to butt into the sands and mud that line the floor of Delaware Bay. On the shore there are several beached crabs lying on their backs, waving their legs at the sky. They bend almost double along their middle hinges, but their best efforts still fail to turn them over (it would be different under water). If they remained on their backs, greater black-backed gulls would soon come along to peck them to pieces. Before setting them aright, I have a chance to see how delicately each of the paired jointed limbs under the head-shield carries a set of pincers at its tip. I am reminded of the manual toolkit owned by the eponymous hero of the movie *Edward Scissorhands*. They are indeed picky little tools. Nearer the front end of the head-shield, where the carapace is doubled back from the top surface into a blunt point, a very delicate set of pincers at the centre of the animal and close to the mouth looks just the thing to feed titbits towards the innards. The bases of the legs are really quite stout and equipped with blunt spines that face one another along the midline of the animal: they can be used like nutcrackers to crack shellfish if needs be. I begin to understand how these creatures can grab a living from the waters of Delaware Bay. Behind the legs are a few pairs of flattish flaps that cover up intricately folded book lungs. Like every marine animal, the horseshoe crab needs to breathe dissolved oxygen, and as long as this breathing apparatus can be kept moist under its protective covers the crab can survive on land. Hence the female can endure her risky excursion to lay her eggs in the sand. The shore may be an unwelcoming nursery, but might still be preferable to a sea where every cubic metre holds a thousand twitching antennae sensing free food. It is time to turn our crab over to allow it to trundle away. It heaves itself along like a battered tank: slowly and undignified, as if to signal "I have survived endless battles, and survival is all." As it performs its lurching exit to the sea, it leaves a track behind on the muddy sand surface. The paired imprints of the limbs are prominent; even the tips of the pincers leave their doubled marks. And the tail, dragged behind, leaves a groove between, as a child might scribe with a stick clumsily trailed across the strand.

Horseshoe crabs are not really crabs at all; indeed, they are only very distantly related to crabs in so far as both kinds of animals propel themselves through the sea on spindly jointed legs. Animals with useful appendages of this articulated kind are known as arthropods (from the Greek: jointed legs). They are classified together in Phylum Arthropoda, a vast animal group that includes all the living insects, as well as spiders, millipedes and a host of marine "bugs" of all kinds. Crabs are crustaceans, along with lobsters, shrimps and woodlice (pillbug to some). Horseshoe crabs are no more crustaceans than are butterflies. They do not have the flexible antennae or "feelers" adapted to sensing the environment that are a common property of Crustacea and insects: these delicate organs both feel and touch, and smell. Instead, in the horseshoe crab the head appendages are modified at the front into a pair of useful pincers, or chelicerae, which I had observed in my stranded animal lying on its back. The significance of this apparently small feature will become apparent. The scientific name of the horseshoe crab is *Limulus polyphemus* (I shall need to use scientific names throughout this book). By day the beach throngs with feeding wading birds: thousands of them skitter nervously away from human intruders in animated, piping, fluttering waves, always beyond reach. Like most waders, many of them are dressed in shades of brown and grey, but the different statures of several species are obvious even to an inexperienced birdwatcher. Small, short-billed sandpipers throng on short legs; slightly larger pale-bellied sanderlings dash along the water's edge; taller, long-billed dowitchers elegantly stride among them. The iconic species for the area is the red knot, which has a dramatic cinnamon-coloured belly when in breeding plumage. All wader species—and there are many more in the crowd—are united in rapt attention along the shoreline, pecking and probing incessantly at the ground, like chickens fed in a yard on the best grain. They are undoing the work of all those heaving masses of horseshoe crabs the previous night, gorging on the green, millet-seed-sized eggs the female crabs sought to sequester beneath the sand. For the red knot the eggs provide vital refuelling, as this particular population started its migration near the tip of South America. By the time they arrive in Delaware Bay on their way to the Arctic, the birds may have lost half their body weight and they are starving. The crab eggs must taste like the best caviar. The birds would not survive without those countless horseshoe crabs performing their spectacular mass mating ritual. These inelegant invertebrates are completely unaware of the gift they are providing to an animal many millions of years their evolutionary junior.

Birds always attract devotees, and naturalists' concerns for the welfare of the red knot probably accounts in turn for their anxiety about the state of the horseshoe crab population. If that were to fail, then so would the long migration of the attractive waders. A recent census estimated that there could be as many as 17 million horseshoe crabs in the Delaware Bay area, and that concerns about their decline may have been exaggerated. Since the horseshoe crab has a range that extends along the shore north to Maine and south to the Yucatán

Peninsula in Mexico, the population is assuredly larger still, although the densest concentration of individuals is probably where I saw the heaving multitudes at Kitt's Hummock and Pickering Beach. Delaware Bay is also where the mature crabs grow largest at maturity. Since there has, indeed, been a decline in red knot numbers, the cause must lie somewhere else in its complex migration story. The weakest link in an ecological chain is always the critical one.

It had always been a dream of mine to see throngs of jostling horseshoe crabs reach the climax of their life cycle. For more than three decades at the Natural History Museum in London, I studied fossils of trilobites. This once important group of sea animals went extinct something like 260 million years ago, when the world was a very different place. Trilobites had once swarmed in all the ancient oceans, but now their remains have to be patiently collected by splitting open the rocks that have entombed their shelly remains. Like horseshoe crabs, trilobites are arthropods: animals with jointed legs and all the muscles and tendons tucked inside an exoskeleton. However, unlike horseshoe crabs, trilobites did not survive the mass extinctions that redesigned the biological face of our planet. It is astonishing to learn

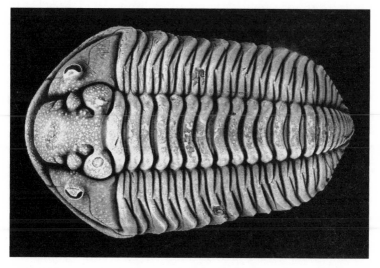

A Silurian trilobite, *Calymene blumenbachii,* from the limestone quarries of Dudley, West Midlands, England.

from unchallenged fossil evidence that relatives of *Limulus* were con-temporaries of trilobites. That nocturnal scrimmage on the beach in Delaware might have happened many millions of years before; I might even have been listening to sounds that had been rehearsed in Palaeozoic times. There were relatives of the horseshoe crabs in the sea long before other arthropods, such as insects and spiders, had ventured onto land, or before crustaceans—shrimps, crabs and lobsters—had taken up the central roles in the ecology of the ocean they enjoy today. So it would not be incorrect to describe the animals thronging along Delaware Bay as primeval. Indeed, many scientists believe that *Limulus* is the closest living relative of trilobites them-selves. Would the head-shield of the giant Cambrian trilobite *Para-doxides,* a fossil 510 million years old, have felt the same to the touch as the beached *Limulus* I restored to the sea in eastern America early in the twenty-first century? Like a horseshoe crab, a trilobite would surely have contemplated me through compound eyes set within its head-shield; its eyes are preserved in detail as fossils. Trilobite legs would have scraped against my mammal flesh with just the same spikiness as *Limulus.* It would have crept and it would have crawled, brother under its external skin to the hordes on Delaware Bay.

So a visit to Delaware is to me rather like a visit to the holy city of Rome to a Catholic. Naturally, I had to meet the Pope. The Pontiff of horseshoe crabs is Carl Shuster, in his tenth decade still a giant of a man: craggy faced, walking without the aid of a stick, with lively eyes beneath towering eyebrows, memory and curiosity undimmed, and only betraying his years by his deafness. Like all field biologists he wears a coarsely chequered thick shirt and blue jeans, hitched up with a stout belt. He was brought up during the Great Depres-sion, when he had to run a farm, so he is himself a survivor, like the horseshoe crabs to which he has devoted his life. His father was a mathematician who gave succour to penniless intellectuals dur-ing the tough years, while young Carl raised asparagus, chickens and strawberries. He brought together all the current knowledge about his favourite animal in his book *The American Horseshoe Crab.* He is accompanied in Delaware by his former student, a man of boundless enthusiasm, Glenn Gauvry (himself an implausibly youthful sixty years), who coordinates much of the research on *Limulus* around the

Bay area. Those volunteers who help with counting the crabs during their nocturnal orgies receive a handsome little pewter lapel pin as a record of their collaboration in the conservation project. Naturally, the pin features a horseshoe crab. It signifies membership of one of the more exclusive clubs in the world.

Carl Shuster and his colleagues established the biological facts about horseshoe crabs that allow naturalists to understand how they fit into the ecology of the Atlantic coast. *Limulus* is a typical arthropod in that it must moult in order to grow, shedding its old coat as a kind of pale ghost of its former self, and growing a new and larger external covering. With a little vigilance it is possible to find one of the cast "shells" on the beach: they are almost as light as tissue paper, for the animal recycles what it can. The newly emerged horseshoe crab is capable of moving immediately. In this it has the advantage over other marine arthropods in the area; freshly moulted blue crabs, for example, are virtually motionless until their new "shell" hardens. They often hide away. Younger horseshoes resemble the older ones apart from being a little spinier. The surprisingly tough but flexible exoskeleton is made of a chitinous material, similar to that forming the wings of beetles. A typical horseshoe crab takes ten years to reach sexual maturity, after which it does not moult again, but heads to the beach for reproduction. Only these mature animals partake in the littoral orgy, which is why one does not see any little limulids scuttling between the adults. They are not demanding animals: a fully mature animal might go for months without eating. When her time comes, a female may well lay 80–100,000 eggs, and enough of these survive the depredations of wading birds to secure the future generations. The greenish eggs are laid in the sand in batches of 4,000 to 6,000 in spheres about the size of a golf ball; the female makes repeated visits on successive tides to complete her duties. Females can be recognised by the scars left behind by the mounting males: up to fifteen males may have a chance of fertilising the eggs of any female. Nonetheless, only about thirty-three eggs out of a million survive to adulthood. This means that at various stages of its life *Limulus* provides a lot of food for other animals. Loggerhead turtles are an important predator on the crabs, even when they are adults. Turtles, too, are animals with a long geological history, so they may have had eons to make something nutritious out of horseshoe crabs, which now seem all sinew and horn and little enough meat. The crabs themselves can survive on molluscs and carrion and almost any kind of scraps, but the strong inner parts of the legs can also crush thick shells if needs be. Although they seem to lurch in an ungainly way on land, under water the crabs are more streamlined and can move quite fast, even sculling on their backs. They can easily right themselves if they need to. In short, they are tough, jack-of-all-trades kind of creatures, built to last. They remind me in a way of a Volkswagen Beetle that I once owned (a beetle being another arthropod, of course) that carried on carrying on even though its coachwork was full of rusty holes, its suspension was down almost to the tarmac, and it often fired on only three cylinders.

This analogy particularly came to mind when I saw a badly damaged horseshoe crab still trundling gamely onwards, even with a great hole punched right through its head. Looking over the beach more carefully, I noticed a lot of these war veterans: lumps out of the thorax, broken tail spikes—clearly, it must take a lot to finish these creatures off. Glenn Gauvry pointed out to me what a great advantage for reproductive success this resilience would furnish. Such endurance is possible because the blood of *Limulus polyphemus* has exceptional clotting powers; the animal does not bleed to death because its blood coagulates and "walls off" damaged areas. And the blood is blue. Does not the horseshoe crab begin to seem "curiouser and curiouser," as Alice would have said? The blood of *Limulus* is blue because it is fundamentally different from that flowing in red-blooded creatures, like you and I and the kangaroo. Whereas we have haemoglobin as our oxygen-carrying pigment, which includes the element iron as an indispensable component, the horseshoe crab carries a copper-based molecule called haemocyanin to do a similar job. In nature, copper often comes with such a blue colour tag. The molecular structure of both these vital molecules is now known in detail, as is the way they move oxygen through the tissues, although this is not directly part of our story. But the *Limulus* narrative would not be complete without exploring the extraordinary coagulating properties of its blood a little further, because this affects the very survival of the species.

This discovery was made in 1956 by Fred Bang of the Marine Biological Laboratory at Woods Hole. He noticed how *Limulus* blood clotted dramatically when infected by a particular bacterium. Subsequent research showed that the crab's blood had an extraordinary sensitivity to a vast range of micro-organisms that are found almost everywhere in nature—known as gram-negative bacteria. A few cubic centimetres of seawater may contain hundreds of thousands of these tiny organisms. Since some of these bacteria are also agents of disease in humans, this property was of immediate interest. It seems that a hypersensitivity to microbial enemies helps to protect the crabs in their natural habitat—as soon as the bacteria enter a wound their defences were up. *Limulus* has a very diffuse blood system compared with ours, with the interior of the animal bathed in blood, and lacking the defined circulation of veins, arteries, and capillaries we are used to in humans and other vertebrates. In horseshoe crabs the job of defending the works of the animal is given to one type of protective cell, the amoebocyte, which contains a mass of granules capable of promoting clotting. When a gram-negative bacterium is in its vicinity, an amoebocyte will react by rupturing and then the granules are released. A clot follows and the infection is sealed off. Now we know why dented and holed crabs can totter on regardless. They have had hundreds of millions of years to come up with an effective response to some of their most dangerous and invisible enemies.

Twelve years later, in 1968, Bang and his colleague Jack Levin had managed to prepare and extract the active principle "*Limulus* amoebocyte lysate" (known as LAL) that clots human blood plasma when exposed to gram-negative bacteria. This is an extremely useful substance in medical diagnosis, readily used for the detection and measurement of the poisonous toxins (called endotoxins) belonging to the appropriate bacteria. Poisonous endotoxins are released into the host organism (you, me or a horseshoe crab) when the bacterial cell wall ruptures. LAL is a highly sensitive chemical able to detect minute quantities of the offending substances. The LAL test is now widely applied, having been sold on to pharmaceutical companies for commercial manufacture. It has to be prepared close to the *Limulus* populations, but is then exported around the world. This means that there is a tremendous demand for the horseshoe crab's blue blood. What had saved it from harm for millions of years now made it a desirable commodity.

The influence of this new industry has been the subject of some debate. Carl Shuster told me that recent estimates say that there may be as many as 17 million adult crabs in the Bay. By 2003, some 3 million crabs were harvested for the pharmaceutical trade—an unsustainable quantity. The bird people were worried about the fate of the red knot and its fellow waders. Now the number has been reduced, and the technique for obtaining blood has been adapted so that it does not require the animals to be killed: they have become blood donors! Four companies have the right to what is called "bleed and release," and this method has been applied to some 600,000 crabs per annum. Maybe it is possible to get the LAL and not threaten the horseshoe crabs, after all. Meanwhile, some ornithologists are sceptical, and feel that the companies, as beneficiaries of the local species, should put something back into safeguarding the regional marine habitat for the benefit of all parties, crabs included. Given the long time taken for the crabs to reach sexual maturity there is always a lag before the effects of a population slide can be observed when the summer high tide hits the beach. But there are certainly some very vigilant people on the case.

There was another threat to the welfare of the horseshoe crabs. They were harvested in great numbers to use as bait to catch the giant marine snail, or whelk, known up and down the Atlantic coast as conch (*Busycon*). Big, pinkish conch shells are a familiar item in every seaside knick-knack shop; put to the ear, they allow the listener to "hear the sea." Roadside stalls sell the shells for a modest price in the small villages around Delaware Bay. The molluscs inside the shells have a loyal following among connoisseurs of seafood. The only time I tried them I found the flesh very tough. A crab split in two is a pungent treat for the big snails, however, and fishermen along the coast from Delaware to New Jersey are well aware of the power of this lure. The U.S. Fish and Fisheries Commission limited the number of crabs to be used to 100,000 in New Jersey, and later introduced a moratorium, but some fishermen moved northwards to Massachusetts or even Maine (the northern limit of the crabs' distribution) to continue their trade. A method of sustainable fishing must be worked out, and there are hopeful signs. On Delaware Bay, for example, the use of a different kind of trap employing mesh bags has cut bait use by 50 per cent. One cannot help feeling that the horseshoe crabs deserve to prosper unhindered. There may seem to be endless crowds of them jostling for a space on the sand, but the example of the American passenger pigeon comes to mind: countless millions of the birds were slaughtered in the nineteenth century until the species finally became officially extinct in 1914. How dreadful to contemplate the thought that an animal that has survived in readily recognisable form from the early days of the dinosaurs might become extinct in the cause of being a special garnish on a plate of *fruits de mer*. It is fortunate that its blue blood is so valuable.

Limulus polyphemus is not alone. There are three additional, Asian species of living horseshoe crabs, but none of them can be found in anything like the profusion of the North American form. *Tachypleus tridentatus* has been given protected status in Japan, where it lives on the Seto Inland Sea. Its shallow-water habitat there is under threat, and increasing industrialisation on the Seto Sea seems unlikely to spare it. Various attempts have been made to conserve the crab, but none of them has solved the problem of making this ancient animal at home in the modern world. Nonetheless, it has an important place in Japanese history, as its form is believed to have inspired classical samurai masks, and brave warriors were supposed to be reborn in the guise of horseshoe crabs. The contemporary artist Takeshi Yamada has made a modern interpretation of masks based upon the ancient icon. A related species, *Tachypleus gigas,* is fairly widespread in eastern Asia. I saw the fourth species, *Carcinoscorpius rotundicaudatus,* while I was on fieldwork in Thailand more than a decade ago. As its second, species name would indicate to those with a smattering of Latin, this particular type has a rounded tail spike compared with other horseshoe crabs. When I first set eyes upon *Tachypleus,* the poor crab was in one of those tanks in a restaurant where delicacies are displayed before consumption, along with several sad-eyed fishes and a dying lobster. I was puzzled by what there could possibly be to eat on the horseshoe crab, because I knew from dissecting its

A growth series of juvenile *Tachypleus tridentatus,* the Asian species
of horseshoe crab, collected from tide strand lines in Deep Bay
estuary, Hong Kong. (The scale is a 15 cm/6 in ruler.)

American relatives that they are not exactly meaty. If the species in
the tank was indeed a relative of trilobites, this might be my first and
last chance to taste something resembling my own speciality. When
the cooked item finally arrived, I felt a twinge of conscience, for it
transpired that large, yolky eggs hidden under the head-shield provided the delicacy, which was only available while the gravid females
came inshore. I felt that I was consuming the next generation just to
satisfy my curiosity. The eggs came served in a thin sauce on a mountain of noodles. They had a rather overwhelming rancid-fishy taste; I
am not anxious to repeat the experience.

As for the place of the horseshoe crab in the tree of evolution,
the Latin name of the horseshoe crab I saw in southern Thailand
might offer a clue. *Carcinoscorpius* means "crab scorpion," and that
is what these curious creatures are: superficially crab-like relatives
of the scorpions. They are the most primitive living examples of a
great group of arthropods that includes all living spiders and mites
as well as scorpions, and several less familiar kinds of animals. This
group is referred to as chelicerates, after those special appendages—
chelicerae—located on the head-shield, which they all share in one
form or another. All the relatives of the horseshoe crab I have mentioned live on land. But the cradle of life was the sea, and *Limulus*
and its relatives take us back to the far, far distant days when the land
surface was barren of larger organisms.

In the darkness along Delaware Bay the scratching percussion
of the crabs provides an unmusical accompaniment on an imagi-
nary journey backwards in time: to an era well before mammals
and flowering plants; a time before the acme of giant reptiles, long

before *Tyrannosaurus*; backwards again through an extinction event 250 million years ago that wiped nine-tenths of life from the earth; and then back still further, before a time of lush coal forests to a stage in the earth's history when the land was stark and life was cradled in the sea; a time when a myriad trilobites scuttled in the mud alongside the forebears of the horseshoe crabs. The trundling, heaving, inelegant not-so-crabs along Delaware Bay are messengers from deep geological time.

A palaeontologist would naturally want to track the history of the horseshoe crabs back into the distant past. A few years ago I visited the famous quarries in Bavaria, southern Germany, where the Solnhofen Limestone of Jurassic age, 150 million years old, had been excavated. Great opencast pits scour the gently rolling countryside revealing thin slabs of cream-coloured limestone, where each bed represents the former sediment surface. The limestone provides the perfect fine-grained stone for the manufacture of lithographic printing plates; this is still a popular medium with artists today, but had an even greater use in the past for graphic illustration. The Germans called this kind of rock "plattenkalk," which is an appropriate name, because if a fossil turns up it will be laid out on the surface of the slab like a fish on a very flat plate; and some of the fossils are, indeed, those of fishes. The most famous fossils from the Solnhofen Limestone are skeletons belonging to the early bird *Archaeopteryx lithographica*, complete with feathers, but they are very rare—only one turns up in an average decade. Some other fossils are quite abundant, like those of little sea lilies (*Saccocoma*). The Solnhofen Limestone is thought to have accumulated in a warm lagoon, or a series of lagoons, not far from a biologically diverse land habitat, but with periodic influxes of waters from the sea. From time to time the lagoon became salty enough from evaporation to poison living organisms, and its deeper parts were depleted in oxygen sufficiently to deter scavengers. The result is the outstanding preservation of delicate animals. When sticky mud was exposed, animals could get trapped upon it, such as delicate flying reptiles or dragonflies. Operating together, these special conditions preserved a huge cross section of Jurassic life. One of these animals is a horseshoe crab called *Mesolimulus walchi*. It really is remarkably similar to our living *Limulus polyphemus*. At first glance it looks as if it had just wandered in from Delaware. One has to look hard to notice that its marginal spines are longer than in our living blue-bloods, and there are a few other minor differences. Nobody could doubt that this species, too, trundled through the shallows, nor that it carried its eggs under its head-shield. To that showy new upstart—a feathered bird—it may already have seemed archaic.

Up to this point I have avoided describing the horseshoe crab as a "living fossil." This is not only because I am chary about using a phrase that is a paradox and an oxymoron rolled into one, but also because it is a misleading description. Charles Darwin himself was cautious when he introduced the term in the phrase quoted at the start of this book. Despite what I have just said about *Mesolimulus*, it is *not* exactly the same as *Limulus*. Consider everything we have learned about our living horseshoe crab. It is woven deeply into an ecology that is utterly different from that in the Jurassic. Millions of birds of many species depend on the horseshoes' eggs every year, whereas its old relative was probably irrelevant to the life cycle of what is often called "the first bird." *Limulus* has adapted to many changes of circumstances: new predators, new climates and now humankind. It is a winner in the lottery of life, and not just because of its long family tree. "Living fossil" seems to imply a negative judgement somehow, as if the poor old organism was just about tottering along on its last legs, having hardly changed in tune with a changing world, awaiting an inevitable end. A similar misplaced judgemental tone is often applied to dinosaurs. "We mustn't be dinosaurs! We must change with the times!" is a mantra of commerce. The dinosaurs were actually superbly efficient animals, and their extinction was most likely a combination of external factors (a drastic meteorite impact is favoured by many) that had nothing to do with either their virtues or lack of them. They were animals of the wrong size living in the wrong places at the wrong time. Bad luck! Meanwhile, the living fossils trundled on through the crisis because . . . well, we will come to that.

Modifications are happening at the genomic level all the time. There really is no such thing as "no change"; the very flexibility of the DNA molecule is what has kept natural selection on its toes for thousands of millions of years. Nor is change in DNA *necessarily* related directly to any change in the appearance of an animal. Many muta-tions accumulate in the large fraction of the genome that apparently does not do much work in the specification of proteins, or initiating developmental changes, or any of the other vital, active stuff. These mutations might well be irrelevant to the kind of changes in shape or colour that indicate the appearance of new species. A living fossil may indeed have accumulated many changes at the molecular level that have not even been expressed in its surface appearance, which is the phenotype that has to face the world. Fluctuations in gene fre-quency are the stuff of life, but they don't map one-to-one on skel-etons and limbs, which are the usual stuff of fossils. So a little caution in terminology is wise.

There is also a temptation to think of the living fossil as if it were a true, surviving ancestor. When the coelacanth fish was discovered, it was presented in the popular press as "old fourlegs," as if it were just about to march onto land on its stumpy fins as a thorough-going tetrapod. Not only does this scenario happen to be wrong, but the likelihood of any such ancestor surviving unchanged to the pres-ent day through many millions of years is also exceedingly remote. Time, chance and competition will see to it that change is inevitable. What can be said without demur is that the ancient survivor and its other living—and more evolutionarily advanced—relatives will have shared a common ancestor, and that the features of the living fossil will be close to those of that ancestor. The discovery of ancient fossils more or less similar to the survivor will date the appearance of the whole animal group to which they belong, and point up the changes that must have happened through geological time along the subsequent branches of the evolutionary tree. The survivors from the early days carry with them a package of information revealing primitive morphology, development and biochemistry that can illu-minate histories that would otherwise be hidden from us. Fossils never preserve blue blood. The "living fossils" may not be *the* ances-tor, but they are survivors carrying a precious legacy of information from distant days and vanished worlds.

Hence *Limulus* allows us to understand something about deep branches in evolution. It is far from unique. If every descendant spe-cies had simply replaced its predecessor, the history of life would be like one of those patients described by Oliver Sacks who live perpetually in the present day, constantly erasing the memories of yesterday. Fortunately, life is not like that. Deep history is all around us. In the life of the planet, the latest model does not always invalidate the tried-and-tested old creature. Groups of organisms that originated long, long ago, in very different worlds, have been able to evolve and adapt alongside their more recent cousins and second cousins. The story of life is almost as much about accommodation as it is about replacement. To look at a living horseshoe crab is to see a portrait of a distant ancestor repainted by time, but with many of its features still unchanged. This book reflects my interest in living survivors from the geological past and what they can tell us about the course of evolution. I have spent the last few years seeking out animals and plants that have helped to illuminate our understanding of the history of life. Wherever possible, I have visited these organisms in their natural habitats; none has proved less than fascinating. Observing how they survive today has allowed me a glimpse of their biology and provided clues about the reasons for their longevity. I have carefully selected the old timers I visited because I wished to understand their biology in depth; I have had passing encounters with several more. A few organisms proved too rare or inaccessible for me to discover personally—the coelacanth comes to mind—and then I have relied on the accounts of others. I shall relate many of these case histories to those of their fossil relatives, which is only to be expected of a palaeontologist. This will illuminate the vital fourth dimension—time. I soon discovered that there were too many potential candidates for inclusion, and I am obliged to mention some of them only briefly. I believe it is better to deal with a smaller number of organisms in detail than swish around vaguely with a broad brush. My specialist friends will probably complain that I have left out their particular favourite beast or weed, and my answer is that these survivors have lasted so long that they will almost certainly still be around for someone else to champion in the future.

Consider scorpions, for example. In some ways they are as impressive as horseshoe crabs as survivors. I have met them several times in my fossil collecting career, usually hiding beneath a log or a rock, for many of them stalk their prey at night and stay out of sight by day.

In the Oman desert I once disturbed a huge, black knobbly scorpion that came running at me with its tail-sting erect, while I backed off in the opposite direction, gibbering foolishly. My Omani companions laughed and told me that its sting (in the tail of course) was relatively mild. A few hours later I lifted up a rock slab and nestled beneath it was a small yellowish scorpion with a flattened, side-wound sting. I was about to poke at it with my geological hammer when my companions tugged at my arm. "Don't touch, it's a real killer!" The smaller, insignificant-looking creatures can often be the most deadly. The spike on *Limulus'* tail and the sting on the scorpion's are closely related structures, and indeed both animals belong to the same great arachnid group. The scorpion learned the trick of arching over the very end of its body to dart poison into enemy or prey. Encased within an external skeleton (cuticle) that prevents evaporation with outstanding efficiency, the scorpion has been able to live far away from water; some species specialise in surviving in the driest places on earth. Scorpions started out as sub-aqueous creatures like

Gustav Vigeland's sculpture of a survivor, the scorpion,
Frogner Park, Oslo, Norway.

Limulus, and only later did they acquire the skill of living on land. Back in Devonian times, 400 million years ago, their relatives, the sea scorpions (eurypterids) were the largest invertebrate predators ever to have lived, some as long as a man. There are fossils from the Carboniferous age that really *do* look like living scorpions, at least to the non-specialist. They, too, have faced out extinction events that have blasted greater and more glamorous animals. The scorpion is built into mythology as a sign of the zodiac; it features on Roman mosaics, and in the Bible (1 Kings 12:11): "my father hath chastised you with whips, but I will chastise you with scorpions." So one could argue that the scorpion's connection to human culture is more pervasive than that of the horseshoe crab.

My choice of organisms has been guided by the place they occupy in the tree of life, rather than by their innate charisma or significance in folklore and culture. The horseshoe crabs earn a special place in this natural history because their relatives root down to the beginning of the diversification of animals. The early Permian *Palaeolimulus* is clearly a horseshoe crab, for all that it predated the great extinction that put paid to most species at the end of the Palaeozoic Era, 250 million years ago. *Limulitella* is present in 242-million-year-old (Triassic) strata dating after the great trauma, evidence of their survival. Those distinctive tracks left by the tips of the legs, and the trail of the tail, that I observed in Delaware have been found as fossils even in the absence of the body itself. There were horseshoe crabs crawling among the coal swamps of the Carboniferous (Pennsylvanian), a little better segmented perhaps than the beasts on Delaware Bay, but carrying the distinct signature of their ancestry. In 2009 my Canadian colleague David Rudkin announced the discovery of the oldest typical horseshoe crab in rocks of Ordovician (approximately 450 million years) age, thus taking these simple arthropods back before *all* the major extinction events that have rocked the Phanerozoic biosphere. Whatever the magic ingredient for survival is, the horseshoes clearly have it in spades. "The meek shall inherit the earth" may be an appropriate motto for their longevity.

Let me describe these early days in more detail. I have already remarked that the horseshoe crabs had set out upon their distinctive path before the first land plants had advanced upon harsh and barren shores; although recent discoveries suggest that a few simple plants may have already ventured onto mud flats. These had probably not yet been followed by insects or spiders. There were fishes already, of a primitive cast, but they had no ambitions then to invade the land, although a few species may have nudged into waters that were not fully marine. The seas abounded with trilobites, which occupied every ecological zone from shallow shores to ocean deeps. These prolific arthropods must have been many times more abundant than the early relatives of the horseshoe crabs. They evidently had an advantage at the time. This may have been the evolution of their robust dorsal "shell" of calcite, which allowed them to develop spines, armour and a tough anchor for muscles, as well as an ability to roll up into tight, impregnable balls when threatened. Trilobites soon learned an array of different feeding habits; some were predators, some ate soft mud, others swam in the open seas. They died out some 255 million years ago. By contrast, the relatives of *Limulus* may have stayed conservatively on the sea floor as scavengers and predators. The horseshoe crabs on Delaware Bay have an exoskeleton of chitin, which is a natural polymer that is quite tough and flexible, although no substitute for stony calcite. But the horseshoe crab has turned what might have been a weakness to advantage by developing an exceptional immune system. Survival in the long term may depend on more subtle features than armour alone.

It is possible to trace the horseshoe crab story still further back, into the Cambrian Period more than 500 million years ago, to a time when many of the major types of animals converge towards their common ancestors. The Cambrian was an interval of unprecedented evolutionary activity, and I shall describe its special features in more detail in the next chapter. Early relatives of *Limulus* have been identified in Cambrian strata, but they include some species that look a little different from their hardy survivors. Some of them also look more like early trilobites, such as *Olenellus,* a form with a big head-shield surrounded by a narrow rim; one might expect a family resemblance if they are indeed closer to a common origin. There are important differences, too. Where the limbs of trilobites are known, they are similar all along the length of the animal: the paired limbs are each split into a walking leg carrying a comb-like branch near its base that in all likelihood functioned as a gill. They are not subdivided into different "packages" in different parts of the body, separating walking and feeding appendages in front from gills behind, as they are in *Limulus*. To add to this, all trilobites had typical "feeler" antennae near the front of the head, and none had the strange chelicerae. This may not be so important, since having antennae seems to have been a general property of primitive arthropods. It might just be that the trilobites still retain this one characteristic more primitive than *Limulus* and its allies, but they could yet have descended from a common ancestor. Trilobites are abundant fossils on account of their easily preserved calcite hard parts. Fossils of unmineralised animals are altogether more unusual. Spectacular recent discoveries of fossils of soft-bodied animals preserved within Cambrian rocks have been made in China and Greenland. These have revealed an almost embarrassing variety of undoubted arthropods early in the Cambrian. Some of them might seem to bridge the differences between *Limulus* and its allies and the trilobites, but for every feature that points one way, there seems to be another that suggests something else. For more than a decade now palaeontologists have argued about how these fossils should be classified, and about the only thing they all agree upon is that the Cambrian threw up many animals with curious combinations of characteristics that were probably winnowed out by subsequent evolution. It is not, perhaps, so surprising that "mixed up" animals lived at this Cambrian time, because *all* the arthropods were not genetically far apart then—they would have had the subsequent 500 million years to box themselves into more separate evolutionary compartments. In these early days the destiny of one animal to become a crustacean, say, and another a chelicerate was not easy to anticipate.

When scientists are confronted by conundrums of this kind, they usually turn to computers. There are now sophisticated computer programs that deal with the problems of determining relationships between animals. They work by identifying the particular arrangement of the creatures analysed on a branching tree that most succinctly accounts for the features they share with their fellows. The most significant resemblances in morphology should result in organisms being classified together on a single branch. Like so many computer methods, the inner workings of the process are staggering in their complexity, so that for a big problem like analysing the Cambrian arthropods, millions of potential arrangements of trees embracing the animals under study will be inspected and rejected. My own appreciation of what goes on inside these machines is thoroughly naïve, and I cannot suppress a vision of thousands of cards being shuffled into piles like a supercharged game of Patience until the answer "comes out." The end product is a diagram tree (technically known as a cladogram), which often provides the basis for portrayals of the more familiar "trees of life" showing how the major groups of animals and plants are related. Like all computer methods, the latter are subject to the familiar caveat of RIRO (Rubbish In Rubbish Out), but the fact that they have been so widely used indicates that they have helped with thorny problems. According to the analyses to date, on balance the trilobites indeed do still classify within a group that also includes the horseshoe crabs. Despite all the confusion of the Cambrian, it seems my crusty-shelled friends and the dogged, eternally trundling horseshoe crabs are sisters under the external arthropod skin.

They do share special features. The larva of the horseshoe crab is a pinhead-sized object long known as the "trilobite larva," because it does resemble the tiny larva of many trilobites.* Both kinds of animals grow larger with each moult in similar ways, casting off their old external housing and re-growing larger premises. Then there are the compound eyes. In both trilobites and horseshoe crabs, the eyes are included as part of the head-shield, rather than sticking out separately at the front on flexible stalks as they are in the majority of crustaceans. Most of us will have looked a lobster in the eyes before popping him into the pot. The lenses of the trilobites are unique in the animal kingdom, since they are made of the mineral calcite. Hard calcite makes up the hard parts of the trilobite, providing the crusty shield that covers the back of the animal known as the dorsal exoskeleton. Calcite has also been recruited to provide the material for the lenses of the eye—so they have become "crystal eyes" if you

---

* I dealt in detail with the growth of arthropods in general, and trilobites in particular, in my book *Trilobite! Eyewitness to Evolution.*

will. The individual lenses are minute in many trilobite eyes (they can have several thousand), but each separate lens presumably responded to an external light stimulus, and then an optic nerve conveyed the information to the brain. Eyes with many small lenses are usually thought of as particularly sensitive to movement: a moving image progressively impinges on different lenses within the field of view. Both trilobites and *Limulus* have eyes that look predominantly sideways, scouting around over the sea floor where they live.

Strangely enough, the eye of *Limulus* has been very intensively studied. Haldan K. Hartline of the University of Pennsylvania used the eye of the horseshoe crab as his experimental material to investigate the physics of animal vision. In the 1930s he was the first scientist able to record the activity of a single optic nerve fibre attached to a lens (ommatidium). *Limulus* has about a thousand such fibres in the eye, and we might well imagine that trilobite eyes had at least a comparable sensitivity. He later showed how different fibres in the optic nerves respond to light in selectively different ways. This opened up the route to a whole new field of physiology—and earned Hartline the Nobel Prize in 1967.

Robert Barlow and his colleagues are now building further on Hartline's research. They have attached miniature video cameras onto living animals in order to scrutinise exactly where the horseshoe crabs are looking. The eyes seem to exhibit an unsuspected sophistication. There is apparently a natural, or circadian, rhythm in the sensitivity of the ocular system, which combines with other dark-adaptive mechanisms so that their sensitivity at night may be as much as a million times more acute than in the daytime. Crabs are particularly attuned to recognising potential mates, which, given the frenetic activity along Delaware Bay, is not altogether surprising. The ability of the *Limulus* eye to eliminate visual "noise" is quite extraordinary (think of our own faltering attempts to really see very faint stars on a dark night), and Dr. Barlow is currently trying to understand how this works right down at the molecular level. It is probably the case that we know as much about the visual system of this ancient arthropod as about that of any other living creature. But the more we know, the more we might wonder whether this particular survivor is primitive or just exquisitely adapted. Did the trilobites have blue blood? There is no final proof one way or the other; nor can there ever be with such perishable stuff as blood. However, there are many examples of trilobites that have been severely bitten and yet have survived. They usually show a sealed-off gouge on one side. Even in the early Cambrian there were predators such as the lobster-sized *Anomalocaris* and its relatives that might have regarded a trilobite as a crunchy snack. *Anomalocaris* was a strange but evidently raptorial arthropod with two long grasping arms and a mouth surrounded by plates. In those days of accelerated evolutionary change, natural selection would rapidly have favoured any mutation that stopped a wounded trilobite from bleeding to death, and the same would have applied to any of its relatives. Since the circulation system of *Limulus,* and doubtless of a trilobite, is diffuse compared with our own—it more or less fills the open spaces between the other internal organs—a general clotting agent would have been at a premium. It does seem possible that the alternative way of making blood—the copper route—could have had a very long pedigree, and that the blue ichor's ability to seal wounds and its sensitivity to infection could have helped both trilobites and horseshoe crabs to survive in a newly vicious world. This is one of those moments when palaeontologists wish they could circumvent the rules of the space-time continuum, and go back and see for themselves. As it is, we have to make do with more or less plausible guesses, in the process trying to persuade our fellow scientists that we have undoubtedly arrived at an entirely logical conclusion. History, of course, does not necessarily have to follow our own human logic, and may have surprises of its own.

Could a scene like that witnessed at the beginning of this chapter been played out by trilobites in deep geological time? It is possible. To see evidence, I must take you with me to the small town of Arouca in northern Portugal. It lies at the end of a very winding drive into the hills from the old seafaring city of Porto. The prevalence of hillsides covered with eucalyptus trees in some parts of this landscape can be depressing, as these antipodeans are out of place here; but their contribution to the local economy pushes all ecological niceties to one side. Most go to pulp for paper, for these efficient trees grow faster than native species. So in another sense these eucalypts, too, are natural survivors. Every now and then bush fires flare up uncontrollably, fuelled by the volatile oils of the "gum trees"; black swathes along the hillsides record their ugly legacy. In the higher hills, pretty valleys contain ancient mills and farmhouses built of crudely squared-off large blocks of the grey granite that makes up the highest bare ground in the region. In geological terms, obstinate granite is probably the longest survivor of all. Little has happened to the face of these sensible buildings since medieval times other than a dappling of face-paint provided by lichens. On the bleak granite moors nearby are burial chambers that have seen much of human history pass, but still endure. Since the time of the trilobite, whole mountain ranges comprised of this most persistent stone have been worn away grain by grain by the inexorable forces of erosion, and rendered down to sea level. Life outlasts even mountains, for the greatest survivor of all is DNA.

Arouca must be the only town in the world with a trilobite monument, which is a tall spike sitting a little uneasily in the centre of a roundabout. The small hill town is bidding to achieve European Geopark status, and part of its claim is as the home of giant trilobites, which figure prominently on the monument. To see the real thing I head off to the slate quarries above the town near the little village of Canelas. Mining has been a part of the culture in the region for a long time. The Romans were in the hills seeking gold, and old workings excavated into tough Ordovician sandstones can still be seen atop a local high spot, where a dark and slippery stairway leads down into a ferny crevice. The same sandstones preserve fossils of burrows made by trilobites digging into an ancient sea floor, providing yet another type of treasure. Nowadays, the booty is roofing slate. A mass of Ordovician slates known as the Valongo Formation overlies the sandstones and runs across country. The slates are nearly black, and split into flat sheets that can be further split again until they make usable roofing slabs. Once prepared this way, they are very durable commodities. The slates are extracted from large quarries by blasting out huge chunks of rock, which are then carried away to a factory for further working. The flat planes along which the slate cleaves also furnish a record of Ordovician sea floors, albeit now turned almost vertically as a result of convulsions of the earth. Every now and then a slab covered with trilobites is discovered. They are, as my Spanish colleagues exclaimed without overstatement, *¡espectacular!* The fossils often show up pale greyish against their dark background. Under normal circumstances it is a rare event to find trilobites much bigger than a small shoe, but in this locality they are often as big as tureens. The largest trilobite in the world, perhaps a metre in length, may be lurking among unstudied collections, but specimens 70 cm long are already familiar. Not only are they large, they are also numerous. Because whole bedding planes (which represent former sea floors) are extracted, an extensive view of a tragic moment in time is occasionally recovered: it is a fossil graveyard, with bodies laid out at the moment of death, a community of cadavers. A scenario like this could not be extracted by the usual tapping of the geologist's hammer upon a rock: it requires activity on an industrial scale. Fortunately, the quarry owner is aware of the importance of his slates in exposing a sea floor perhaps 470 million years old; time enough to erode three mountain ranges, granite and all. Many of the trilobites are preserved on site in a private museum that the owner has generously dedicated to conserving these fossil remains. A dozen different species are on display there; to one in thrall to the past like me, it is an extraordinary experience to see these grey bodies covering every wall. It has something of the feel of a picture gallery, and I have to remind myself that these were once scuttling animals as intent on their business as any living horseshoe crab.*

Many of the large slabs show assemblages of just one species together, and the individuals are all large and of similar size. They are often complete bodies, which suggests they are entire animals that have been killed rather than, say, the "cast-offs" left behind after moulting, when bits and pieces might be expected, arranged

* I am lucky enough to have one of the giant trilobite species named after me: *Ogyginus forteyi.* This might offer me an excuse to explain the rules of nomenclature and classification. The generic name *Ogyginus* is spelled with a capital letter (as are all genera), while the species name is conventionally spelled with a small initial letter, whether or not it is named after a person. The island of Ogygia was where, according to Homer, the nymph Calypso detained Odysseus for seven years. Many fossil names have similar classical roots. *Limulus* is derived from the Latin meaning "to look sideways," which I surmise refers to the field of view of the lateral eyes.

higgledy-piggledy. There are examples where the bodies partially overlap. All this is very like the mating congregations of horseshoe crabs along Delaware Bay. Imagine if some catastrophe had killed and preserved the crabs at the height of their nuptials. A volcanic eruption might fit the bill; this would bury and kill the animals at exactly the same time. After eons passed, the sediments would have hardened into rock and the crabs would be fossils; compaction of the sediment would also have flattened down the buried beasts. Some fossil specimens would still partly overlap their partners, frozen in the act. The younger animals would have been elsewhere, so they are not represented among the fossils. It is a plausible scenario, even a tempting one. We have to add an extra complication, because something additional had happened to the Portuguese trilobites during their long sojourn in the rock. The whole mass of slates of which they have become a part has been squeezed in a tectonic vise that has twisted some specimens out of true until they look a little lopsided. Others have been stretched somewhat, and as a result claims about the "longest ever trilobite" have to be treated with caution.

A more critical examination of the evidence identifies some important differences between Delaware Bay today and Ordovician Portugal. The most obvious of these is that the giant trilobites were clearly not gathered upon a beach. They were overwhelmed and killed on the sea floor. Local Portuguese geologists believe that the Ordovician animals lived in a marine basin with poor oceanic circulation, so that deeper layers could become stagnant. The congregated trilobites might have been overcome by a phase when oxygen dropped to lethal levels: after all, even trilobites needed to breathe. Such anoxia is not much of a problem in Delaware Bay. The trilobites could still have been gathering for reproductive purposes, of course. They might have even been safer from predators in the deeper basin. But then it makes us uneasy to think of the trilobites depositing their eggs in such an inhospitable place—unless anoxic events were so rare as to have little effect on their long-term survival. Then there is the fact that a number of the slabs seem to show a *mixture* of species. Could they have been gathered together for some purpose other than mating? Unlike *Limulus,* a freshly moulted "soft shelled" trilobite would have been vulnerable until it grew a new hard carapace. Maybe these congregations were huddled together for mutual protection away from the prying eyes of predators. With these ambiguities in the picture, the case for a direct comparison with the behaviour of modern horseshoes begins to seem weaker. A sceptic would say that it is simple-minded to expect similar habits to endure for hundreds of millions of years. Perhaps so, but common problems in nature often come up with comparable solutions, the more so if the organisms concerned are related. Those trilobite examples with marginally overlapping bodies might merit further examination, since they do recall the struggle for mating among the horseshoe crabs, and it is more than a guess that eyes in these animals were particularly attuned to seeking out mates. I still like to think that the crystal eyes of trilobites may have had similar lustful intent.

2

The Search for the Velvet Worm

New Zealand is a country that beguiles but deceives, for much of it is dressed in false colours. Although there is still some almost untouched forest on the South Island, human hands have transformed much of New Zealand in the service of forestry and sheep.

The story of these islands is one of isolation. Their origins lie within the great and ancient vanished land of Gondwana, from a time when peninsular India, South America, Africa and Australia were united together as a "supercontinent." Something like 100 million years ago, the nascent New Zealand separated from its parent, as Gondwana began to fragment progressively into its individual plates. These eventually forged the continents of the southern hemisphere that we would recognise today. Unravelling this story was one of the great achievements of modern science, and it is linked to some of the stories of biological survivors in this book. New Zealand may be just a small part of that story, but its own narrative is geologically complex. To a kernel of old Gondwana rocks, newer rocks have been added piece by piece because the islands have sat in a tectonically active, though isolated, zone for millions of years. Volcanoes have made their fiery contribution in ash and lava, other igneous rocks have been intruded into the Alpine range as it grew, and then sediments eroded from the young mountains completed a dynamic rock record. It could be said that the geography of New Zealand has been under constant revision. But animals and plants were also carried onwards into the growing New Zealand from the ancient Gondwana days, a persistent legacy of an old continent bequeathed to a future land. Sometimes the evolutionary signal of an organism betrays a far-distant past in surprising ways.

The ancient coniferous podocarp forests that once covered much of the North Island have all but disappeared. Little patches of it hang on almost by oversight. They are dark and mysterious within; silent, but for melodic tweets from birds high up in the canopy feeding on the little fruits the trees produce. Podocarps are southern hemisphere conifers of several species that make superb and stately trees if they are allowed centuries to grow to maturity. This is too long for a healthy profit. The original forests were felled for their good timber, but were replaced in many areas by quick-growing conifers such as Californian pines deriving from the other hemisphere. Huge areas of the North Island are covered with conifer plantations. Periodically they are felled en masse, and then the rolling hills are scenes of devastation, with nothing green left standing but wrecks of stumps and unwanted branches in rough piles everywhere, and small fires smouldering as if shells had exploded not long before. When I drove through such an area, I was torn between recollections of battle scenes from World War One and J. R. R. Tolkien's descriptions of the ghastly land of Mordor in *The Lord of the Rings*. I suppose the latter might be more appropriate, since splendid alpine New Zealand has been repeatedly used as the location for the movie version of Tolkien's saga. The sheep country looks like steep sheep country everywhere, and reminded me of Wales and Scotland, even to the extent of carrying scrubby patches of brilliant yellow-flowered gorse—which, of course, is a troublesome introduction from Europe. There are so many other Europeans on these islands, not just Smiths and Joneses in suburban villas, but oaks, sycamores, elderberries and implacable ivy. They compete for space with other native trees, including the New Zealand red beech, *Nothofagus fusca,* with its delightfully delicate little leaves and graceful habit. I could not help feeling that a coarse and unthinking hand has been at work, interfering with the landscape, scrubbing forests out, planting weeds. This is grossly unfair to the New Zealanders, the kindest people on earth.

Podocarp trees are in a sense "survivors" from the time of Gondwana. These trees are found in Australia, New Caledonia, South America and Sub-Saharan Africa—one or two genera are even in common between New Zealand and the Andes. Gondwana may have split into its separate pieces, but the identity tags of its former inhabitants were not redesigned so easily. These Gondwanan coniferous trees, with their relatively large leaves and bright berries, do have a very special appearance, at least to a European accustomed to pines and firs with their dry-looking cones. A botanist would remind me I should really describe the berries as "fleshy peduncles" because they carry exposed seeds at their tips. On the wet west coast of the South Island near Karamea, I walk into a podocarp forest where dampness rules. Everything that could be is covered in moss, epiphytes or filmy ferns. They clothe the trunks and branches of trees in a creeping, delicate and close-fitting cloak of tiny green leaves. Inconspicuous orchids are there somewhere, perched on branches, sporting small yellowish flowers, the antithesis of tropical showiness. Where light breaks through the canopy, tree ferns erupt like green fountains perched on shaggy stems, adding ebullience to the primeval atmosphere. Little brown birds with bright little eyes—tom-tits, New Zealanders call them—pipe tamely from exposed twigs, hoping that these clodhopping visitors might disturb insects for their supper. Trunks of the podocarp *totara* tree soar upwards, while the *rimu*—the most elegant of its family, with weeping, cypress-like branches—breaks through the canopy like drapes. The wood of this tree is so hard that the heart is still sound for working from trunks lying on the ground years after the outer layers have rotted. The more familiar southern beeches (*Nothofagus*) are unsuitable for major construction since they rot from the inside out, but they also have a Gondwanan signature, following closely the pattern of the podocarps. I recall that Charles Darwin observed how the natives in Tierra del Fuego ate a curious fungus looking like a cluster of yellow golf balls that grew on southern beech branches. The fungus was named *Cyttaria darwinii* by Miles Joseph Berkeley, the great nineteenth-century mycologist who worked out the fungal cause of the Irish potato blight. Further species of the same fungus were discovered, but they only grew on southern beeches: fungi can be choosy. The Gondwana legacy even applies to soft, edible fungi that would never stand a chance of being preserved as fossils. Biologists must have their wits about them if they are to understand the complexity of the past.

New Zealand took away its package of Gondwanan plants as the continent broke up. Later, it was colonised by birds, and they evolved in isolation to produce a host of endemic species. Some are almost comical, like the kakapo, a ground parrot of remarkable stupidity (and now a threatened species), and the kea, a mountain parrot of legendary intelligence and a fondness for eating the windscreen wipers of cars exploring the Southern Alps. It is said that keas can be found solving crossword puzzles left behind by tourists. Other birds became intimately involved with perpetuating the podocarp forest by swallowing and distributing their seeds. Some still remain, singing sweet songs high in the canopies of the stately stands that survive. Many scientists believe that at some stage in New Zealand history the sea level rose to a point where mammal species could not endure and breed. Today, it has no endemic terrestrial mammals. Whatever happened, nobody could question the fact that this antipodean island represents the acme of avian evolution in the absence of serious mammalian competitors. The loss of the ability to fly is common— why bother to take to the air when you can safely amble about in the bush? The kiwi is the amiable emblem of the country; a variety of kiwi species show the fecundity of this ground-dwelling option. None is safe for the future. The largest flightless bird that ever lived, the moa, lived in huge numbers in New Zealand. A Brobdingnagian ostrich, it was meat on legs for the first human invaders, who undoubtedly hastened its extinction.* In the Karamea forest I see the dark entrance to cave systems perforating Honeycomb Hill, from which dozens of moa bones have been recovered, and marvel at a sudden vision of an island swarming with the giant birds. If only we could turn back the clock. So many New Zealand bird species are

---

* The scientific description of the moa was prepared by the first Director of the Natural History Museum in London, Darwin's antagonist, Richard Owen. I have used a photograph of the ageing Owen standing by a much taller moa skeleton in my history of the Natural History Museum, *Dry Storeroom No 1.*

either extinct or threatened. The new generation of New Zealanders are almost neurotically aware of what human interference has done to the natural environment. The introduction of the possum from Australia was a particular disaster, since these aggressive vegetarians seem to particularly relish New Zealand tree flowers. They threaten the livelihoods of all the nectar sippers and honey eaters among the bird species. The restocking of offshore islands with native birds in a rat-free, possum-free and cat-free environment seems to be the best option at the moment. It is at best a despairing attempt to store away from further trouble a remarkable history running into millions of years.

I have to understand New Zealand's long history before my search for an animal that has survived from a period even earlier than the first appearance of the horseshoe crabs. My quarry is the velvet worm. This creature will help us climb downwards to a still lower branch of the evolutionary tree. George Gibbs from the University of Wellington is my guide. He knows the secretive ways of these elusive animals. We drive out along Route 1, west of Wellington on the southern edge of the North Island, prior to walking up the Akatara Ridge along a small country track. The whole area was milled in the 1930s and 1940s, so the mature podocarp forest has all gone, but there is secondary growth of tree ferns and *rimu* and *Protea* in a dense thicket. Some of the common native birds have adapted to the new circumstances. We hear the distinctive whistle and churr of the *tui* as we park the car. New Zealand birds usually have a distinctive and attractive song, even those that are unspectacular to look at. As we walk up the track, I notice another survivor, the lowly herb *Lycopodium,* growing on the bank, a plant we shall meet again. It is a steady climb, though hardly taxing. Towards the top of the track the landscape opens out into gently rolling, wooded farmland. A scattering of cows and white sheep graze on the cleared, grassy hillsides, and dotted among them are Californian pines. The wind blows through the trees with a sound like the gentle crash of waves. The "old homestead" proves to be an antique wooden building in the bottom of a small hollow surrounded by a circle of ageing pine trees. George locates the bleached remnants of rotten pine logs lying on the ground nearby. For some reason they had not been tidied away after felling, so they have had the opportunity slowly to break down *in situ*. Selecting one log, it soon becomes apparent that inside its pale exterior the decaying wood is rusty red and fibrous. George starts beating at it with a small mattock brought along especially for the purpose. I cannot help leaning expectantly over his shoulder. Each hack of the instrument beats away 10 million years of geological time. Can the velvet worm be hiding inside this curious sarcophagus? Where is its time capsule?

But the first log yields nothing. A second log is soon under attack. It seems softer somehow, more decayed. As the wood splits easily apart, tiny white termites are exposed to the air, looking something like pallid ants, almost transparently delicate. They move slowly, as if stunned by being exposed suddenly to bright light: they are creatures of habitual darkness. Their little antennae can be seen waving furiously. Termites are wood eaters hiding deep inside the log, living in chambers they make running along its "grain." We had opened up their secret world. And then we see there's something else, something caterpillar-like, hiding in the termites' tunnels. It shrinks away as if it does not want to be seen, or as if light is somehow an embarrassment to it. George coaxes it expertly into full view: it's the velvet worm!

This is the creature we had come all this way to find: *Peripatus novae-zealandiae,* to give it its scientific name. Because it does not move very fast, it proves relatively easy to catch and bring out into the light. It is indeed about the size of a very large caterpillar, light brownish and with a stripe running down its back. I gingerly touch it and find it soft and giving—if hardly velvety. George soon finds a second worm hiding away inside the log, and then a third; they evidently do not mind one another's company. They attempt to twist away from us in a most peculiar fashion: they seem to be capable of drastically changing their length. It looks as if they can stretch or squash like concertinas. They are highly flexible, too, and one of them turns into a tight "S" shape with no trouble. "That's not like a caterpillar," I say to George. He grins back at me, sharing my pleasure in the discovery. They clearly have a front and a back, for at the forward end are a prominent pair of antennae—which lead the way the animals want to flee. Their movement is not worm-like at all, despite their name. It is accomplished by means of little conical stumpy legs on either side of the long body. On the hand these make an oddly prickly-tickly sensation. Velvet worms are clearly very odd invertebrates.

The *Peripatus* animals evidently live alongside the termites inside rotting pine logs; indeed, they feed on the little insects, pursuing them through the chambers inside, doubtless detecting them with their sensitive "feelers." They trap their prey by means of a sticky slime produced in special glands. Nothing else in nature feeds in exactly the same way. One of George's students proved that the slime only entraps termites of the right size—not too big to escape, not too small to be uneconomic—after all, slime is protein, and that is expensive for the creatures to make. I try out the feel of it; it is distinctly tacky, and it must be like glue to a termite. Both the velvet worm and the termites shun the sunlight with good reason. They lose water very rapidly through their thin "skins." The velvet worm is little more than a bag of fluid surrounded by a membrane. In bright sun it would soon dry to a crisp. Inside the hermetic and lightless world of a decaying pine tree, the relative humidity is nearly always 100 per cent and it is perfectly safe.

Poking about some more in the rotten wood, we make another discovery: baby velvet worms. They are only about one centimetre long, and pale in colour, but they seem to be exact small versions of the large ones. I presume they must eat suitably diminutive termites. The worms grow continuously to achieve adult size, blowing up like balloons. The velvet worm actually gives birth to live young, and the ones we saw may have been newly born. This is unusual among invertebrates, and even among vertebrates is only characteristic of mammals (and a few specialised reptiles). The eggs of this particular velvet worm are few in number, and large and yolky, thus allowing for further embryonic development within the female; only three or four young are born at one time. There are two "litters" a year. Since the animals live for three years, they only have about twenty offspring, which is an extraordinarily small number when one remembers that most arthropods, for example, lay thousands of eggs: recall our horseshoe crabs. The most prolific velvet worm species produces no more than forty young a year. It seems that *Peripatus* is an animal with a personality all its own.

Looking a little more closely at the velvet worm, the first thing one notices is that the body seems to be made out of many rings that encircle it completely, even the legs and antennae. They remind me of the Michelin Man, supreme advertising logo of the famous tyre company, all dressed up in his bands of rubber. It is this distinctive structure that accounts for the body's elastic properties. Muscles circle the body cavity inside the skin. Then it is obvious that this is a segmented animal rather like a trilobite, with lots of similar units repeated along the length of the body. Each body segment carries a pair of those stumpy legs. Among living species of velvet worms, the number of segments varies quite widely, but, biologically speaking, that is only a matter of tacking on extra identical units, and does not require massive tinkering at the genetic level. To prove this, there is even one velvet worm species that can have between twenty-nine and forty-three body segments. The short, stumpy legs propel the animal along by working in sequence in waves, a common feature among segmented animals. From the side, it looks as if one leg hands on a motion to its neighbour progressively in a common direction. Forward movement would obviously not be possible if legs pushed forwards entirely at random; cooperation is required. The legs remind me of the limbs of a child's stuffed toy, rather like those belonging to Piglet as illustrated by E. H. Shepard in *The House at Pooh Corner,* but they work well enough to catch up with termites. After all, one does not need a Maserati to overtake a donkey. Looking more closely at the surface of the "skin," one can see that each of the body rings carries a line of protuberances, giving the external surface a knobbly appearance, especially on its upper side—these are known as papillae (they may have even smaller secondary papillae upon them). The patterns of the papillae vary between velvet worm species, as does the overall colour. There is one magnificently blue species elsewhere in New Zealand.

The head of *Peripatus* is most obviously identified by its pair of antennae. But close to the front on the underside is the mouth, which is provided with sickle-like jaws to either side, each equipped with a pair of blades at the tip that are produced by a local thickening of the skin, or cuticle. They are simple but efficient shredders. The ducts for the slime glands open at the side of the head. There are no eyes. As for the legs, they are little more than stumpy projections off the body equipped with muscles internally to swing them backwards and forwards. Their feet carry two sickle-shaped claws at their tips, which are much like the jaws in structure; this may indeed provide a clue to the evolutionary origins of the more specialised jaw. Males and females are similar, except that the former are usually a little smaller and are less common.

Inside, they are pretty simple, too. The major part of the body is taken up with the stomach, which runs along the length of the animal to the anus at the end. Between the gut and the mouth there is a short oesophagus and a muscular pharynx, which is used for initial food processing. Oxygen absorption is achieved through tiny tubes inside the body called tracheae, which have their apertures located in depressions between the papillae. There are no special gills or lungs, because animals of this size can get all the oxygen they need through thinned parts of the cuticle. The heart is another simple tube, positioned at the top of the body above the stomach. The rest of the vascular system is much as in the horseshoe crab *Limulus*, distributed rather diffusely through the internal cavity. *Peripatus* gets rid of its waste products by means of nephridia, kidney-like organs, located in the legs along with small excretory openings. The nerve cord is a double structure running along most of the length of the animal, with cross connections that make it look somewhat like a ladder: nerves extend from this into the segments and limbs. A larger ganglion in the head is all that this basic creature can display as a brain.

Simple though it may be, the velvet worm functions perfectly well. For a moving animal, there is quite a short list of vital functions: sensory equipment to find a source of food and tools to help eat it; a method of locomotion; a way to breathe and distribute oxygen to internal organs; a system of waste disposal; a reliable way to propagate the species. *Peripatus* would be the kind of creature one might put together from a "how to make an animal" kit, except that like almost everything else in nature it has some tricks all its own— its gluey trap, its ability to produce little peripati by live birth. It is a simple creature in many features, specialised in other subtle ways; but it is also another old timer, a messenger from the distant past.

Its more recent history is not very different from that of the podocarp trees. *Peripatus* and its relatives number about two hundred living species (placed together in Phylum Onychophora, informally known as "onychophorans").* They also have a distribution over the areas that once formed Gondwana: Australia, New Zealand, South Africa, South America and Assam (India). There are also velvet worms in Irian Jaya and New Guinea, where it is very likely that further species still remain to be discovered in mountainous and inaccessible areas. All of them carry the long memory of the vanished supercontinent as they tramp their unadventurous way on their stubby legs. Velvet worms had once wandered over Gondwana, but, like the podocarps and southern beeches, new species arose on the separate pieces of the progressively fragmented continent; for evolution does not stand still. I could have gone in search of the velvet worm in any one of these other regions. The New Zealand species I happened to pursue is particularly interesting because it has developed a relationship with a special kind of termite that is regarded as the most primitive of its kind (of the Family Kalotermitidae), among which most individuals finish up as flying insects. The other termites are noted for their extraordinary caste system, with specialist workers and soldiers that never change their roles. It seems possible that a whole primitive ecology was transferred to New Zealand when Gondwana broke up, and there it endured, virtually unchanged, encased in logs. But something did change. I found *Peripatus* inside a pine log belonging to a species that was not native to New Zealand, so at some recent date the velvet worm must have followed its termite prey into a new habitat. You can teach old worms new tricks.

Since the velvet worm has a body as soft as dough, it is most

---

* The hierarchy of classification includes many more inclusive categories above the genus and species (family, superfamily and so on), but in this book we are principally concerned with the greatest groups of related animals, classified together as phyla (singular phylum) and the major divisions within phyla known as classes. The Phylum Arthropoda contains animals with jointed legs: horseshoe crabs are placed within the Class Chelicerata with relatives like scorpions; trilobites comprise Class Trilobita; crabs, shrimps and their relatives form Class Crustacea; and Class Insecta consists of butterflies, beetles, ants and hordes of other small terrestrial animals. These groupings all represent major branches on the evolutionary tree.

unlikely to be preserved as a fossil. Shells and bones leave behind their hard evidence, but can we expect a shy, soft package of flesh to do the same? Even the Solnhofen Limestone fails to preserve a single fossil of a *Peripatus*. Fortunately, there is one example preserved in Cretaceous amber from Burma (Myanmar), perhaps 100 million years old, a contemporary of the dinosaurs. Amber preserves the most evanescent of creatures: flies, beetles, even mushrooms. This fossil species is very like the living *Peripatus*, and there is no question that it lived in a similar fashion. It provides the proof that velvet worms of modern type were alive at the break-up of the Gondwana supercontinent—which is good to know, although we might infer it anyway from their distribution today. But we want to go back much further than this, 200 million years earlier. A remarkably preserved impression from the Carboniferous called *Helenodora* tells us that in the swamps of the coal measures, distant relatives of the velvet worm—but still eminently recognisable—were wandering their deliberate way through the damp undergrowth. Their contemporaries at this stage in the evolution of life were inelegant amphibians and very early reptiles, accompanied by the first flying insects. The velvet worm was terrestrial then, just as it is now. It may even have developed its special slimy-gluey glands, although at this early date it must have fed on something other than termites: for these insects were not even a twinkle in the eye of evolution in the Carboniferous. The velvet worm is beginning to look at least as ancient as the horseshoe crab. The velvet worm likewise survived the great extinction at the end of the Permian, and then it slid through the major event that secured the removal of the dinosaurs from our planet; like *Limulus'* ancestors, *Peripatus* is made of sterner stuff, not to be seen off by mere global catastrophes. But now there comes a surprise. When we go back yet another 200 million years all the way to the Cambrian Period, to the time of "explosive" evolution at the beginning of complex animal life, there, too, were relatives of velvet worms—they prove to be more common as fossils in Cambrian rocks than they are in rocks laid down in later geological periods. They began their history under the sea, in the cradle of life, like everything else. And they proved to be survivors. They shared their early world with trilobites, and the first relatives of horseshoe crabs, and the distant ancestors of scorpions. So much in biology seems to converge back more than 500 million years ago to the Cambrian ancient sea floor. The ancestors of the velvet worms were yet another kind of animal that later moved onto land—and this happened at least 300 million years ago. Because of their rarity as fossils, it is not possible to say whether velvet worms got onto land before or after scorpions; we shall probably never know. Unlike scorpions, they needed to stay in wet, or at least humid, environments, but just like those venomous arachnids, none of their close relatives managed to survive to the present day beneath the sea. For *Peripatus* and its relatives, going on land was arriving at some sort of haven.

Probably the best-known onychophoran from the Cambrian is called *Aysheaia pedunculata*. It was named a century ago by the renowned palaeontologist Charles Dolittle Walcott of the Smithsonian Institution, Washington. It occurs in what is probably also the most famous rock formation of that age, the Burgess Shale of British Columbia, Canada. A locality near Mount Field in the Rocky Mountains discovered by Walcott yielded the first known, diverse fossil fauna of "soft bodied" organisms, that is, those lacking hard mineralised shells, which are the kinds that give us "regular" fossils. The Burgess Shale allows us to see something of the whole panoply of marine life at a seminal time—although admittedly it only samples the larger organisms. The fossils are preserved as silvery films on the surface of the black shale, so that they are subtle casts made by fine minerals before the animals could be scavenged or they fell apart. The exact circumstances of their preservation are still being debated, but it is certain that quick burial and protection from normal decay played an important part. Whatever the cause, *Aysheaia* is preserved in extraordinary detail.

Comparing *Aysheaia* with *Peripatus* reveals that they are of similar size and shape, the former reaching about 6 cm in length. The fact that differently sized animals of *Aysheaia* retain the same form as they get larger implies a simple growth pattern like that of the modern velvet worms. In *Aysheaia* the fine rings encircling the body are clearly visible, and little prickles are much like the papillae of the living animal; add to that their stumpy conical legs look very alike, and at the tips in the fossil little sickle-like claws can be clearly seen. But

Cambrian lobopod fossil *Aysheaia pedunculata* from the Burgess
Shale in the Canadian Rockies, British Columbia.

there are some differences between this most ancient animal and the
creature I helped to dig out of its woody habitat—it would have been
astonishing if there were not. Most obviously, there is a pair of gill-
like structures on the head end of the fossil. This is hardly surprising
since the animal was living under water. There is also no sign of the
special slime glands in the fossil. This must have been a later development, which presumably would also have been acquired after the terrestrial invasion. But it would take a hardened sceptic not to believe
that these animals were related. Of course, in science there are always
such sceptics, and the special features of *Aysheaia* were emphasised
by some at the expense of its many similarities to *Peripatus,* but I
believe most students today would accept the onychophoran tag on
the Cambrian creature.

The story got interesting when a second, and much more
peculiar-looking, Burgess Shale species was assigned to the onychophorans. This animal had been named in 1977 *Hallucigenia* by the
Cambridge palaeontologist Simon Conway Morris, but his original
description of the fossil was upside down. *Hallucigenia* carried paired
spikes on its back which Conway Morris had originally interpreted
as legs (he later acknowledged his error with good grace), while the
true legs were more spindly affairs than those of living velvet worms

or, indeed, *Aysheaia.* The spines arose from hardened plates, which had been found separately as fossils in early Cambrian strata, but had been unfathomable up to that time. The mystery was not fully elucidated until much better preserved, soft-bodied fossils began to be found over the last decade or so in strata cropping out around Chengjiang in Yunnan Province in China (these are known as the shales of the Maotianshan Formation). The new fossils were up to 10 million years older than the Burgess Shale examples, and have now proved even more diverse. They include at least six animals that can be assigned to the same group as the velvet worms. One of them carries spikes on its back and was an additional species of *Hallucigenia;* another one (*Paucipodia*) was an altogether slimmer affair than its distant living relatives, with only nine pairs of slender legs. One fact was now becoming clear: the relatives of the velvet worm were much more varied in the early days. There were lots of them of several distinct kinds, but they did all share those lobe-like legs, often tipped by little claws. An appropriate term for the whole group, both living and fossil, achieved wide currency during the 1990s—they were "lobopods." Thanks to the special preservation of these Cambrian fossils it was possible to see surprisingly varied and delicate lobopod animals in unprecedented detail. Living velvet worms began to seem more of an evolutionary afterthought.

The plot thickened still further at this time, for up in Greenland Dr. Graham Budd and his colleagues were finding yet more soft-bodied animals in the early Cambrian Buen Formation. These showed certain similarities to onychophorans, like the rings along the body, but the animal named *Kerygmachela* by Budd had a pair of grasping appendages at the front and was obviously a hunter capable of grasping prey. The lobopods were clearly going to spring yet more surprises.

The question now arises as to where this curious bunch of animals fits in on the tree of life. I have already described how Cambrian fossil faunas included many kinds of jointed-legged animals or arthropods, such as distant relatives of the horseshoe crab. All these arthropods would have had a tough chitin covering over the body that made the "invention" of hinged joints necessary. Without them, the animals would have been as helpless as a medieval jouster whose articulated armour had rusted into immobility. But *with* hinges added, arthropods were equipped with a versatile covering that could be recruited to be armour, jaws or toolkit as the occasion demanded. The future walked on spindly legs. Like arthropods, velvet worms and their relatives were, and are, segmented animals. Unlike arthropods, they did not have a strong coat made of chitin: no hinges were possible. Their lobopod legs were effective enough in their own plodding way, but they could not be extended into the attenuated pins of a daddy-long-legs. That requires serious mechanical engineering, and the stiffening support of a hard skeleton. On the other hand, some features of internal anatomy seem to be very similar between living onychoporans and arthropods. I could mention the diffuse circulation system and the arrangement of the nerve cords, and some scientists are impressed by the presence of antennae in both kinds of organisms. At least one of the Cambrian lobopods shows evidence of simple eyes. The musculature is differently arranged in lobopods and arthropods, which actually allows the lobopods greater bodily flexibility.

Their fundamental similarities make it likely that *Peripatus* and arthropods share a common ancestor. The arthropods seem to be more advanced in several respects: the jointed legs could only have been added when the "skin" acquired its hard outer layer, and sophisticated compound eyes like those of *Limulus* must surely have been a later development. This is another way of saying that lobopods are probably sited on a lower level on the great tree of life, likely to have been around before the arthropods evolved. There are some scientists who would claim that they are the true ancestors of the arthropods, or even that different kinds of lobopod gave rise to different kinds of arthropods. Partly, this depends on the interpretation of the jawed animal *Kerygmachela* from Greenland that seems to display something of an amalgam of lobopod and arthropod characteristics. Whatever the final interpretation, these recent discoveries of Cambrian fossils provide another case of neat categories of animal classification blurring at the time of the "explosive" phase of animal evolution. The story also takes us back further in time than we have been before.

1. Horseshoe crabs (*Limulus polyphemus*) on the beach in Delaware Bay, gathering in millions for mating. The females are mounted by males intent on fertilizing the eggs.

2. (BELOW) A mass gathering of fossil trilobites (*Dikelokephalina brenchleyi*) from the Ordovician strata of Morocco, 470 million years old. These large trilobites are similar in size to living horseshoe crabs, and may have gathered together for similar reasons.

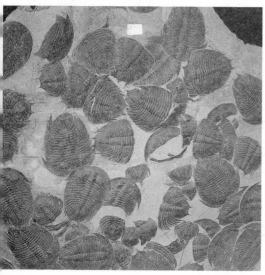

3. (ABOVE) The moulted carapace of a small horseshoe crab from Delaware Bay perfectly showing the limbs, and behind them the book lungs.

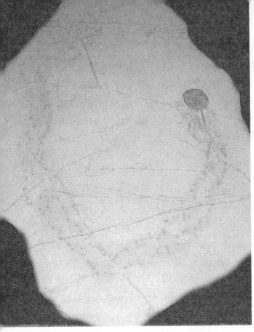

5. *Mesolimulus* from the Jurassic Solnhofen Limestone in Germany is obviously very similar to the living horseshoe crabs.

4. Fossil horseshoe crab *Mesolimulus* from the Solnhofen Limestone, Jurassic in age, died at the end of its tracks. Similar fossil tracks have been found in even older strata.

6. (LEFT) The author, right, with Juan Carlos Gutierrez-Marcos in the quarry of the giant Ordovician trilobites in Arouca, Portugal.

7. (ABOVE) *Anomalocaris* from the Burgess Shale, Canada (Cambrian), a large predator that lived more than 500 million years ago.

8. A lycopod with spore-bearing structures, by a path in New Zealand. Carboniferous relatives of this herb were tree-sized.

9. Native podocarp forest in New Zealand, Paparoa, South Island. Podocarps and tree ferns clothe the riverbank. The *rata* tree on the left is a New Zealand endemic.

10. Breaking open logs in New Zealand in search of the velvet worm, with George Gibbs on left, author on right.

11. The velvet worm, *Peripatus*, from New Zealand showing antennae at front and stumpy legs. The flexible body is made out of fine rings carrying low tubercles, and is capable of stretching considerably. The poison glands that allow the animal to capture termites are situated in the head end.

12. (ABOVE) One of several fossil lobopods from the Cambrian Chengjiang fauna of Yunnan, China (*Hallucigenia fortis*). Relatives of the velvet worm were marine during their early history and much more diverse than they are today. Soft-bodied fossils are preserved as coloured films on the surface of a fine-grained shale. Exceptional preservation like this helps greatly in our understanding of the origin of the animal phyla.

13. (LEFT) Large animals appearing before the Cambrian: fossils of the enigmatic *Fractofusus* lying on the strata surface at Mistaken Point, Newfoundland. Coin for scale.

14. Strata at Mistaken Point dip gently to expose a series of ancient sea floors (the notice has partly blown away in the customary winds).

15. (LEFT) Fossils from the late Precambrian, Ediacara, Australia, *Charniodiscus* to show frond and holdfast (half life size).

16. (ABOVE) The curious animal *Spriggina*, whose affinities have caused much debate in the palaeontological community.

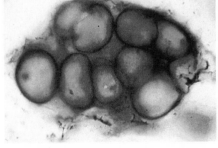

17. (ABOVE) A vision of the early earth: stromatolites growing at Shark Bay, Western Australia. Each mound is about the size of a large cushion.

18. (LEFT) Fossil stromatolites, showing columnar structure, with fine internal laminations caused by their slow growth layer upon layer. Precambrian ("Sinian"), Yangtze Gorge, China.

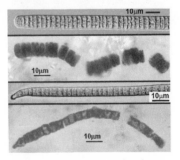

19. "Blue greens": two examples of *Cephalophytarion* fossil threads from the Bitter Springs Chert (850 million years old) shown below living examples with their characteristic hue. The longest survivors—note their small size.

20. Minute colony of chroococcacean cyanobacteria from the Bitter Springs Chert, central Australia (Neoproterozoic, 850 million years old).

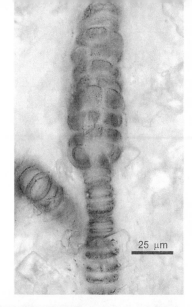

21. (LEFT, TOP) This *Bangiomorpha* fossil is a red alga closely similar to the living *Bangia* (see pp. 84–85), and is described from 1.3-billion-year-old rocks in Canada.

22. (ABOVE) *Porphyra umbilicalis*, a red alga growing on rocks in Devon, western England. "Seaweeds" are survivors from the Precambrian.

23. (LEFT, BOTTOM) Primary colours, ancient life: Crested Pool, Upper Geyser Basin, Yellowstone National Park. Early life tolerated high temperatures.

24. (RIGHT) Runoff from hot springs in Porcelain Basin, Yellowstone National Park. Different thermophile microbial species are fine-tuned to changes in temperature, painting the waters in different hues.

25. Calcareous springs at Mammoth Springs in Yellowstone National Park produce exquisite terraces of travertine, as here at the Minerva Terrace, where archaea and bacteria flourish.

26. (RIGHT) Iron Spring Creek, Porcelain Basin, Yellowstone National Park.

27. (INSET, RIGHT) Some of the minute threads of *Zygogonium*, a filamentous green alga growing in the streams, which are coloured purple by pigment. They grow well at 35–40°C, forming mats and streamers.

28. The Dragon's Mouth in Yellowstone National Park, where sulphur-loving microorganisms thrive in near boiling conditions.

29. (INSET, ABOVE) The minute archaea *Sulfolobus* that is characteristic of this habitat. It thrives where most life would die immediately.

30. Brothers black smoker, a haven for ancient sulphur-loving organisms, in deep water north of New Zealand on the Kermadec Arc.

31. Digging up the past. Paul Shin and his assistant prospect on the shore in the New Territories, Hong Kong, in search of survivors, especially *Lingula*.

32. *Lingula anatina*, a brachiopod survivor, freshly dug from the sediment to show its long stalk, or pedicle.

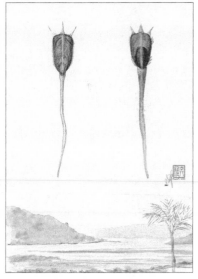

33. Sally Bunker's painting of the locality and living fossil. The bunches of hairs on the valve margin serve to separate inhalant from exhalant feeding currents.

34. (ABOVE) The clam *Solemya velesiana* from Stradbroke Island, Queensland. The pink areas discernible inside the shell are the bacteria-filled lungs of this survivor. Twice life size.

35. (ABOVE) The cephalopod mollusc *Nautilus pompilius* living on the Queensland Plateau, showing eye and tentacles.

36. (ABOVE) Fossil Jurassic nautiloid *Cenoceras*, about 12 cm across, similar in size and shape to *Nautilus*, sectioned to show internal chambers and tube-like siphuncle in the middle.

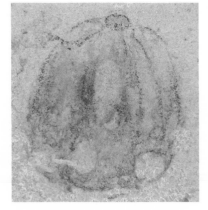

37. (LEFT) Fossil comb jelly *Maotianoascus octonarius* from the Cambrian Chengjiang fauna; this fossil is the size of a gooseberry, and even preserves ciliate edges.

38. (ABOVE) The recently recognized "living fossil" sponge *Vaceletia*, from the Osprey Reef in the Coral Sea (approximately life size).

39. (LEFT) Elegance and radial symmetry: living medusae of the ancient phylum Cnidaria.

40. Glass sponges have a 600-million-year history. *Caulophacus arcticus* is a mushroom-shaped species pictured at 2,400-metre depth off Svalbard.

41. *Huperzia*, a lycopod forming upright growths 10 cm high, here growing north of Oslo in Norway.

42. A tiny tetrad spore from the Ordovician of the Arabian Peninsula (diameter 30 microns). Spores of this kind are air dispersed and provide evidence of the earliest movement of plants onto land, probably in something resembling a liverwort.

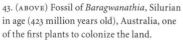

43. (ABOVE) Fossil of *Baragwanathia*, Silurian in age (423 million years old), Australia, one of the first plants to colonize the land.

44. (LEFT) Liverwort growing over a damp rock surface in a chalk stream in Oxfordshire, UK. Small photosynthesising pads like this probably were the first colonisers of the land.

45. (RIGHT) *Amborella trichopoda* in flower. This most primitive, living flowering plant is a native of New Caledonia, and has relatives from the early Cretaceous.

46. (BELOW) The horsetail *Equisetum* is ubiquitous, but it is a survivor from the Carboniferous. Here it grows in the New Forest, Hampshire, UK.

47. (BELOW) Fossil ginkgo leaf from the Eocene (58 million years old) has a fan-shaped leaf very similar to that of the living tree *Ginkgo biloba*.

48. (ABOVE) Dramatic twilight falls on one of the ancient surviving ginkgo trees in the Tianmushan of China.

49. (RIGHT) A stately *Araucaria* species growing in Brisbane, Queensland.

50. (INSET, RIGHT) The male cones of another monkey puzzle tree. Araucarias have a typical Gondwana distribution.

51. (ABOVE) Cycad showing fleshy covering of seeds, City Botanic Garden, Brisbane. These Mesozoic survivors can be grown around the world.

52. (BELOW) The extraordinary plant *Welwitschia* in its natural habitat in the Namibian desert. "Flowering" structures at centre.

53. Basal chordate, the lancelet, *Branchiostoma belcheri*. Arrangements of musculature, nerves and presence of a notochord all prove its relationship to higher vertebrates.

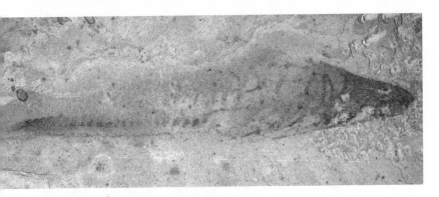

54. (ABOVE) Cambrian fossil more than 500 million years old, *Myllokunmingia fengjiaoa*, evidently related to the lancelet. Chengjiang fauna, Yunnan Province, China. The fossil proves that the history of our own phylum extends to the Cambrian.

55. (LEFT) The author holding a fine specimen of the Australian lungfish, *Neoceratodus*. Among the living fauna, this fish is probably closest to the evolutionary line that became terrestrial, leading to tetrapods.

56. *Latimeria chalumnae*, the living coelacanth fish, "Old four legs," painted by Gordon Howes. Perhaps it could be considered the type example of a "living fossil," because extinct species were known to science long before the living species was caught in the Comoro Islands.

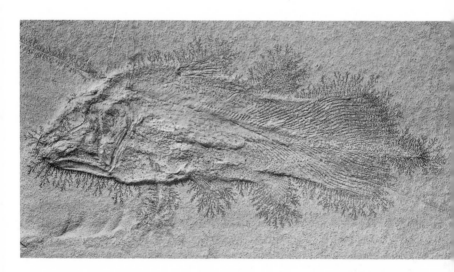

57. Fossil Jurassic coelacanth fish *Undina* is very similar to *Latimeria*. The feathery-looking surroundings are a mineral growth in the enclosing rock and have nothing to do with the original animal.

Recently, additional evidence for the velvet worm's place on the tree of life has come from the genome of the living species. Ancient fossils do not preserve DNA, which is a large and delicate molecule, readily fracturing into pieces. But by studying the molecules of living survivors from deep branches in the tree of life we are afforded a kind of telescope to see back in time. For the genetic code of DNA records another kind of history; it retains the accumulated narrative of all the changes at the fundamental molecular level that have built up slowly over time. Mutations that have been incorporated in the genome provide a kind of ancient fingerprint. But the code of life is famously huge—which means that the investigator may be obliged to seek out the particular piece of the genome that contains the information he needs. Although, as this is written, more and more organisms are having their entire DNA sequenced, this is still the prerogative of a privileged few—unsurprisingly, those like wheat or influenza that have a particular importance to *Homo sapiens.* For many organisms, it is more feasible to use a particular chunk of its genetic code to compare with the same chunk from a range of its potential relatives. This might be a particularly suitable gene or series of genes, for example, that do not change too rapidly to be useful through long periods of geological time. Obviously, the chosen gene has to be present in all the organisms under study. Other workers favour sequencing parts of the RNA molecule in the ribosomes that are present in the cells of all living organisms as the centres for protein synthesis. Comparing the similarity of gene sequences is one way of assessing how closely (or not) organisms are related to one another. The results can be drawn up as another kind of tree, with branches drawing together the closest related species, and deeper patterns of branching inferred from still more fundamental inherited similarities. This is not as easy as it might sound from this bald description, as various kinds of "noise" can obscure the signal the investigator seeks, and there are always genes that change too fast to retain meaningful signals from deep time. I need hardly add that computer programs have been designed to help out.

The technical problems are not part of our story, except in so far as they have produced different "trees" of relationships between organisms since the methods were first developed. Indeed, early attempts sometimes look quaint or improbable. But recent studies seem to have stabilised, and produce trees that appeal to prior knowledge and common sense, mostly by lumping together evidence from many different genes and finding the best fit. These then make a meaningful contribution to the summary trees of evolutionary history. The latest molecular analyses to treat the velvet worm and its relatives show interesting results. It places our chosen survivor as the bottom branch of a tree that includes all the arthropods above it—which must therefore have arrived later. Another name appears *between* the lobopods and the arthropods. This is Tardigrada (water bears), a group of tiny creatures that often live between sand grains and in other cryptic habitats. They are interesting in their own right, but they have but one known fossil, so they will not be described in detail here. Many tiny animals have no fossil record at all, but that does not mean that they did not exist in the past. The important point for us is that the molecular evidence supports the idea that lobopods are a branch even lower on the tree of life than arthropods. Those stumpy legs have walked on and on from a time even before the Cambrian. The very earliest Cambrian strata contain the traces of animals, but not their bodies. This is probably because those early animals lacked readily fossilisable hard parts, and the special conditions required to preserve the slightly younger Chengjiang fossils were not present at this particular time. No matter, for some of the tracks and trails that *are* preserved as fossils show clearly the traces made by arthropods of normal size digging their way into soft sediments with their numerous paired legs. It is even possible that these could have been tracks left behind by soft-bodied "proto" trilobites since they are similar to tracks made by the same animals higher in the geological column; at the moment we simply do not know. But we now do know that there must have been lobopods on that same sea floor, too, stomping ever onwards. More than that, they must have been present even earlier, before the first arthropods, because both the molecules and the anatomy of the animals tell us that they preceded the jointed-legged organisms. This takes us back into the mysterious world of the Ediacaran, a period whose remains lie above the Precambrian, and below the Cambrian, before the time of abundance and variety of marine life and before the appearance of shells.*

The story of the lobopods now disappears. There are no velvet worms or indeed any kind of lobopods in strata of Ediacaran age. There has been no shortage of attempts to find them. Geologists and palaeontologists have been cracking open likely rocks for decades now. The fact is that there are no trilobites, no early horseshoe crabs, nor any old familiar biological friends to be found in Ediacaran age strata. As in *The Hunting of the Snark* by Lewis Carroll, searchers vowed: "To seek it with thimbles, to seek it with care; / To pursue it with forks and hope," but to no avail. Even big hammers did not work. Instead a whole series of fossil animals have been recovered which have proved as enigmatic as they are exciting: not snarks but boojums. They are not small—some of them are bigger than a dinner plate—and neither are they uncommon if the searcher goes to the right place. The Ediacaran Period takes its name from the Ediacara Hills in the Flinders Ranges in South Australia, where a diverse selection of these remarkable early fossils was first collected. They appear as impressions on fine sandstones, many looking like strange leaves or fronds. Most of them show evidence of divisions or compartments dividing up the body, but they are not simple segments, because they are usually offset from one side of the animal to the other. Similar fossils are now known from more than thirty localities all over the world: from Arctic Russia, Canada, America, Newfoundland and Great Britain. Everyone agrees that these fossils lacked skeletons, but otherwise the experts disagree on almost everything else. Most of them would now concur that the Ediacaran animals were not obvious ancestors of the animals we know from the Cambrian onwards; they were genuinely inhabitants of a former world that did not survive. It seems only fitting that in a book about survivors I should also go to visit a world that failed to endure. The journey took me back

---

* The Ediacaran (635–542 million years ago) is the latest geological period to have been named, only added to the official list of geological periods in 2004. I should note that a small number of skeletal fossils are now known in strata earlier than the base of the Cambrian, such as the enigmatic little shell *Cloudina*.

to Newfoundland, where I had spent a year at Memorial University in St. John's when I was a young scientist. So I was travelling into my own past as well as towards a far, far deeper time.

Newfoundland is an island at the tip of eastern Canada and is itself something of a survivor. Built on the fortunes made from codfish on the Grand Banks, it has survived the great crash in the population of its most important crop. It is the textbook case for the effects of over-fishing. In the thirty years I have known the "rock" (as the natives call it), I have watched with bewilderment as fishermen have laid up their boats, and an apparently endless resource has all but disappeared. The codfish has not become extinct, of course, but the decline of this otherwise unfussy fish does prove that nothing in nature can be assumed to be unassailably fecund. High-tech factory ships from outside the island indiscriminately scooping up huge quantities of fish are mostly to blame. The Newfoundlanders, ever resourceful, have now taken to oil. The name of the Come-by-Chance refinery is somehow appropriate to their persistence in the face of setbacks not of their making. The little fishing villages along the coast are known as "outports," and ever since they have been required to eschew the cod, those young outport men who have not gone to Come-by-Chance have left to find work at Churchill Falls, the huge hydroelectric plant in northern Labrador, or even to become hands on the extraction of the Athabasca "tar sands" on the other side of Canada. They are a breezy bunch, despite their peripatetic life, and have an unusual accent: Irish with added stretched vowels, and wheezy interpolations of interjections like "Jeez, my son." The outports are all freshly painted these days, with wooden houses in cheery colours scattered up the hillsides. For the few who stay behind, there is nothing much to do except repaint the picket fences.

The drive south along the Avalon Peninsula from the capital St. John's passes several sheltered coves tucked away inside a coastline of magnificent cliffs. The geology is laid bare all along the rim of this island: the only problem is reaching it. Inland, the opposite is true; an endless forest of short conifers interspersed with scattered birch and aspen trees is interrupted only by shallow lakes called "ponds" hereabouts, which are a legacy of the last ice age; the bedrock is hard to see among the scrub. As we approach the end of the Peninsula, the trees get shorter and shorter, planed off by the fierce winds. Finally they crouch against the ground, as if terrified to poke up a twig. Usually the whole of this exposed area is swathed in fog, so the landscape supplies a passable setting for a vampire movie starring Vincent Price. But the day we visit it the weather is clear and sunny, with a few fluffy white clouds in a faultlessly blue sky. My companions are astonished; it was the best day they had seen in the last decade. The warden of the Reserve came from Wales, and remarked ruefully that he had chosen to work in the only place in the world with worse weather than Ffestiniog. One of the Newfoundlanders mumbles to me under his breath that the warden will be betrothed before Christmas. "Not a lot of single men around here," he says, with a wink.

At Mistaken Point, a path leads for a mile across a bleak coastal heath, which is less forbidding examined closely. Berry-bearing plants hidden in the close sward bear blue-black or scarlet fruits, and bright yellow tormentil flowers smile at us along the way. Patches of *Sphagnum* bog support pitcher plants whose leaves trap flies and mosquitoes to compensate for the poor nutrition offered by the damp wilderness. Even wild roses are tucked into natural hollows. As we approach the sea, grasses take over to make a natural lawn. Fulmars wheel in and out, just to have a look. The path leads onto the cliffs, which are quite comfortable to clamber over in this part of the Avalon Peninsula. The sedimentary rocks of which they are composed form a series of ledges that dip at a gentle angle into the sea, forming steps that we can climb up or down to explore different strata. The rocks are dark in colour, and the more resistant beds have made natural groynes that project out into the ocean. Waves break continuously over the ledges, throwing up foam—and this on a calm day. When winter storms are raging, salt spray must blast all the exposed surfaces. It is not hard to imagine how Mistaken Point got its name. The bones of fifty ships lie offshore, waiting to be fossilised.

Each of the flat surfaces exposed on the ledges is an ancient sea floor. In 1967, a graduate student geologist called S. B. Misra at Memorial University of Newfoundland discovered the most extraordinary organic remains preserved on these stretches of petrified sediment surfaces. Only a year later, an account of the finds was published in the most prestigious scientific journal *Nature,* jointly with Mike Anderson, also of Memorial University. The rocks were recognised as being late Precambrian in age (this was long before the Ediacaran had been named). There was palpable excitement in the scientific community at finding such large fossils in rocks of this great antiquity, although it was not known at the time just how old they were. Misra subsequently described the original conditions under which the sediments had been deposited. There were some special features about this discovery. First, the fossils could not be safely collected. They were impressions on the exposed surfaces of a very hard but brittle rock, shot through with cracks, and often located in the middle of a great uncompromising slab. The best way to study the remains was to pour a latex solution onto the surface of the rock, allow it to dry—even that might be a challenge with the Atlantic hard by and fog always lurking in damp banks—and then take the hardened cast off to somewhere nice and warm. For scientific description, it is usual to have an actual specimen on which to found a scientific name, and this should be kept in perpetuity in a public museum. This was obviously going to pose a problem, unless a public museum was constructed over the cliffs. Second, with such unusual material it is rather hard to know where to begin, since most of the usual biological pointers are absent. How does one describe an enigma, except as "enigmatic"? Perhaps it was a combination of these factors that stalled a full account of these remarkable fossils. Anderson took over the material when Misra went back to India, and when I met him in the late 1970s, he seemed to be crippled into inaction by these admittedly difficult problems. At the same time, he put his marker down upon the fossils so that nobody else could study them. The result was that most of the Mistaken Point fossils did not receive proper descriptions and the respectability of scientific names for several decades. Guy Narbonne and his colleagues from Queen's University, Ontario, are making good this omission even now. It is a strange fact about science that until an object or a phenomenon receives a name in some way it does not exist. Names really matter. They retrieve something from an endless chaos of anonymity into a world of lists, inventories and classification. The next stage is to understand their meaning.

A notice at the top of the cliffs points the way (a quarter of it had blown away in the last gale) accompanied by a pinned-up sheet of paper instructing visitors to "remove footwear before visiting fossil bearing surfaces." I confess that the idea of taking off one's boots in a howling squall to safeguard fossils that had survived since the Precambrian had its funny side. In the event we are provided with a pair of rather fetching blue over-socks. Visits to the famous fossils are now strictly supervised, as the site is now part of the Mistaken Point Ecological Reserve, and quite right too. Canadians are strict about protecting their national natural heritage. There is an architect-designed Visitor Centre to explain all to those who have made the trip. I climb down onto the best surface, in my special socks, and it takes a while to identify what to look for, but once they are pointed out the fossils are obvious. Any doubt that they were of organic origin was immediately banished from my mind. The fossils are strewn over the black surface of the gently dipping former sea floor almost as if laid out for the convenience of future inspections: one here, one there. The most conspicuous look like leaves or fronds, and are about the same size as a domestic *Aspidistra* leaf or some other showy tropical pot plant. They are pleated within, and the closer one looks the more subdivisions inside the "leaf" one begins to see. Such spindle-shaped fossils are the commonest type. There are more than a thousand of them on display under the Newfoundland sky. They were named *Fractofusus misrai* in 2007, four decades on from their original discovery, thereby commemorating the discoverer in perpetuity in the species name.

The name *Fractofusus* is quite descriptive—the "fusus" part refers to the fusiform (spindle-like) shape of the whole organism, and the "Fracto" part to the fact that it appears to have a fractal structure. Fractals, those intriguing mathematical entities recognised by Dr. Benoit Mandelbrot in 1980, are shapes that seem to repeat themselves precisely when the scale is focused down to a smaller level. So, the largest primary divisions within *Fractofusus* are subdivided into identical-looking smaller frondlets, and those in turn into identical-looking "sub-frondlets," and so on. It seems that these Precambrian organisms favoured this kind of structure; indeed, Martin Brasier of Oxford University has shown rather ingeniously that several of the organisms at Mistaken Point can be understood as a kind of three-dimensional origami played out by folding such fractal objects in different ways. But there are also some frond-like organisms that seem to be attached to the former sea floor by a kind of disc-shaped holdfast. *Charniodiscus masoni* was perhaps the earliest Ediacaran species to be recognised—from Charnwood Forest in Leicestershire in England, as the generic name should make clear (like *Misrai,* the species name is after its discoverer). The same "frond" is known from a very large number of Ediacaran localities, including several in the Ediacara Hills themselves, so it is almost totemic for this early and vanished marine world. The disc is thought to have held the organism in place while the frondose part was maintained aloft in the water current. There are several additional forms from Newfoundland that have their counterparts in Leicestershire, but since the latest reconstructions of the later Precambrian world place these areas quite close together geographically, this is not as surprising as it may seem at first. Some other oddities are pointed out to me; one is a kind of plate with tumid blobs arranged all over it. It was called informally "the pizza." The name reminded me that in my excitement I had not yet eaten lunch, so there I sat on an Ediacaran sea floor eating a cheese sandwich, looking out to sea on a perfect day while fulmars wheeled past on a light breeze. For a palaeontologist, it doesn't get much better than this. I realise that whatever we eventually make of these strange fractal beings, it cannot be doubted that there was a lot of conspicuous life in the later Precambrian, but apparently no relatives of velvet worms. These special fossils position a time line in our story; they offer a calibration for evolutionary invention.

I wonder what lucky circumstances account for the preservation of the fossils. After all, they are soft bodied. They could have vanished leaving no trace. My guides tell me that the area now so often coolly fog-bound was volcanically active in those distant days. Periodic ash falls cascaded into the sea and rapidly killed off and buried the Ediacaran fauna. They point out the *Charniodiscus* bending over in a common direction, flattened by the incoming volcanic Armageddon. I should have noticed this before. Each fossil-bearing sea floor is the record of one tragic moment for the Ediacaran animals, though it is no less than a miracle for us intelligent primates. Volcanic rocks have another property in addition to their role as natural undertakers; they yield minerals that can be used to obtain a radiometric age for the eruption. They both write the obituary and record the date. A time label of 565 million years ago has been obtained recently from an ash layer immediately above one of the best fossil-bearing beds. This is more accurate than can be achieved with many younger deposits, because datable volcanic rocks are not commonly interleaved with fossil-rich sedimentary rocks. Given that the best date for the base of the Cambrian Period is 542 million years ago, the Newfoundland rocks are only 23 million years older. I use the word "only" advisedly; although this might seem like a long time, it is a short span in the history of the horseshoe crab or velvet worm. Even if we went back 23 million years from the present day, we would readily recognise a world of mammals, birds, butterflies and flowers; and our own distant ancestors were already in the trees. But the world of Mistaken Point seems to have nothing to do with the marine world familiar from Cambrian strata, with its arthropods like trilobites, together with molluscs, brachiopods and echinoderms, ancestors of today's sea urchins and feather stars, not to forget the distant relatives of velvet worms.

It is no wonder that an attempt to understand the Ediacaran world has attracted the attention of researchers around the globe. Some facts have become quite well established, but there remain many disputes, which is hardly surprising when considering scientific forays into such mysterious and ancient environments. In fact, the stuff of science *is* disagreement. If there were no disputes, there would be no incentive to drive scientists out (without shoes) onto exposed Atlantic shores in order to crouch over cold wet rocks for hours on end. They want to get one step ahead in the race for the truth. However, most specialists do concur that the Ediacaran sea floor was very different from the seabed on the continental shelves today. The surface was coherent, even rubbery, due to a thin-skin veneer composed of bacterial mats. Sediments were almost cling-film wrapped, and holdfasts probably got a good purchase on this kind of surface. There is also a less universal consensus that the reason for this skin-like surface was that a range of burrowing organisms had not yet appeared to churn up the sediment. The sea floor nowadays is often a mass of so-called infaunal animals that live in the silt of the seabed and have a vital role to play in the food chain. Think of the huge flocks of waders that strut around on muddy estuaries when the tide is low, pecking down into the mud—not every dunlin has to rely on horseshoe crab eggs. Little churners and burrowers, especially marine polychaete worms, oxygenate the lower layers of the sediment as they work away. In the absence of such activity, an anaerobic layer soon develops beneath the surface, which can be recognised by the preservation of fine, horizontal layers when the sediments eventually harden into rock. Many Precambrian strata do indeed look like this—though by no means all. Sometimes the more fine-grained sedimentary surfaces betray a wrinkly skin, which is finely puckered, almost like the skin of an elephant, enabling us to visualise the gummy bacterial surface, although the minute organisms that made it are not preserved. These curious sea conditions have been ingeniously invoked to explain the preservation of many Ediacaran soft-bodied fossils. After a sudden overwhelming event—it could be a sudden slurry of sediment or a volcanic ash fall—the organisms are entombed, and a new mat then quickly grows on top of the grave sealing the dead animals in the sediment. Then the reducing conditions that inevitably ensue in the absence of wormy disturbance help to mobilise iron in the sediment in a form that migrates to make a kind of "death mask" around the potential fossils before they have decayed away. The endurance of so many soft-bodied organisms certainly implies a lack of those scavengers that make short work of dead bodies in today's oceans. As for the texture of the Ediacaran organisms, they may have lacked shells, but they seem to have been membranous, possibly even quite tough. Some scientists believe that they were divided into chambers rather like an old-fashioned quilted eiderdown. Their apparently fractal structure is probably a reflection of a particular style of growing, whereby the same set of rules are repeated over and over. It may just be a simple way of growing big. However one looks at them, these organisms do seem irredeemably strange.

My visit to Mistaken Point convinced me that it was possible for whole groups of organisms to disappear from the biosphere. There are some scientists who claim that the organisms preserved there—they have been called Vendobionta, among other things—are a kingdom (like Animalia) that has become extinct; a kingdom of "quilted" animals that many of the same scientists also think may have harboured bacteria in their body compartments in some kind of symbiosis. The somewhat younger fossils from the Ediacara Hills in Australia also include a variety of "quilted" organisms, but some of these seem to show a clear front end—a head. One of these, a creature called *Spriggina*, has been quoted as a kind of soft-bodied trilobite precursor. The more I look at *Spriggina*, the more I doubt it. The numerous "segments" seem to be out of step on either side of the animal, and the head end looks like a boomerang and not really like the forerunner of a head-shield. In fact, when you examine it impartially, it looks more like another apparently quilted and very un-trilobitic animal called *Dickinsonia*. But there is no question *Spriggina* is an intriguing animal, and I would love to be proved wrong. An Australian school of palaeontologists identifies soft-bodied ancestors of a few, living types of animals among a group of strange Ediacarans that are *not* quilted. An odd, radially symmetrical creature called *Arkarua* is claimed as an ancestral echinoderm, for example; a thing that looks something like a snowshoe called *Kimberella* has been claimed as a mollusc. Every one of these animals courts controversy. But at least some of these Australian Ediacaran animals, including *Kimberella*, are symmetrical about a line running along their midriff. This may not seem much, but it does show that below the Cambrian there were animals that could be placed in Bilateria—that is, animals with left and right sides that are mirror images (or bilaterally symmetrical). The common ancestor of arthropods, molluscs, annelid worms, and flatworms, not to mention the ancient relatives of velvet worms, would have been bilaterally symmetrical. We shall return to the interesting questions of the early days of animal evolution.

Vendobionts (or call them what you will) seem to have colonised all the seas of the world before the Cambrian Period. They were the first large organisms, and the younger and more advanced ones were certainly animals. Explaining exactly what they *were* has taxed the ingenuity of many clever people; but they have in all likelihood vanished from the world (the organisms, I mean, rather than the clever people). Some of the quilted animals that lived in shallow water may, possibly, have housed symbiotic algae or bacteria in their tissues, and basked in the sunshine, like prostrate reef corals. On the other hand, the Mistaken Point fauna appear to have lived in too deep and too turbid an environment for this to be a plausible option. It is perhaps not surprising that such strange creatures have inspired strange explanations. One worker even claimed that the vendobionts were not animals at all, but lichens, the living symbiotic collaboration between fungus and "alga"* that coats trees and rocks almost everywhere in the world. Lichens are the ultimate biological survivors in the simplest sense, because they seem to relish hardship and the tough life. However, none of them is adapted to life in the sea. The fact that some lichens have a flat and foliate form, as do the Precambrian "spindles," indicates no more than a broadly similar way of growing over flat surfaces. Life's history is as full of repetition as it is of endless inventiveness.

The waves surge and retreat from the stacked-up sea floors that once built Mistaken Point. This continually punished land will inevitably succumb to erosion, and the record of ancient life buried by chance so long ago beneath clouds of volcanic ash will be returned to the sea as a billion tiny particles. In the end, only the sea endures, it is the greatest survivor of them all. Even the continents mutate and remake themselves, driven by the internal engines of the earth powering slow but inexorable movements of tectonic plates. Mountain ranges are elevated and then reduced to rubble, but life can outlast mere Himalayas. *Peripatus'* relatives once walked upon Gondwana when Africa was united with Australia and the Americas. The memory of that vanished geography still lingers under rotting logs, or whispers through the leafy boughs of podocarp forests. Briefly, at least geologically speaking, all the continents were united together in the supercontinent called Pangaea (Greek: "all earth") some 270 million years ago. But that mighty entity, too, was just a phase, just one configuration

---

* In many lichens the fungal partners are more correctly referred to as dinoflagellates.

Pangaea, where the continents of the world were united as one "supercontinent" 270 million years ago. The southern mass (South America, Africa, India, Antarctica, Australia) is Gondwana.

of the earth's ever-changing physiognomy. For earlier still there was a time when continents were dispersed once more, making for a geography that looks still odder to our eyes. Science tries to reconstruct this former world map: it is like cutting a jigsaw puzzle into a set of new pieces, and then attempting to refit them into another picture altogether. By the Cambrian Period some 500 million years ago, these scattered continents were naked with their rocks unclothed by plants. The distant relatives of the velvet worm were there, though, living beneath the sea among a host of other creatures: some strange, some familiar. The lobopods were more diverse then than they have ever been since.* The branches of the tree of life were drawing closer to a relatively few common major limbs, but there was still a great variety of crawling, swimming, floating, burrowing creatures. There were

---

* Nor are they unique. In Chapter 5 I will visit two rather obscure groups of living marine "worms," Sipunculida and Priapulida—informally known as "peanut" and "penis" worms, respectively—which fossils have revealed were also more varied in the Cambrian than they are today. They, like the lobopods, have declined in variety since their early days.

livings to be earned: prey to hunt, hideaways to construct, plankton to be filtered, mates to be found. But then we must go back further, still further, into the Ediacaran. The surf at Mistaken Point washes over an even earlier, but alien world, a vanished world of soft-bodied, fractal things. There may have been no predation then, no burrowing, no grazing, no evidence of "nature red in tooth and claw." It was a different biosphere, and its mysteries still elude us. And the fossils of Mistaken Point prove that not everything survived.

The search for the velvet worm leads to unsuspected places and puzzling worlds.

3

Slimy Mounds

Shark Bay is a long way from anywhere. In Australia, distance soon acquires its own curious rules. Within the suburban strip that lines favoured parts of the coast, there are traffic jams and shopping malls like anywhere else, but away from civilisation the outback country stretches onwards forever. Far from the mountainous east, much of the country is flat. No doubt connoisseurs of the horizontal find infinite entertainment in its small variations, but for me a bemused puzzlement sets in after a few hours apparently rehearsing the same piece of landscape numerous times. Time begins to stretch in odd ways. After a snooze, I wake up unsure whether I have been asleep for ten minutes or two hours. Small eucalypts line wandering creeks while sand dunes are covered with scrub, occasional scruffy fences mark obscure ownership, and there are groves of taller gums or isolated she-oaks stocked with the noisy parrots known as galahs. Then the sequence repeats, but not necessarily in the same order. The landscape is utterly distinctive, like that of nowhere else in the world, with a stark beauty under a clear pale blue sky, but it is also relentlessly repetitive. Anyone foolish enough to leave the marked track will find it is easy to get lost. Bush stories are full of sticky ends and grieving widows. I know that maps do not really work in a landscape that repeats like an old tune whistled over and over.

Route 1, running up the west coast of Western Australia towards

Shark Bay, seems never to end. The Greyhound bus runs onwards through the dark, with nothing really distinguishing the passage of miles except sporadically a startled kangaroo picked out in the headlights. Occasional vehicles pass the other way, and each one seems something of a surprise. What can they be doing out here? I have to remind myself yet again that I am en route to see one of the holy relics in geology; it will be worth the effort. After countless hours, the Overlander Roadhouse welcomes me—a neon-lit marker set down in the endless landscape; a gas station, with a rudimentary restaurant, a place to loaf about until the next bus arrives. Aboriginal people wait there desultorily for relatives who have been off to Perth or somewhere to make a few dollars. Flies buzz about, with irritating persistence; there must be *something* else for them to do than endlessly return to drink from the same sweaty brow, or so one would think, but round and round they go. Backpackers loiter, waiting to embark on the next section of an adventure planned in theory, but now measured out in sweat and flies. It is a kind of end-of-the-world place, on nobody's list of "must-sees," but an essential stopping point before negotiating the wilderness. This is a place where timetables mean something to somebody, a place where I can get the next bus to see the stromatolites. Not far from the Overlander Roadhouse is a place that tells us of the transformation of the very air we breathe, a window opening into remote Precambrian times.

Though the outback may look pristine, in this part of Australia the wildlife has been transformed by human introductions. Feral goats have degraded the natural bush, and cats have culled the nocturnal mammals that were once numerous. The big-eared marsupial bilby, with its back legs like a miniature kangaroo and improbably long tail, is such a charming animal that it has become a kind of mascot for the conservation movement hereabouts. It would indeed be tragic if its only permanent memorial were in one of those perfectly photographed wildlife television programmes. Conservationists in Australia have taken to referring to the "Easter bilby" rather than the "Easter bunny" (bunnies being voracious introductions, too). It is already too late for many small marsupials in the eastern states of the country; their only record now being watercolour drawings made by the early naturalists. These harmless creatures could not outwit intelligent feline and canine hunters, and they failed to survive. Australia is full of poignant paradoxes. This land has many ancient biological survivors, yet it is also, much like New Zealand, a place where the extinction of species is still in progress. This is despite the efforts of a generation of Australians many of whom treasure their unique fauna and flora. Almost every town boasts dedicated people concerned with "bush regeneration," and in Western Australia new species of beautiful indigenous plants are still being discovered regularly, even around Perth. It is a very biodiverse region, despite the challenges of the climate, and not yet fully known. While I was there, Tropical Cyclone Hubert turned the sky black, and sections of the main road were closed. The species that live in this tough land must be natural survivors to be able to negotiate fluctuations between flood and swelter. That description, of course, also includes feral cats.

Shark Bay is a huge and ragged bite into the profile of the west coast of Australia. It has now become a World Heritage Area, which brings more money and more tourism. Much of the former, and nearly all of the latter, is directed to the beach resort of Monkey Mia, where "swimming with dolphins" is on offer. When I flew over the Bay and its clear waters in a light plane, I saw an undulating submarine prairie of sea grass, dark emerald green, broken into banks like meadows. A tenth of the world's dugongs—250 kilograms of peaceable herbivorous sea mammal—graze in leisurely fashion upon this luscious expanse, many living to seventy years or more. Juicy fishes doubtless account for the name of the Bay, since they attract fourteen species of sharks, including species like the tiger shark that command respect. From the air, I saw how Shark Bay is divided into two large lobes by a median peninsula; the aboriginal name for the Bay is "cartharrgudu" ("two bays"). The top of the peninsula is now the Francois Peron National Park, and a serious attempt is being made to clear this sandy area of feral goats and predators for the benefit of the native fauna and flora. Dirk Hartog Island provides an outer barrier to the Bay, which protects the coast from storms cutting across the Indian Ocean. I never knew before visiting Western Australia that this island was the first landfall for any European. The Dutchman Dirk Hartog landed here on 25 October 1616, beating Captain Cook to it by 152 years, and leaving a pewter plate nailed to a post as evidence. That plate is still preserved in the Rijksmuseum, Amsterdam. William Dampier, "the buccaneer explorer," spent a week there in 1699 and gave the Bay the name we use today. The aboriginal fishermen were plying their trade at the time, but there is little evidence of them now. I conclude that it is not only small and shy marsupials that failed to survive.

My quest is for something altogether more *recherché* than shark or dugong. At the tip of the eastern bay, the edge of the sea provides a prospect of life 2 billion years ago . . . I am travelling incredibly far back in time. The journey to the old telegraph station at Hamelin Pool takes me through undulating, intensely green scrub interspersed with a few mallee trees, interrupted only occasionally by flat-bottomed depressions carrying scrappy salt-scrub and patches of white gypsum—the aboriginal inhabitants called these clay-pans *birridas.* I missed the flowering season, and now all the bushes seem to bear black nuts. Next come low dunes made up of startlingly white tiny shells. I crunch my way across the dunes, and beyond lies a very shallow arm of the sea. This is where the stromatolites grow. I am approaching the famous site, where living analogues still flourish of the most ancient organic structures on earth. They ought to have disappeared long before the first velvet worm or horseshoe crab, but here they linger on, a marvel of anachronism.

Back at the highway must be the only road-sign in the world that points to "STROMATOLITES," and no geologist or palaeontologist could fail to follow its bidding. Here they are growing by the shore while the sea beyond shines an almost improbable ultramarine. Is this luminous vision the time warp I sought? Some part of me expects the stromatolites to be green, but they prove to be darkish umber brown. I confess I am momentarily disappointed. They comprise flat-topped cushions and low pillars, or even giant mushrooms expanding upwards like plush stools, with sandy gullies between them. They are regularly disposed along a seaboard more than a hundred yards wide; seawards they disappear beneath the barely lapping waters. It is a scene of perfect calm. A little walkway has been built over the strand so that visitors can get close without damaging the organic structures. I touch one of the hummocks. It is actually quite hard (why did I expect it to be soft?) and slimy or tacky to the touch when moist, but almost crispy when dry. In the bright Australian sun it is even a little warm. Now that I get closer I can see other kinds of surfaces along the shore, particularly sloping stretches of dimpled microbial mats, a fruity brown colour, running down to the glittering sea. They make stretches of the shoreline resemble wrinkled skin. Stromatolites growing at the water's edge look less like cushions and more like knobbly cauliflower heads. The inevitable flies are buzzing about my head, and some antipodean swallows chirrup cheerfully about the platform. I hope they are after my flies.

Up on the shore are some dead stromatolites, left behind by the sea maybe a thousand years ago. By now they have decayed into iron-stained ruins, but where they have broken open they show the internal structure of the cushion-shaped columns. It is clear to me that the columns are layered internally parallel to their top surfaces, rather like filo pastry. They seem to be built up layer by layer—a little like those giant stack pancakes an unwary visitor gets offered in New York for breakfast. The columns were evidently living things, self-made towers. A little museum on the site of the old telegraph office nearby provides more explanation. I peer closely at a stromatolite kept in a glass tank; its enveloping seawater must be refreshed every month. I see that when water covers the column its surface is slightly fuzzy—no doubt, it is still *alive*. A lack of crisp definition is somehow a proof of metabolism in action, life blurring the edges. Little bubbles fizz upwards off the top in a steady stream, none bigger than a lentil: they are bubbles of oxygen. So the column is evidently more than a brownish crust; it is something altogether more potent and dynamic, and it is breathing out oxygen, the element that babies and bilbies and bunnies all need to stay alive. Everyone has had nightmares about suffocation, when fighting for breath becomes fighting for life, so we all know in our bones how quickly we would perish without oxygen. The exhibition reminds me of the demonstration of nature in action at my very first school, when us kids looked at water-weed in a full glass beaker, and saw the same little bubbles of oxygen rising to the surface. This was the first time I heard the word "photosynthesis."

The survival of the stromatolites on the beach is another measure of their toughness. On the foreshore I see two broad grooves carving their way through the stromatolite grove. These are the persistent traces of a former industry. In the late nineteenth and early twentieth centuries, camel trains brought bales of wool here to Flagpole Landing. These were then carted off the foreshore to lighters that sailed 190 kilometres to a boat waiting in deep anchorage off Dirk Hartog Island. Then the wool was transported to Fremantle and finally to the United Kingdom for manufacturing. We are fortunate that these activities did not destroy the mounds completely. But it is also a measure of the slow rate of biological activity hereabouts that the old tracks are still visible after a century.

Sea conditions in this part of Shark Bay are quite particular. The shallow seawater evaporates fast under the relentless sun. It is the basis of a salt industry at Useless Loop nearby. The very clear water has an elevated salinity and is very poor in nutrients. Hamelin Pool is backed up behind a sand bar known as Fauré Island, lying about forty-seven kilometres out to sea, so it is almost a lagoon. Only specially adapted or tolerant organisms can survive under these conditions. One of those animals is a little clam called *Fragum hamelini*, which, as the name implies, is special to this locality. It is so abundant that its snow-white shells, none bigger than a walnut, make up the dunes that line the Bay. After some decades the shells harden into a shelly rock—it would be an exaggeration to call it a limestone. An old quarry above the shore records the employment into which this curious white stone has been pressed. Cut into blocks the size of large bricks it made a serviceable, if hardly robust, building stone. Some of the older edifices made of it still stand. The stone was used to build the walls of the Pearler Inn in the town of Denham, eighty kilometres distant. This pub looks as if it were constructed from a mass of white peas. In order to survive the testing conditions in the Bay, *Fragum* has incorporated photosynthesising algae into its tissues: sunlight is the ultimate source of its food, just as it is for the stromatolites. But *Fragum* is an evolutionary newcomer, whereas the stromatolites are very, very ancient.

Stromatolites are mounds slowly built up by microscopic organisms, layer by layer. The mounds are not composed of a single organism: they are a whole ecology. The tacky or slimy skin that caps the stromatolites is the living part. This very thin layer is composed mostly of cyanobacteria, organisms that are often called "blue greens" (or, formerly and incorrectly, blue-green algae) on account of their characteristic hue beneath the microscope. This may well explain why I expected the stromatolites to be green. The conditions in Hamelin Pool suit their growth. There are many organisms in nature that like to graze on "blue greens." Think of those finely scalloped trails wandering over moist rocks by the seaside, made by the rasping action of the sea snail's feeding apparatus as it scrapes away the thin nutritious bacterial layer that paints the rocky surface. This is not inappropriately compared with grazing by herbivores on terrestrial environments. Like grass, the "blue greens" grow back, and the molluscs move on. But these micro-organisms never have the chance to build complex or elaborate structures like mounds or "stagshorns" because the constant assault of herbivores renders their best attempts at architecture futile. Everything is eaten back before it can grow too big. However, in the special, warm world of Hamelin Pool the grazers are kept at bay. No snails sully the sticky surfaces of the stromatolites; the fish there don't nibble away the "blue greens" for supper; in fact, nothing much ventures into the almost unnaturally limpid seas. Some authorities believe that the very low nutrient levels in the Pool are as important in growing stromatolites as the absence of grazers. Whatever the reason, the simple organisms have it all their own way for once. And when they do, they reconstruct the Precambrian world. This is how life was before marine animals chomped and scraped away ancient biological constructions that had covered much of the sub-aqueous environment since life began. In Shark Bay a prelapsarian age can be restored to view, a time before velvet worms or even vendobionts, or anything that crawled upon its belly in the mud. I have seen dozens of artist's reconstructions of ancient seascapes that owe a debt to the prospect at Hamelin Pool. So when I saw the living stromatolites, I was not unprepared for the experience. However, I recall seeing Picasso's *Guernica* for the first time; just because an image is familiar does not diminish the impact of the real thing.

Cyanobacteria are simple organisms that often make long, green and narrow threads with organic walls which can be as thin as a few thousandths of a millimetre, but which often occur in sufficient profusion to make green slime. Other species are tiny round cells that grow by fission—essentially splitting in two, to double up as identical twins. They are ubiquitous. When a glass of water is left in the light on a window ledge, cyanobacteria will usually appear as a green smudge. They have been wrongly called "blue-green algae" in old textbooks, but as we shall see the algae are altogether more complicated organisms. When raindrops wash over rocks in a desert, these tiny organisms will soon take advantage of the opportunity to grow, and the rocks will shortly glisten with microscopic life. In the sea, their numbers occasionally erupt into "blooms" of billions of cells that can poison fish, or even humans, if they eat the wrong kind of shellfish too soon after one of these events. In biological jargon the "blue greens" are described as prokaryotes. They are both the smallest and the simplest-looking cells—often no more than a sphere or a sausage—but there are hundreds of different species. They lack an organised nucleus surrounded by a membrane that is present in every cell in what are termed eukaryotes. Every organism that has been mentioned so far in this book (including the author) is a eukaryote, which is another way of saying that our narrative has now arrived back to a simpler way of organising a living entity. Prokaryotes came before eukaryotes in time, which also means that they are closer to the main trunk of the tree of life. So there was a world before eukaryotes where the cyanobacteria were state-of-the-art and where the prospect before us in the shallow waters in one corner of Shark Bay would have been typical of much of the world, rather than a special survivor. I should flag up at this point that this prokaryote–eukaryote division is itself an over-simplification, and this topic will be revisited in the next chapter.

Modern seaweeds are both plants and eukaryotes, to emphasise the point again, and do not build stromatolitic mounds. In Shark Bay, the majority of such "advanced" organisms are discouraged by the low levels of nutrients available there; hence they leave the dominant cyanobacteria to cooperate in making different kinds of mounds. In the typical stromatolite the mode of growth is cumulative. The living "skin" is a thin layer of growing threads matted and twined together. The technical term for it is a "biofilm." The cyanobacterial mats are positively attracted to light and grow upwards. Any blown dust and other fine sedimentary material becomes incorporated in the surface layer and maybe provides the modest nutrient required. The slimy surface layer of the bacterium encourages the precipitation of calcium carbonate from its dissolved state in seawater, thus making a thin "crust." A new living layer grows on top of the one beneath, and may be able to extend a tiny bit further laterally: this is why some of the stromatolite mounds are wider at the top than at the base. Naturally, the "blue greens" are only able to grow in the sunlight that gives them nourishment, and are quiescent at night. Some scientists at the University of California even claim to have recognised daily growth increments. The overall rate of growth is extraordinarily slow, however, and certainly less than 1 mm a year (and possibly as little as 0.3 mm). It has been stated that some of the Hamelin Pool structures could be a thousand years old; that is, they grow more slowly than the slowest-growing conifer on land. The life and death of the wool industry would register as no more than a hand-depth on the height of a stromatolite column. Time can be ticked out in microscopic laminations, and history reduced to a measuring stick made by timekeepers invisible to the naked eye.

Stromatolites vary in form according to where they are found on the shore. It is easy for me to see that ones at the edge of the sea are little more than pimply mats. At least to this unschooled observer, some of them superficially do not look very different from some of the mats that covered sediment surfaces in the Precambrian at Mistaken Point. They are made particularly by one of the spherical, or coccoidal types called *Entophysalis,* and the internal layering is not well developed. Further down the shore in Hamelin Pool, the stromatolites that I tentatively touched represent the dominant kind in the intertidal zone, with a typical columnar-cushion shape. This kind of column is constructed particularly by a filamentous cyanobacterium called *Schizothrix,* which under the microscope is an intense emerald-green colour. It has lots of apparent partitions that make the organism look something like an old-fashioned tube of circular cough sweets. These particular stromatolites are very well laminated internally, so that the mechanism of being built up layer by layer is particularly patent. It has been proved that the cushions "lean" a little to the north, each component filament attracted preferentially to the sun (but on such a minute scale) in this, the southern hemisphere; the god *Ra* evidently ruled in the prokaryotic shallows. Further out to sea again, to a depth of a little more than three metres, there live the lumpier, bumpier, lobed and somewhat rounded stromatolites that are a collaboration of many different microbes. These include cyanobacteria of the genera *Microcoleus* and *Phormidium*; the latter is another concatenation of delicately segmented threads, while the former comprises microscopic "ropes" made up of bundles of a kind of entwined green spaghetti. The different species collaborate to grow together, like a confederation of medieval guilds, with each tiny specialist contributing to the function of an integrated community. True algae—diatoms—may chip in as part of the community among the deeper water stromatolites, but this group of eukaryotes probably did not evolve until much later. Beneath the surface skin of the growing mound, bacteria of a different kind from cyanobacteria process waste products and can cope with low, or even no oxygen; they are like artisans that moved the dung from the streets of the medieval village and made it a trade. Life encouraged specialised habits and habitats from the first.

Stromatolites are the most ancient organic structures, and their recognition as fossils transformed the way we understood the endurance of life on earth and the evolution of its atmosphere. I admit that viewed with complete impartiality when it comes to visual impact, the Shark Bay mounds are not on a par with the Empire State Building or the pyramid of Cheops. But stromatolites are one of the wonders of the world. Rationalists are not permitted to have shrines, but if they were then Shark Bay, where stromatolites were discovered alive, might be high on the list. Although many more living stromatolites have since been discovered, those in Shark Bay have been most thoroughly studied. From their initial recognition in 1954, the fame of these living stromatolites spread, until by the late 1960s they were finding a place in textbooks. As so often happens in science, the discovery of these living mounds happened just when palaeontologists were making major finds of microscopic fossils in rocks of Precambrian age, opening up debates about the biological history of the earth. The strange creatures of the Ediacaran, like *Fractofusus* and *Charniodiscus,* took the record of life back tens of millions of years before the great burst of familiar fossils such as trilobites that appeared in the Cambrian, 542 million years ago. But there remained more than 3 billion years of the history of life on earth in the Precambrian still to account for. This was the era of the stromatolites.

It is necessary to have a digression on geological time at this point. The age of the earth had been established at close to 4.5 billion years by the time Shark Bay was becoming known to the scientific world. The precision of this figure was largely a consequence of refinements in dating techniques, using the slow radioactive decay of naturally-occurring uranium isotopes into other isotopes of lead: turning rocks into clocks, one could say. The samples collected from the moon by the Apollo Mission- were first unpacked on 25 July 1969. I recall the excitement of seeing a small black piece of the earth's barren satellite when samples from the collection made on the Sea of Tranquillity were distributed to major museums, including the Natural History Museum in London. Like the stromatolites, it was not so much the thing itself; it was what it implied that made it so special. After the moon rocks were dated using the best technology of the day, the question of the antiquity of the green planet to which the moon was partnered was finally laid to rest: 4.55 billion (plus or minus 0.05).

The geological time period before the Cambrian was simply known as Precambrian for more than a century—after all, that is what it undoubtedly was, "before the Cambrian." But when this time period was recognised as so vastly long, it became necessary to divide it into several named chunks to help us order events in the earth's history. The Archaean Era is that part of deep geological time that ends at 2,500 million years ago, or 2.5 billion years if you prefer. After this came the Proterozoic Era, which, in terms of strata, lies above it and extends to the base of the Cambrian Period 542 million years ago (the Cambrian is the first subdivision of the Palaeozoic Era).

The Ediacaran, the latest addition to the roll call of geological time, begins at 635 million years ago and is slotted into the top of the Proterozoic. The Proterozoic covers a very long period, and these days is usually divided into three, which used to be known as Lower, Middle and Upper, but are now formally known as Palaeo-, Meso- and Neoproterozoic, respectively. The Neoproterozoic begins arbitrarily at 1,000 million years ago, and the Mesoproterozoic at 1,600 million years ago (1.6 billion), so the Palaeoproterozoic occupies the time period 2.5–1.6 billion years ago. Names really do help us get a grasp on the immensity of geological time, though phrases like "Palaeoproterozoic digitate columnar stromatolites" do not exactly trip off the tongue. But it is as well to get our labelling sorted out.

This modern classification is the end point of a long scientific battle. The intellectual classes had been debating the question of the age of the earth since the Comte de Buffon's estimate of 75,000 years in 1774 based upon the idea of the planet cooling down from a molten state. Apart from a purblind few who insisted (and indeed still insist) upon a biblical timescale based on totting up the generations mentioned in the Bible—Bishop Ussher's 4004 BC estimate for the Creation—the time available to "evolve" the earth increased fitfully throughout the early days of geology as a science. The bolder savants soon speculated in more and more millions. The longer time got, the more questions were raised about life's early days, because of the apparent absence of "organic remains" in Precambrian rocks. Charles Darwin famously fretted about it. Geologists of his time were beginning to explore large areas of the world comprised of Precambrian strata as the rocks of countries such as Canada were mapped for the first time. It was soon evident that sedimentary rocks, much like those found in younger geological formations, were widespread over these ancient lands. The seas were apparently barren in these ancient worlds; seas not so different in their physical properties from those that gave succour to the trilobites and snails that could be so easily collected from younger strata. In 1883 the American palaeontologist James Hall found some intensely layered Precambrian rocks apparently "growing" upwards incrementally from a narrower base to which he gave the name *Cryptozoon* ("hidden life"); however, their organic nature still remained controversial. Nonetheless, with the mere application of a scientific name, the biological virginity of the Precambrian had been breached. It was time for the stromatolites to be recognised as organic constructions.

The first time I saw fossil stromatolites in the Precambrian was as an undergraduate at the University of Cambridge, when I took part in an expedition to the Arctic island of Spitsbergen in 1967. My doctoral thesis was to be on rocks of Ordovician age exposed along the cold remote shore of Hinlopen Strait on the eastern side of a northerly peninsula called Ny Friesland. The small boat that took

us to our field area dodged between ice floes stained with the droppings of countless seabirds, for the Arctic summer is a brief period
of plenty for animals that live off the ocean. On land the scenery
was bleak: a succession of cobbled beaches raised above the present
sea level, across which the occasional polar bear or Arctic fox wandered in search of a feathered snack. It was not an inviting prospect,
although it was to be my home for several months. On the way to
reach my Ordovician rocks, and before passing the Cambrian strata
that lay beneath them, the boat had to chug past a great thickness of
even older Proterozoic limestone and shale, piled up layer on layer.
There are few places in the world where a young scientist can cruise
through such a momentous stretch of geological time, let alone along
an outcrop that has survived so unaltered by subsequent earth movements. These ancient rocks were in almost pristine condition. On
one occasion we landed to make a temporary camp and pick up fresh
water. I wandered over to the nearby rock outcrop just to have a look.
The rocks were not horizontal; instead they had been tipped gently
(but less so than the rocks at Mistaken Point). I could easily make
out the flat bedding planes breaking up the shore into a series of
steps that recorded a succession of former sea floors. The rocks were
a mixture. Most were very pale grey, sometimes almost pearly, and
hard looking. Others were yellowish, in patches somewhat sugary
and brightly tinted. The latter were dolomites, a calcium magnesium
carbonate rock that at the present day particularly forms in areas surrounding the more arid tropical regions. The off-white rocks were
limestones, that is, made of calcium carbonate in a finely crystalline
form. Looking closely, I could now observe that several of the limestone surfaces were finely scored. Many years of erosion had picked
out subtle differences in hardness within a single bed of limestone,
so that lines even a millimetre or so apart could be clearly discerned.
A comparison with layered pastry came to mind. I imagine that a
blind man could have read the rock surface like Braille, just by gentle
palpation. Where the rock surface was weathered at right angles to
the bedding surface, providing a natural cross section, these finely
layered rocks were arranged in a series of undulating columns, widening a little from their base. They were stromatolites, or to be more
accurate, sections through stromatolites—with a thousand years or

more of slow growth preserved in a fossil grave of fine limestone. *Cryptozoon* proved to be not so cryptic after all; it was the stony legacy of cyanobacteria. When I followed the stromatolites over onto the bedding plane beyond to get a vision of the ancient sea floor, they converted into balloons or pillows stretching away from me, each one showing the top of a stromatolite head. This was the fossilised version of the view at Shark Bay I was to admire decades later. It was bleached to white limestone by the passage of a thousand million years, perhaps, but it was still recognisable, a picture petrified from a former earth. This was the nearest thing I will ever experience to being in a time machine, even if my appreciation of it were countermanded at once by the shrill cries of Arctic terns above my head: chicks to feed, human business to be done, and the earth has moved on. Nothing remains exactly the same forever.

Had I looked more widely along that unwelcoming shore, I would have observed a greater variety of shapes carrying the telltale laminations of stromatolites. I would also have noticed some occasional shiny black patches within them; these are made of chert, a very hard, flinty rock composed of the mineral silica. Andrew Knoll, who is perhaps the doyen of Precambrian palaeontology, visited the same rocks in Spitsbergen a few years later, as he has described in his book *Life on a Young Planet*. From those cherts he recovered remarkable small fossils, which helped to make his reputation. He also recorded a whole range of different rock types produced by ancient microorganisms; the most general term for these rocks is "microbialites," which is a term that I trust does not require further explanation. Subtly mottled microbialites can present the appearance of ornamental marble, or the interior of a sponge cake, or the dimpled mats that I saw on Shark Bay; they can all be attributed to the work of bacteria and their relatives. Evidently, ancient microscopic communities did not just manufacture columns, though these do display several different shapes. Over much of the vast compass of Precambrian time, it was a dominantly microbial world, and there was nothing to prevent tiny organisms from constructing a variety of edifices.

Stromatolite fossils are not at all rare if the right rocks are explored. It is not surprising that many rocks have been altered by heat or pressure if they have sojourned on the earth for billions of years. The great motor of plate tectonics has been continuously in operation, building mountains and moving continents around. It is a lucky rock that escapes unscathed. Most of those that have successfully dodged being crunched or heated up are found around the edges of the most ancient and stable continental cores often known as "shields." These bits of earth's crust stabilised early on, and have been pushed around the earth during successive phases of continent building rather like counters being shoved around a draughts board. They survive to play another game. If there are patches of stromatolites preserved upon them, they are handed onwards. Perhaps the Canadian Shield is the best known of these ancient areas, but parts of southern Africa and Western Australia are equally familiar to geologists. However, the list of stromatolite occurrences is much longer than that of ancient shield areas, the rocks Andrew Knoll and I examined in Spitsbergen being a case in point. It is obvious that these strange organic structures were almost ubiquitous at least in the shallower parts of ancient oceans. Cyanobacteria would have needed light to grow, so the particular stromatolitic structures made by them must have been confined to comparatively shallow water. The early Precambrian ocean depths are unknown to us, since ocean floors are consumed in the inexorable slow dance of the plates. But it is more than likely that there, too, were structures made by different bacteria that flourished away from light. After all, life never misses a trick.

Stromatolites are found way back into the Archaean. The oldest ones of all are almost miraculous survivors found on the scraps remaining today of the most ancient continents. Fossils dated at 3.5 billion years old have been found in the Apex Chert in Western Australia,* and in Swaziland. It is hardly possible to imagine such antiquity. I have the same trouble trying to grasp the number of stars in the Milky Way, for the mind soon loses its normal frame of reference when the figures get so large. I can probably do no better than echo the words of the pioneer geologist John Playfair in 1788, when he became convinced of the reality of the vast age of the earth: "the

---

* I should point out that doubt has recently been cast on the authenticity of the Apex Chert fossils by Martin Brasier of Oxford University, but these questions have been vigorously countered by the original describer, Bill Schopf of the University of California in Los Angeles. More of this controversy in the next chapter.

mind seemed to grow giddy looking so far into the abyss of time."
Nonetheless, it is important to at least get a feeling for this "abyss,"
an intuition of its magnitude, because it shows just how long it has
taken for life to arrive where it is today. The two stories of life, and
the earth itself, have been intimately intertwined for billions of years.

Stromatolites began in the Archaean as relatively simple domes,
but later they evolved into a number of different forms. Some of the
more distinctive shapes have been dubbed with Latin names, just as if
they were organisms in the conventional sense (*Pilbaria perplexa* and
the like). As we have seen, they are actually collaborations between
several organisms, so such an approach does not fit in with normal
biological procedure. However, it is useful to have a way to refer to different shapes and forms, and some of the names have achieved wide
currency. In the far reaches of the Precambrian, stromatolites could
grow in a wider range of marine environments than they do now, and
this may partly account for some of their different shapes. In deeper
or calmer water, for example, it was possible for relatively delicate,
branching, even candelabra-like forms to grow. In complete contrast,
one of the most distinctive varieties produced massive cone-shaped
bodies that could grow to be tens of metres high. These microbial
behemoths have received the appropriate name *Conophyton* ("cone
plant"). They have been memorably described as making outcrops
in the field look like a series of rocket launchers placed side by side:
they must have taken many centuries, even millennia, to grow. The
vocabulary used to describe different kinds of stromatolites gives
some indication of their variety of form; they have been compared
with fingers, fists, cauliflowers, columns, spindles, trees, mushrooms,
kidneys. Given time enough, these most simple organisms could produce an art gallery's worth of shapes. Nature was patently a sculptor
from the first. The view from Hamelin Pool was, it now transpires,
only a partial glimpse of a richer, but now vanished, stromatolitic
world.

The controls on stromatolite growth were probably quite simple.
The growing surface film was attracted towards the sun, while the
supply of calcium carbonate from seawater dictated the dimensions
of the layers produced. A group of Australian physicists have developed computer models that "grow" stromatolites by playing around

with these simple elements. *Conophyton* emerges naturally as a shape in response to strong solar attraction; prokaryotic life, it seems, simply could not help building regular structures. Where there's life, there's architecture. But there is also good evidence that the variety and complexity of stromatolites increased during their extraordinarily long tenure of the earth's seas. The few kinds of simple domes and cones that dominated their first billion years, during the Archaean, were supplemented by dozens of additional shapes during the Proterozoic, when branched structures and pleated columns on many different scales appeared. The first occurrence of particular stromatolites has even been used broadly to subdivide this long period of time. They probably achieved their greatest variety about a thousand million years ago, but still long before the emergence of large animals, even the strange Ediacaran ones. Many early stromatolites were fully submarine, rather than living between the tides. Their living analogues have been found in the Bahamas near Exuma Island, hidden in marine channels. Here, these large, lumpy columns rising from a lime-mud sea floor probably provide a closer match to many Proterozoic environments than does Shark Bay. The biofilm forming the living skin is known to be a complex microbial community, and much more than just a photosynthesising surface. Several other kinds of bacteria have their homes there, some with the capacity to "fix" nitrogen, like the little nodules harbouring bacteria that grow on the roots of beans and contribute to soil fertility. These kinds of bacteria work at night, when the cyanobacteria are "sleeping." Once again, the mat is a whole ecology, a world measured in millimetres.

As for the fossils of the organisms that made the Precambrian mounds, the apparent absence of which so perplexed Charles Darwin and his contemporaries, well, they were lurking there all the time; it is just that they were very small. The cherts, like those tucked among the limestone rocks on Spitsbergen, held the secret. In some cases such siliceous rocks were formed early enough to petrify the fine threads and other cells making up the ancient biofilm. The process is somewhat analogous to that involved in making artificial resin souvenirs in which butterflies or scorpions are preserved, colour and all, which then lurk on the mantelpiece forever. Silica petrifactions were already well known from higher in the geological column, even preserving tree trunks down to the last cell. Considering that the dimensions of the Precambrian fossils are often measured in a few thousandths of a millimetre, the preservation of their cell walls is remarkable, almost miraculous. However, when very thin sections were made of the right Precambrian cherts they became transparent; these preparations were then examined under the microscope and revealed the unmistakeable imprint of life. The discovery was reported in detail in 1965 by the resplendently named American scientist Elso Sterrenberg Barghoorn Jr. based on fossils obtained from the Gunflint Chert, a rock formation exposed along the northern shores of Lake Superior. Barghoorn's co-author, Stanley Tyler, had previously recognised fossil stromatolites preserved in rather beautiful red jaspers (an iron-rich form of silica). At the edge of the Canadian Shield, the Gunflint Chert was one of those special survivors that had escaped the subsequent adventures of our mobile planet, fortuitously frozen in its own ancient time. At 1.9 billion years old, the fossils of the Gunflint Chert lie well down in the Palaeoproterozoic. Among the organic remains seen in thin sections of the chert, the commonest are probably thin threads not unlike those so abundant in living mats and biofilms. Some of these show the kind of transverse striping that are typical of some "blue greens"; interestingly, the threads are narrower than they were later in the Precambrian (and narrower still than they are today). They are accompanied by a range of other tiny organisms, some generalised rod-like bacteria, others more distinctive, like the spherical *Eosphaeria* with its cell walls apparently divided into compartments, and the enigmatic *Gunflintia*. Palaeontologists continue to argue about the biological identity of some of these fossils, although it is beyond doubt that "blue greens" were certainly present among them, but the important point is surely that this is an early community, already divided into different biological "trades." The kind of prokaryote collaboration happening today was already happening then. Stromatolites were indeed true survivors.

But back in the 1960s, the fossil search was on! The world was scoured for younger, older, similar or, in particular, new and unnamed Precambrian small fossils. Africa, especially Namibia and Swaziland, was mapped and investigated; Australia, especially Western Australia, was crawled over; the Old World was looked at again, and much of the New World was looked at with new eyes. Precambrian fossils turned out to be very widespread, and new discoveries were nearly always heralded by someone spotting stromatolites in the field, which hinted at what might yet be found at the microscopic scale. Geologists' boots tramped up *wadis* in deserts, their hammers whacked at Arctic cliffs, and their hand lenses focused on limestones outcropping deep in the Siberian *taiga*. These last devoted geologists bore the scars of marauding mosquitoes for weeks. Then by dissolving Precambrian shale in hydrofluoric acid still other microfossils with organic walls were extracted, to be studied in detail on microscope slides. University departments hired staff, and the growth in knowledge was exponential. Many of the famous names in early evolution were students of, or collaborated with, Elso Barghoorn. Andrew Knoll was among them. Bill Schopf, an equally grand figure at the University of California, Los Angeles, is now the elder statesman of the Barghoorn disciples, and did much to push the record of life and its fossils further back, into the Archaean.

Lynn Margulis may be the most luminous name of all those scientists associated with Barghoorn. She espoused and championed an idea that has transformed our way of understanding the history of life. There is a lot of hype in science nowadays, the more so since big claims often result in further research funding. I have never heard anybody announce "a minor discovery" or "a modest advance." I have also become allergic to the media's phrase "the textbooks will have to be rewritten," since it conjures up an inaccurate vision of textbooks being hurled with a curse into the waste-paper basket on a regular basis. Textbooks *are* rewritten, but most scientific discoveries are passed from one edition to another, since science generally works by piling bricks of knowledge one on another to make a solid edifice. It is very unusual to scrap a whole chapter and start again. However, this does happen on occasion, and one such occasion was when it was claimed that eukaryote cells originated by a kind of piracy. The vital organelles within eukaryotic cells—things like mitochondria and chloroplasts—were originally free-living prokaryotes. The more complex cell was a result of a hijack, whereby former free-living bacteria were summarily tucked away inside the swag bag of a bigger descendant cell. Unlike the human hijack, though, all parties benefited: the scientific term is symbiosis. The formerly "free" bacteria proliferated in their new habitat, sequestered away from harm. The newly enhanced cells took advantage of the novel vital functions tucked away inside them. For example, in plants the captured chloroplasts concentrated photosynthesis into special sites within the safety of a eukaryotic cell. Plants could now prosper using the energy of the sun. By contrast, mitochondria localise the "furnaces" providing the chemical energy for life, which is essential for organisms to feed and grow. Malfunctions in these organelles are usually lethal, so deeply are they embedded in the "works." Variegated varieties of garden plants can have white patches lacking chloroplasts, but such varieties are selected by gardeners for their appearance, not by nature for efficiency. Such an origin for complex cells is called the "endosymbiont theory"—"endo" meaning "inside" symbiosis. Complex cells arose by incorporation of simpler ones, for mutual benefit. It altered our understanding of life's early history to such an extent that not only had the textbooks to be rewritten, but there also had to be new books to replace the old ones in their entirety. The theory was triumphantly vindicated when the DNA of chloroplasts was investigated and found to be similar to that of free-living photosynthesising prokaryotes. The organelles tucked inside complex cells were, indeed, closely related to free-living, simple prokaryotes. This is the kind of confirmation that most of us can only dream about. It was almost as good as getting into a time machine and travelling back to the Precambrian. The phrase "endosymbiotic events" was soon incorporated into foundation biology classes. Like all science, the story has got much more complicated since the initial insight, as more and more acts of cellular piracy have been detected, but all the complications serve to reveal yet more events deep in geological time.

I should mention that none of these distinguished scientists conforms to the common preconceptions of the "geeky" white-coated specialist. Bill Schopf is a *bon viveur* and *raconteur,* with a very persuasive laugh, and a relentless drive for discovery. Andrew Knoll is the kind of American who makes most of us poor Europeans feel as if we were only given half a ration of energy at birth. He has projects spanning most of the world, and a stable of exceptional students to take the work forward. He manages everything with a kind of urbane

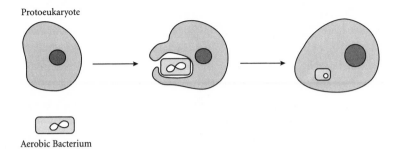

Protoeukaryote

Aerobic Bacterium

Endosymbiotic/endosymbiont theory: a complex cell
"in the making" engulfs a formerly free-living prokaryote, which
is then retained as a symbiont rather than being digested.
This process happened several times, thereby introducing
many more possibilities for life.

good humour and insouciance that defuses any possible resentment
at his omniscience. Lynn Margulis is unique. I know of no other professor who would, or indeed could, quote the poet Emily Dickinson
at length in a supermarket. As a long-term maverick, she is always
on the look out for ideas that will provoke and encourage new ideas
(and completely undeterred that some of them may well prove to be
wrong). Her dress style is equally distinctive, featuring embroidered
waistcoats and pleated skirts, as if she were about to take part in a
folk festival. She has an intuitive grasp of the important collaborations that makes the world work; not just the ubiquitous and versatile
bacteria, but also chemistry, and geology, and politics.

This account of the geological importance of stromatolites and their
discoverers has entailed a distraction from the little bubbles that provide the nub of this chapter. Recall the tiny gas beads rising from
the top of the stromatolite in its tank, arising from a living biofilm
breathing out oxygen. Now combine that scene with the picture I
have attempted to paint of the fledgling earth in the Archaean and
Proterozoic, where shallow seas and lagoons were covered with stromatolites, some of them gargantuan by recent standards. Imagine
thousands and thousands of dimpled miles, exhaling oxygen by day

in a thousand billion tiny bubbles, stimulated by the primeval sunshine. Even the slimy surface of the threads had a part to play in mitigating the effects of harmful ultra-violet radiation; we should all be grateful to slime. Now imagine this process continuing for *billions* of years, six times longer than the history of the velvet worm. The effect was to change the atmosphere, bubble by bubble. The early earth had little or no free oxygen; the "blue greens" changed the air itself. Animals need to breathe oxygen to power their life functions. They could not exist before the slow, relentless preparation of the atmosphere effected by lowlier organisms on the tree of life: no gills, no lungs, no blood blue or red. If some malevolent God were instantly to reverse the work of the stromatolites, we should all be gasping on the ground like beached trout within minutes. So we are, in a sense, the children of the sticky mounds.

The process of photosynthesis transformed the world. It is one of the most important components in the chemistry of life. Balancing oxygen release, the other part of the process is "carbon fixation," whereby this most versatile element, the basis of the molecules that build all life, is taken from the atmosphere and processed through the chloroplasts. The source of the carbon is carbon dioxide gas. I was brought up as a youngster on the naïve notion that photosynthesis used the sun's energy to cleave this molecule in twain, grabbing the carbon, as it were, and letting the oxygen go. The biochemistry is now known in considerable detail and is completely different: for a start, the oxygen is derived from water molecules. The discovery of the Calvin cycle earned its eponymous researcher the Nobel Prize in 1961. The minutiae of the biochemistry of photosynthesis are not our story here, but it is a fascinating tale, well told by Oliver Morton in *Eating the Sun.* An essential part of the process is mediated by an enzyme known as Rubisco, the most abundant enzyme on earth. Photosynthesis turns over about 10 to the power of 11 tonnes of carbon dioxide every year, and it has been doing so for a very long time. Rubisco has made it all possible. The carbon dioxide we humans are adding to the atmosphere right now, through our industrial and domestic use, is undoing the work of cells and sunshine over hundreds of millions of years, to what end we do not yet know, though it is unlikely to be good news. Biochemists tell us that, for all its importance in photosynthesis, Rubisco is a very slow enzyme, and not even a particularly efficient one. Apparently, oxygen competes with carbon dioxide for attention at its important "active site." This fact has led some to speculate that Rubisco may have originated in a world before free oxygen. If so, the enzyme itself is a survivor just like any of the organisms mentioned in this book. I should perhaps have made more of the fact that bits and pieces ranging from the molecular level to vestigial tails are also handed down through geological time. And as Richard Dawkins has repeatedly emphasised in *The Greatest Show on Earth,* not everything in nature and evolution is necessarily the best of all designs in the best of all best possible worlds. Life is a whole lot messier than that, and once an imperfect design is locked into history—well, there it stays, sustained by a little tinkering here and there. And so it may prove with Rubisco.

The early introduction of significant quantities of oxygen really did repaint the skin of the world. It altered much of the earth's surface chemistry. Iron, a common component in so many rocks, was then allowed to rust. Many forms of bacterial life that had been dominant in still earlier times found oxygen a deadly poison and retreated to redoubts where it posed no threat. When a dozen years ago I wrote my summary of the history of life, *Life: An Unauthorised Biography,* the growth in oxygen was portrayed as if it were a regular bubble-by-bubble increase through the Precambrian. The appearance of free oxygen is often dubbed the "Great Oxygenation Event" (GOE) and usually dated at 2.4 billion years ago. Prior to that, the oxygen produced by biofilms was immediately removed by oxygenation of minerals, and the free gas could not accumulate. However, it was not a slow, regular increase because it stalled at 1.8 billion years. Oxygen was "on hold" for a very long time, even as much as a billion years thereafter. The period in question has been labelled the "boring billion," which is one of those alliterative phrases that lodges immediately in the memory since, of course, you wouldn't imagine that anything as big as a billion could possibly be dull. Why oxygenation should have slowed during this Mesoproterozoic Period has been much debated, and I can do no better than recruit the latest explanation published in 2009 by the Knoll consortium. They maintain that the culprits were another group of photosynthesising bacteria—including purple bacteria—that had a different kind of metabolism from the helpful "blue greens." They used sulphur as fuel in their cellular chemistry, and there was a plentiful supply of sulphur during the boring billion. Their reproduction and growth did not release oxygen, and required less energy than normal photosynthesis, so away from inshore habitats they prospered at the expense of the oxygenators. Positive feedback kept the same system turning over and over: the arrival of animals was delayed due to boredom.

The early evidence of eukaryotic plants is ambiguous. A curious and distinctive spiral organism the size of a smallish coin called *Grypania* has been found in several sites where shale has survived in good condition from deep time. The latest report in 2009 is from India, in rocks about 1.6 billion years old. Although it has been claimed as a colony of bacteria, a number of palaeontologists accept *Grypania* as some sort of alga. The oldest examples reported are from Michigan at around 2 billion years old. Although he accepts an early origin for eukaryotic cells,* Andrew Knoll is cautious about attributing much evolution within the eukaryotes until some time after about 1.5 billion years, when the variety of fossils increases.

It does seem that there is indeed a time lag between the first appearance of this major advance in life and its subsequent flowering in the fossil record. Knoll is keen on the idea that the availability of vital elements as nutrients might have been critical; he cites the element molybdenum as one example, for it is essential for advanced eukaryotic organisms to process nitrate. Environmental conditions in the mid-Proterozoic allegedly removed this particularly vital element from oceanic waters. Organisms that needed it could only prosper in coastal waters where it was available. Others might favour sex: Nicholas Butterfield reported a fossil he christened *Bangiomorpha pubescens* in rocks 1.25 billion years old from Arctic Canada, which is extremely similar to the living red alga *Bangia,* and the first sexually differentiated organism known. Outbreeding, so the story goes, leads to more possible genetic recombinations, and ups the ante for

---

* Some of this evidence is chemical. Eukaryotic organisms leave behind durable molecules called steranes when they decompose, and these have been found in some early Proterozoic rocks.

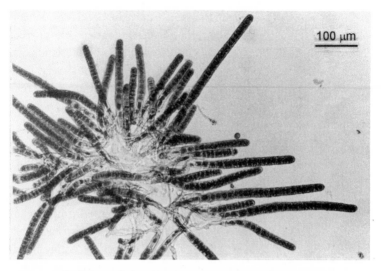

The living red alga *Bangia* is very similar to fossils more
than a billion years old (see image 21).

making a successful new species. Which is a somewhat ponderous
way of saying that sex spiced things up from the outset.

Clearly, I needed to visit a thriving colony of red algae to get a picture of the world in the later Proterozoic. Fortunately, this involved
no more than a short trip to the seaside—a jaunt, as my mother
would have described it. Sidmouth in western England is a little town
tucked into a cleft in the rolling Devon hills, and is flanked on either
side by high, strikingly bright red sandstone cliffs. The sea front dates
from the late eighteenth or early nineteenth century, and retains a
certain elegance, being backed by stuccoed terraces with wrought
iron balconies; a short promenade follows the pebbly beach. Inland,
country houses are tucked into the hillsides the better to enjoy the
view. Peak House on the western edge of town is a substantial grey
Georgian house set back among tall trees, and below it, on the Bay,
red sandstones run out to sea at low tide. This makes the perfect
place for rock pools, which are tucked into crevices in the outcrop.
Exposed surfaces are solid with mussels, which filter out tiny particles of food when the tide is in. This spot must have been where the
great nineteenth-century phycologist Robert Kaye Greville found the

red alga (seaweed) he named *Porphyra linearis*. Greville's pioneer-ing work of 1830 made known the wonderful variety of seaweeds around British shores, especially on the west where the islands face the Atlantic Ocean. He was also a tireless opponent of slavery. I can readily visualise his serious prodding among the fronds in the rock pools: taxonomic man categorising and classifying, and recognising one he hadn't seen before. Now, kids are doing the same thing. With little nets, rapt in concentration, they poke under stones and lift the floppy fronds. There is a natural delight about rock pooling, and I often speculate why so many adults seem to have lost that simple sense of wonder. The pools teem with life, and with water so clear they are like specially planted aquaria. Transparent shrimps twitch nervously under my gaze, then shoot away, and little fishes make themselves inconspicuous against the sandy bottom. A tiny brittle star wriggles away from me; its five flexible arms show that it must be an echinoderm—like our more familiar starfish—but it is another old timer. Fossil brittle stars not so different from this one are found in Ordovician rocks more than 450 million years old. Seaweeds of all colours and shapes speak of still older times: brown, red and green; feathery or leathery; and as elegant beneath the water as they are life-less above it. A coralline alga lines one of the pools, all bright pink and crispy to the touch. I am looking for *Porphyra*, and there is no shortage of it. Although it is a red alga, it is mostly coloured olive green in the summer. It hangs limply off exposed rocks, like dark, slick silk. But under water it makes graceful, leafy fronds crimped at the edges, which grow from a small basal attachment, looking like a carelessly folded lady's handkerchief. There is more than one spe-cies to identify: I have been tutored to recognise the big fronds of *P. umbilicalis* and the thinly purplish ones of *P. purpurea*; whatever their form, these seaweeds are just one cell thick. Longer, thinner fronds probably belong to *P. dioica*, a species named only recently from this locality by colleagues of mine from the Natural History Museum in London. It might seem improbable that a new species of food can be discovered on our own shores. But *Porphyra* is widely consumed around the world. It is known as laver in Wales and *nori* in Japan. I believe it must be the most ancient dietary item; it allows us to munch the Precambrian. It is good for us, too, providing iodine and iron. Laver captures so much iodine from seawater that on hot days the volatile element can escape; this stimulates the formation of water droplets, and has been recognised as a minor cause of sea mists. Despite its soft texture, *Porphyra* is tough, and survives being frozen. This may have helped its relatives survive through the periods in the late Proterozoic* when the seas of the earth are thought to have turned to ice. It is one of the toughest old timers.

From cyanobacteria to seaweeds, we know so much more about the early history of life on earth. It is wonderful to be able now to interrogate the fossil record in detail, when once the Precambrian geological record was regarded as devoid of life. How Charles Darwin would have relished the new discoveries. Another kind of interrogation is being carried out on the genome, which reveals history in a way almost as explicit as fossils. The history is hidden in the base-pair sequences of genes tucked away in the DNA of mitochondria and chloroplasts. Relative similarity in sequences of the same gene is likely to indicate common ancestry: the language of the genes reveals the past as well as directing the present. The language analogy is appropriate: philologists recognise the history of language by identifying common words shared between tongues, but those words don't remain exactly the same; they change through time and with the geographic separation of different peoples. Nonetheless, words can indicate a common source: to take one example, the word for bread ties a thread between European (*pain, pan*) and Indian languages (*nan*) that points to deep common ancestry, and helps to identify the Indo-European language group. Sequencing genes is more definitive. It is now an almost routine procedure, which makes it possible to trace the kind and number of those cellular kidnappings that put together the eukaryote cell. The sequences obtained from chloroplasts from one such event should resemble one another more closely than those derived from a separate kidnap. The "complication" I referred to

* The period 850–635 million years ago when such "Snowball Earths" are claimed to have happened on two occasions (or more, according to some scientists). This is often referred to as the Cryogenian.

previously is that this seems to have happened many times. Even the acquisition of the capacity to photosynthesise came to eukaryotes by repeated symbiotic events. One result of this understanding is to suggest that groups that used to be classified together on the basis of their general similarity probably had been on separate evolutionary trajectories since well down within the Precambrian. In the past all "seaweeds" (algae) might have been considered as one group, but now it is clear that the red algae (Rhodophyta) and green algae (Chlorophyta) are really separate evolutionary "packages" put together by an independent history of endosymbiosis. There are differences in the structures of the membranes surrounding their chloroplasts that reveal a comparable story. They really cannot be grouped together in a classification that reflects evolution, for all that they both include species that are leafy and floppy and live in salt water. Land plants, from modest moss to mighty mahogany, are in their turn related to chlorophytes, but not to rhodophytes: the division is that fundamental.

But whatever their evolutionary origins, all photosynthesising organisms share one thing: they make *food*. They provide nourishment for the lead swinging, Johnny-come-latelys of the evolutionary world, those willing to prosper on the gift of *Ra,* and the hard biochemical labours of others: I refer to animals. "Heterotroph" is a posher word for those after an almost-free lunch: the eater of others. Like photosynthesising organisms, animals started small and got bigger. Single-celled animals still abound everywhere, and their evolutionary story is proving to be very like that of early plants. The old "protozoa" into which they were once lumped is splitting into many different groups. Endosymbiosis also played a big part in their early history. Lynn Margulis could almost be forgiven her hyperbole in describing our world as *The Symbiotic Planet:* symbiosis really does happen everywhere low down in the tree of life, and in a good many places higher up, too.

This book is concerned with visiting fairly conspicuous organisms in the field, and the stories of very small animals must be passed over more briefly than they deserve. I must mention tiny fossils discovered by a Harvard graduate called Susannah Porter, in Neoproterozoic

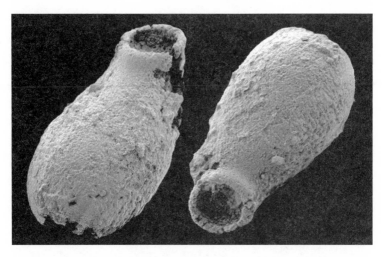

Vase-shaped microfossils of testate amoebae *Bonniea dacruchares* from the late Proterozoic Chuar Group, Grand Canyon.

rocks in the Grand Canyon. This famous gash through the earth's crust is where the magnitude of geological time can be appreciated viscerally as the visitor passes downwards, through deeper and deeper layers of sedimentary rocks. The ripples of fossilised waves, the advance and retreat of the sea, ancient drought and ancient plenty, all contribute to the cliff, bluff and butte as the descent continues. It seems almost incredible that tiny fossils, only about one-tenth of a millimetre long, could be recovered from sediments dated as early as 750 million years, deep in the canyon, but so it is. They are simple things, although distinctive, looking like minute, elongate gourds with a circular open end. They compare closely, I could even say exactly, with the coverings of testate amoebae—a group that could be described as the protoplasmic blob of popular imagination tucked into a tube. I was made aware of these particular single-celled organisms because a former Director of the Natural History Museum, Ron Hedley, was an authority on the group. As a result of their tiny size, the Museum obtained an excellent electron microscope long before many other institutions. Hedley told me that they were ubiquitous: we once shared a Welsh bed and breakfast place while I shot off to look for trilobites and he swabbed trees for testate amoebae. These tiny creatures are heterotrophs; they crawl about the world living off other microscopic cells. The fossils—and there were several species of them—proved that this activity was happening before the strata at Mistaken Point were laid down, and while stromatolites were still practically worldwide. Maybe they even put down a marker for the end of the boring billion. If the prokaryotic world could be portrayed as a peaceful paradise, though like most depictions of paradise a tad dull, then perhaps this was the beginning of the Fall.

Once something was feeding on something else, natural selection favoured anything that offered increased protection. This might be a chemical defence, perhaps, or possibly a toughened surface. Shells would eventually follow. Animal activity even on a humble single-celled scale would set up an arms race between eater and eaten, between grazer and crop, a race that has never stopped. Like sex, animal appetite has provided a stimulus for evolutionary innovation. To judge from documentaries made for television, eating and mating remain to this day the two most interesting activities of anything that wants to call itself an animal.

The most abundant survivors from early seas are probably the seaweeds: red, brown and green. I did not have to go anywhere particularly exotic to see red seaweeds; rhodophytes are common in seas around Britain and just about everywhere else where there are rocky shores. Just because something has an ancient lineage does not mean that a naturalist has to cross the outback to find it. Attached to submarine boulders or bedrock in rock pools, *Porphyra* looks like rich, red and half-translucent paper cut into strips, or folded into delicate fans. It swirls like a saucy Spanish skirt. Porphyry is a red igneous rock from which the ancient Egyptians made particularly durable sculpture, so the long persistence of this red alga in the sea seems appropriately acknowledged in the name (both names are derived from the same Greek root for "red"). I have mentioned that this seaweed is probably the only organism deriving from a branch so deep in the tree of life that provides food for humans on a regular basis. When we eat laver bread, we munch at a Precambrian trough. Red seaweeds have chloroplasts in their cells, and so contribute oxygen to the atmosphere. We must imagine that when waving fields of these eukaryotes were added to the vast dumpy groves of stromatolites, oxygen in the atmosphere must have increased concomitantly.* *Bangia* is supposedly the closest living relative of the 1.3-billion-year-old *Bangiomorpha,* which it closely resembles. *Bangia* is a thin and skimpy thing compared with *Porphyra,* and it hangs on rocks like limp brownish locks of hair at low tide, although it can look more like a bristling haircut when the water is upon it. Individual threads are made of files of piled-up single cells. Despite their apparent simplicity, *Porphyra* and *Bangia* have quite complex life histories. Cells of the "weed" contain only one package of genetic information; they are described as haploid. A second phase in the life history of these seaweeds is called the Conchocelis stage, which makes miniature branching plants, some varieties of which inhabit the borings they make inside seashells. These plants are so different from their "parents" that they were once given the separate generic name, *Conchocelis,* which is now only retained as a label for one stage in the life cycle. Conchocelis plants are the diploid, or the sexual, stage of the red algae. The leafy stage releases gametes that mate with one another, thereby doubling up the genetic content; this produces spores that can germinate into Conchocelis. This, in turn, releases haploid spores that germinate into the more familiar, and possibly edible, seaweed, thus completing the circle. Spores in general are tiny reproductive packages whose size is measured in thousandths of a millimetre but with all the DNA needed for re-growth. Such tiny parcels of genes are widespread among eukaryotes that emerged low down on the tree of life: spores can drift hither and yon in the sea until they find a compatible place to germinate. Much later, the wind would perform a similar distributive role on land. Small spore size was important, because it increased the likelihood of landing on a successful germination site.

Fungi are another Kingdom, as important in their way as plants

* Green algae (Chlorophyta) probably appeared at about the same time and they, too, would have contributed oxygen, although the impact on the atmosphere of both these groups of photosynthesisers was not immediately apparent. Both groups subsequently moved into brackish and fresh water; exactly when this happened is debatable, but it must have been well before the colonisation of land. Not surprisingly with such soft organisms, fossils are rarely encountered.

or animals, although they are much less familiar to the average naturalist. They, too, reproduce by means of spores. Fungi are ancient organisms, but with their small or insubstantial fruit bodies it is not surprising that they have a poor fossil record. They include mushrooms and moulds and a host of less familiar productions of nature, and most of them are, in a sense, heterotrophs like animals. Strictly speaking, as they absorb food rather than actively swallow it, they should be referred to as *saprobes*. Their role as recyclers of vital elements back into the environment is crucial to the health of the whole ecosystem. One mushroom produces millions of spores, each one of which could (potentially) develop into a fine, root-like mycelium that moves through the environment seeking nutritional satisfaction. Fungi break down wood and leaves, cashing in on the photosynthetic labours of others. Many fungi have entered into secondary symbioses with living plants: they coat the roots of orchids and most trees, providing nutrients to their hosts, and receiving sugar in return. As an amateur mycologist, I know the friends of fungi in the UK number a few hundreds; the Royal Society for the Protection of Birds has more than a million members. Yet there are thousands of species of fungi in the UK alone, many of them still poorly known. Potential nourishment that might remain locked away permanently in tough molecules like cellulose is released for the benefit of all life thanks to the vast chemical toolkits of the fungi. They unscrew complex carbon compounds; they re-jig the scaffolding of life. On occasion, however, the versatile mechanic will turn parasite, and devastation can result: I cite the potato blight that caused the Great Famine in Ireland in 1845, or the new varieties of rusts that are affecting wheat crops even as these words are written. Modern studies of fungal DNA have revealed what might have been guessed at all along: the whole group branched off the tree of life close to the *animal* line before taking off on an evolutionary trajectory of their own. They were once considered to be plants, possibly because mushrooms don't move around very much, even though no fungus possesses chlorophyll. If one could magically see the whole fungal system—with the mycelium questing voraciously through the soil and leaves, sometimes making a slowly-moving web many metres across—before the umbrella-like fruit body (which produces countless spores) thrusts upwards into the air, then its ancestry would be revealed in its true colours. And what colours! Fungi revel in pigments: unearthly purples, brilliant reds, luminescent blues—even green, though not the green of grass or trees. They even have sex of a sort, when mating occurs between mycelia of different individuals of the same species.

As I have mentioned, it is not surprising that fossils of such soft things are so rare. One tiny mushroom is preserved in amber, that most exquisite of tombs, but it is only a few tens of millions of years old. Fungal damage, spores and even mycelial threads have been preserved as fossils back to Carboniferous times, more than 300 million years ago, and there is really no question that many ancient trees were broken down and recycled in the same fashion as those in forests today. Dating from about 400 million years ago, a strange fossil column called *Prototaxites* that may grow to several metres high is attributed to the activities of fungi by some authorities, largely because the microscopic details of the tiny tubes make it resemble those of living fungal threads; I remain sceptical. Nonetheless, it is likely that fungi moved onto the land at the same time, or even before green plants, and they have assuredly passed through all the great extinction events. They may even have profited from the misfortune of others at these times of lethal drama, since a "fungal spike"—a massive increase in the numbers of fungal spores—has been identified immediately after both the Permian-Triassic and the Cretaceous-Tertiary mass extinction events; corpses and dead trees to feast upon, one assumes. It is grisly to think of billions of saprobes profiting from hundreds of decaying monsters: rot briefly ruled the world.

The origins of fungi do indeed root deeply back to the time of the stromatolites, when single-celled protists began to diverge into their several kinds. Like almost everything else, the fungi started small and under water. The most primitive living fungi are called chytrids, many species of which tick both these boxes, and a few even live in the sea. The earliest fossil chytrid is known from the Devonian (about 400 million years ago) and it has been assigned to a living genus, which must make it one of the supreme survivors. However, chytrids, or some close relative of them that has not survived, surely appeared in the Proterozoic. Claims about the timing of the first appearance of fungi have fluctuated wildly over the last decade, based on molecular evidence, ranging from nearly 2 billion to 750 million years ago. It is very likely that they were already around when those little testate amoebae now lying deep within the Grand Canyon were crawling around in their unspectacular way (that would correspond with the younger date). At one stage in their lives chytrids even look like tiny animals, since their single cells are propelled by a little whip-like flagellum, which can work for hours and propel the cell towards a reward such as food. This is also a structure shared with several kinds of single-celled animals. When the chytrid eventually forms a cyst full of unreleased spores, it begins to look more like a fungus, a resemblance that the DNA confirms. The important point to underscore is that at the Proterozoic stage of divergence of animals, plants, and fungi our everyday categories begin to blur. Life forms seem to interlock; piracy was everywhere. Genes shuffled sideways in a fashion we now find unfamiliar. But then we would: after all, we are Johnny-come-latelys.

Sunshine glints on the subtly crinkled surface of the aquamarine sea in Shark Bay. The unrelenting light seems to dazzle too fiercely. Mere humans cringe under the sun's merciless glare, but the slimy surfaces of the stromatolites can apparently cope with it; those dumpy, clustered pillars are tough in ways we struggle to understand. Conditions for living organisms would have been far more hostile in the Archaean, before the ozone layer existed, but with the help of slime and appropriate pigments, organic structures learned to survive under the onslaught of untamed light. Life subsequently manufactured its own shield, when oxygen derived from photosynthesis transformed into ozone high in the atmosphere; this made the ozone layer that now absorbs more than 90 per cent of the harmful end of the sun's radiation. The slow transformation of the ocean and air took more than 2 billion years. Life captured and tamed the sun's energy; a planet cannot be transformed except by slow and relentless labour on a vast scale. It is not human nature to acknowledge the work of others, and it is unlikely that many of our species will give due recognition to the contribution of numberless green threads thinner than hair from a baby's cranium. Now, as the visitor stands on the crunchy sand made by a million *Fragum* clams, the scene stretching away to the horizon speaks of a profound calm: uneventful, unchanging, a kingdom of laminae accreted millimetre by millimetre, century after century.

4

Life in Hot Water

In the middle of May in America's Grand Teton National Park, winter still has a grip on the landscape. Not only are the high peaks of this northerly segment of the Rocky Mountains deeply decked in fresh snow, but large patches of ice also cover much of Jackson Lake in its half-hidden valley. Where water shows through, fishermen and birds congregate. The wheeling form of a bald eagle is unmistakeable: this giant raptor still hangs on in the remote mountains, far away from human disturbance. Much of the forest in the valley is a huge swathe of lodgepole pine, a dark tree that stands vertically and in ranks with such a stiff bearing that one feels it might have been specially designed for practical employment. No doubt early settlers thought it had been placed there by a superior power for convenience in constructing cabins and fences. Near the creeks the twigs of the cottonwood trees are delicate and ghostly white, as if spun out of icing sugar. Close to the lake, large patches of swampy willow and osier scrub paint swathes of colour in the middle distance, with their naked twigs tinting patches vivid yellow or orange. Small songbirds hang around here, too, sensing the imminent return of spring. This is the favourite habitat of the moose, and although a sighting of the stately vegetarian eludes us, the prospect stimulates the usual discussion over whether the plural of "moose" is "meese." Native Americans used the whippy willow twigs to weave their baskets, just as the inhabitants do in marshy regions in Europe; similar needs exploit similar plants. On the open mountain flanks I was surprised to observe many small sagebrush bushes, a plant I had last seen in a baking hot desert in Nevada. These scraggy shrubs must be among the toughest organisms in the world; maybe the stromatolites should look to their laurels. Elk pick their way nonchalantly among them, fastidiously nipping off a shoot here and a leaf there. This is a journey to a steaming paradise for the most basal organisms in the tree of life; a place where sulphur can be food and hot acid can provide a home.

The John D. Rockefeller highway cuts along the steep-sided valley, running northwards along one side of Jackson Lake. Although many of the rocks hereabouts are as old as those Precambrian stromatolites, the valley itself is young—speaking geologically, of course. It is a product of movement along a great fault cutting through the earth's crust at the foot of the Teton Range. About 10 million years ago the mountains began to rise to create both the valley and the steep fault scarp delimiting the craggy mountains beyond; snow and ice did the rest. There has been more than 8,000 m (25,000 ft) of elevation along this line, doubtless accompanied by massive earthquakes. Like everything else in this part of the West, the present face of the landscape is the result of deep forces at work. The Pacific Plate has been nudging its way beneath the North American Plate since well before the extinction of the dinosaurs. The great buckling of rock strata that make up the Rockies and adjacent ranges and the active volcanoes and faults that occasionally burst into violent life with dramatic effect are all the result of this sticky collision. One of its least consequences is the course taken by the John D. Rockefeller highway northwards towards Yellowstone Lake. The lodgepole pine forests seem denser now, the country still more rugged. Even in mid-May the road is flanked by high snow banks that frequently obscure the view; rivulets of melt water are the only sign of the upcoming big thaw. Each of the winter snowfalls is marked on the banks by a kind of undulating stratification. My wife and I are determined to arrive at Yellowstone National Park at the moment the gates open to visitors, before the huge influx of tourists follows as the place warms up; just for a moment we want to recapture a sense of wilderness and discovery.

The Old Faithful Inn is itself another kind of living fossil, a vast wooden monument to the "Arts and Crafts" movement, or its American equivalent. The construction is built around a great hall, with an even greater stone chimney as its centrepiece, so there is a medieval feel to the place even though its several storeys reached by wooden stairways around the central atrium are made of split logs, and do not resemble anything I have seen in old Europe. It is an extraordinary building, and the architect, Robert Reamer, was clearly something of a visionary in 1903 in using local materials and making a building that grew out of and fitted into the landscape. I have seen similarly tasteful and thoughtful buildings designed by Mary Colter at the Grand Canyon, so it might be concluded that the concept of the American National Park attracted some prescient "joined-up thinking." All the balconies, struts and supports in the Old Faithful Inn are made from conifer wood in odd elbows and crucks; they resemble the broken bones of some giant animal, and no two are the same. They are called "knurls" around here. Somebody who was less well disposed towards the building than I am might say it was like an enormously inflated woodcutter's cottage. The guest rooms are more like those on an old wooden ship. Writing desks abound on the internal balconies, so for once I can make notes in comfort. The open-air balcony was built to

The Old Faithful Inn, built in a style known as National Park rustic, or "Parkitecture," Yellowstone National Park, Wyoming.

afford an unimpeded view of the Old Faithful Geyser, so I can sit with a gin and tonic and take in the phenomenon. Once a decidedly smart destination, the Inn has regrettably seen better days: having to drink that gin and tonic out of a plastic beaker, while huddled deeply into my overcoat, seemed to symbolise a certain decline. Great coal-black ravens strut about the grounds quothing "Nevermore" (at least they do according to Edgar Allan Poe), but they lie, because Old Faithful erupts regularly every ninety minutes or so. The crowds gather in anticipation, reminded to do so by an old clock. The selection of pre-liminary gurgles and gasps was not something for which I had been prepared. I was reminded of the kind of sneezing attack that threat-ens and recedes, racking up a little each time, until reaching a climax in a most prodigious "Ah-choo!" The great plume never disappoints. As it spouts and whooshes upwards, a breeze catches the droplets and somebody always gets wetter than they expected, but their squeals are masked by the geyser's stentorian exhalation.

We can't approach Old Faithful too closely, but a carefully labelled walkway weaves a trail through the rest of the geothermal field oppo-site the Inn past the other geysers and springs. In the frosty air of the morning the hot water steams into the atmosphere so that the whole field is enveloped in wisps or drifting white clouds of mist. A plume emerges from deep in the forest on the hillside, so another spring must be hidden somewhere there. There's a pervasive sulphurous whiff that is not as unpleasant as I had expected. Duckboards ensure that the delicate crust made by the wash of volcanic waters is not disturbed by human visitors. Buffalo are rather less punctilious, and their hoof prints are everywhere. The naming of these thermal phe-nomena is something of a random process. The Castle is obvious: it is an ancient geyser that has built up a turreted edifice of sinter. The Ear needs no further explanation, but the Lion group is less immediately obvious, unless it was named after the roaring sound it emits. Geysers differ from springs in their periodic eruptions, but their periods can vary from minutes to years. Geyser watchers are thrilled if they man-age to capture the climax of one of the rare ones. One after another, great bowls of unnaturally light-blue water line the track, each one surrounded by a siliceous rim; it is often possible to look down into their plumbing, a profound tube of darker blue from which bubbles emerge in clots or streams. It is hard to gauge exactly how far down the vents extend, as spookily clear water can be deceptive. Some of the pools look innocuous enough but they would boil you alive in a trice if you fell in. Others whistle and spit, and make no secret of the deep energy that drives them. Chinaman Spring is a hot pool that turned into a geyser. An enterprising Chinese laundryman once tried to cut his heating bills by steeping clothes in the spring. All was well until he tipped soap in to help the process along, which triggered the geyser and carried the washing sky high. Geophysical maps have revealed deep connections between springs, so that some will fill while others void, but in other cases even adjacent hydrothermal vents will have separate plumbing from profound depths. Where the boiling water spills over and evaporates, a greyish deposit of siliceous sinter is often left behind, which builds all manner of different shapes. Silica often makes bubbly, popcorn-like, rugose masses that may build into rubbly sprouts resembling some sort of stony broccoli, or it can develop variously pleated growths, or rounded ones like piled-up kidneys. Where a steady stream of water spills slowly from the rim of a pool the sinter forms a series of intricate terraces looking something like a scale model of rice paddies in Indonesia. It is hard to believe that nothing more than the interplay of flow and evaporation sculpts such elaborate designs. All the geysers and springs drain ultimately into the Firehole River, and there are even hot springs within the river bed itself that are betrayed by puffs of steam. Everything here runs hot and cold.

The blue colour of the water is simple physics: it is what remains when all the other wavelengths of light have been removed. But other colours in the thermal field betray the presence of life: it looks as if paint has been thrown into the pools. Where hot water dribbles over the edge of Punch Bowl Spring bright orange stains are flanked by vivid green ones: this is the lurid palette of bacteria. Life decorates with its metabolic pigments, and surrounds the heated outlets, often in the form of bacterial mats. Some of the latter might even be described as stromatolites. In spite of setting a bad example, I found it impossible to resist the temptation to touch one of the mats, and yes, it had the slightly sticky but leathery feel of the Shark Bay mounds. The ground between the hot springs and pools supports more mats and hardy mosses, and the drier parts, tufts of grass and yellow daisies. There is even a tough conifer or two. Small children are admonished not to destroy the fragile ecosystem by treading on it, but a poster showing the fate of one of their number falling into a scalding pool is probably a better discouragement: one small boy seemed more transfixed by the graphics than by the geysers. The ground around the springs does indeed support a particular habitat. Along the runoff channels a specially adapted species of ephydrid fly has larvae that tolerate the difficult conditions (it looks like any other dark fly to the uninitiated), and a closer look will reveal tiny mites and springtails. One species of spider has become adapted to hunt the flies. It makes short dashes onto the hot surface to grab its prey, and doesn't linger. Then an attractive plover known as the killdeer has learned to dip in and out of the wastes to pick up an arthropod snack. The plovers run delicately over the warm crusts. You could not have a better example of a small ecosystem; except for the bird, it might even be an anciently interdependent arrangement. Warnings about the "fragility" of the ecosystem cannot but provoke a wry smile. After all, many of the dominant organisms within and around the springs have probably survived for 3.5 billion years.

A small but colourful geyser encapsulates much of the history of life within its compass. Around its hot centre coloured stains are produced by the most primitive organisms on earth, minute archaea that spill out through the overflow like a lick of blood. Do I recognise the characteristic colour of the "blue greens" on the flat beyond? The age of the stromatolites evidently still leaves its signature. Modern classification separates the archaea as profoundly from the bacteria as the bacteria are separated from the eukaryotes; they are both deep survivors. Then I notice a group of mosses forming low clumps around the cool part of the spill: mosses were among the first plants that helped to clothe dry land long before the vertebrates had designs on terrestrial dominance. On higher ground a cluster of yellow flowers and a stunted tree might be a proxy for the changes introduced into the history of life in the last 100 million years. And one rogue human footprint could very well stand in for our species. So here is most of biological history in an area not much bigger than a backyard. This may not be a big spectacle, but it is what I came all this way to see.

The area around Old Faithful is only a small part of the active thermal system. Hot springs and geysers abound all around Yellowstone National Park. The lake in the centre of the Park is still largely frozen and startlingly white, but West Thumb thermal field makes a lobe at its edge which displays clear patches of open water related to hot sources on the lake bed, or to springs draining down from close on shore. Smart black-and-white diving ducks called scaup make obvious use of the clear patches, and otters employ them as piscatorial doorways during the hard winter when everything else is frozen solid. A well-shaped cone of sinter that has built itself above the lake is full of hot water, and invites the thought that this must be the only place in the world where a fisherman might land a catch and cook it without having to detach the fish from the rod.

The Norris Geyser Basin is in some ways the most impressive thermal site of them all, where a range of volcanic features are laid out around a distinct physical basin, which with its bare flats might be a setting for a science fiction movie, where devastation rules and everything smokes and sweats. The emergent water is exceedingly acid and hot in this basin. Any kind of micro-organism that thrives here has to be some kind of *extremophile*, a term applied to bacteria and archaea that live only under conditions that most life would find impossible; for microscopic living things one cell's poison is another cell's meat. Once again, duckboards allow visitors to stroll nonchalantly over treacherous crusts and bubbling vents that exude a sulphurous whiff everywhere into the atmosphere. It is easy to imagine the awe that the first sight of these hooting vents and malodorous pools must have inspired in their Victorian discoverers. The whole Basin is named after Philetus Norris, Superintendent of the National Park, which was formally declared in 1872, thereby opening up a new concept in our attitude to the natural world. Individual geysers and springs often carry appropriately descriptive labels: Whirligig and Steamboat geysers are easily visualised, and Porcelain Basin is delicately blue. Others require more explanation. The spiny siliceous sinter in Echinus Geyser evidently reminded early visitors of sea urchins. The walkways eventually conduct us away from the pools and terraces and back into the surrounding forests, where songbirds warble away apparently indifferent to the stench of primal fumes. The summer must be a season of plenty for animals and birds, for it allows a very short time to bring up a new generation. The south road into the Park only opens each year in early May and is closed from early November. Fumaroles and pools blast, whistle and bubble regardless of season or spectator. They are in thrall to a deeper control.

It has now been established that much of Yellowstone National Park is the remains of one vast caldera, the scar of the violent eruptions of a "supervolcano" of gigantic dimensions. The comparatively recent eruption of Mount St. Helens to the north in Washington State would have been as nothing compared with this ancient catastrophe. The main eruption 640,000 years ago would have cast a pall around the world. To those who know a little geology, the imprint of these events is everywhere in the Park. Yellow-weathering rhyolite is a dominant rock type on every bluff and road cut (hence "yellow stone"), piled up in layers. A close look at many of the outcrops reveals that the individual "flows" are composed of tiny, glassy fragments that have welded together from the heat they carried onwards from a series of pyroclastic flows and incandescent ashy clouds that would have devastated forests for thousands of square miles. It seems difficult to imagine that trees could re-establish after such a shock, but conifers are resilient. The Park still retains the scorched legacy of "Black Saturday" in 1988, when wildfires blazed out of control during a time of exceptional drought. Twenty years later young trees are starting to overtop the stark, black corpses of dead ones; fire even stimulates some kinds of seeds to germinate. It took a while for geologists to appreciate that the crater extended so far beyond the confines of Yellowstone Lake to embrace virtually *all* the thermal areas that dot the Park. Hot springs and geysers are a memory of that great blast. The deep plumbing that feeds sulphur and arsenic into the vents is locked into a profound geological foundation. Many volcanologists believe that even now a deep magma chamber may be regaining its strength for a future onslaught. The exceptionally cold winters of Yellowstone are one result of the great topographic bowl left behind by the explosion; cold air cannot escape from it. On the other hand, the constant supply of hot or boiling water supplying elements like sulphur helps the survival of communities of archaea and bacteria that root back to the oldest life on earth. Water percolates downwards from the surface through a myriad cracks to refresh the springs eventually becoming superheated at depth, gaining minerals and gases in solution and rushing back to the surface. Geysers often have a subterranean chamber that must fill to a critical level before an eruption can occur. The exit valves of mud volcanoes even sound like deep belches arising from the belly of the earth, exhaling methane, carbon dioxide and steam. The clays are kept soupy by the constant agitation; in spite of their unwholesome appearance, some volcanic "mud packs" are allegedly good for the skin.

Among prokaryote organisms that lack nuclei, one advantage of extremely small size is that reproduction by fission into two daughter cells can be remarkably rapid. It works on the simple doubling series: 1 cell, 2 cells, 4, 8 . . . 2,048 . . . 241,824 cells . . . and so on. It does not take long for a population to rise into billions. All that is required is a ready supply of a nutrient to encourage growth. If the organism in question is an extremophile that has adapted to a special habitat, then not many other species will be able to compete with it.

In the right kind of spring the result will be a churning broth containing countless living cells belonging to just a few kinds of archaea or bacteria. In Yellowstone, Sulphur Cauldron is right by the Canyon Village to Fishing Bridge road, so that passers-by look down on mud pools that are roiling and boiling like a sickly yellow soup—a broth made of split peas perhaps. An intense stench of sulphur poisons the air. It looks like the kind of environment where there should be no possibility of life; a hot, constantly stirred sulphurous pot. But this is the home of a famous kind of archaea called *Sulfolobus* that feeds on the poisonous gas hydrogen sulphide, and sulphur, which in turn oxidises into deadly sulphuric acid; in the process the organism derives energy to grow, prosper and divide. Billions of them thrive down there, growing in a temperature of 80°C. The brew finishes up as extremely acid with a ph of 2 or less—low enough to dissolve iron. Animals wouldn't survive in the broth for two seconds, but this is heaven for the right kind of archaea. These micro-organisms can therefore be described as acidophile thermophilic archaea. Does it get more extremophile than that? The eroding activity of the naturally produced acids is reported as reducing the country rock to mud, and a series of muddy pools on the other side of the road seems to support that idea. The margins of the pools slip like a bowl of porridge shoved by a clumsy giant as muddy, steaming water takes off in slurries downhill. Glooping and bubbling signifies the release of steam and gas into the atmosphere. Nearby is the Dragon's Mouth, a small cave from which issue spooky roaring noises, and from which "breath" spills out and puffs in a most realistic way. A small child asks: "Mom, is it coming to get me?" and hides behind her mother's legs.

These days, bacteria are artificially grown in laboratories for many commercial reasons. Under artificial conditions constant agitation is produced by special machines designed to regularly shake the culture tubes: twitching the fluid keeps the nutrients circulating, and the bacteria dividing. In Yellowstone, exhalation of gas and fluids in

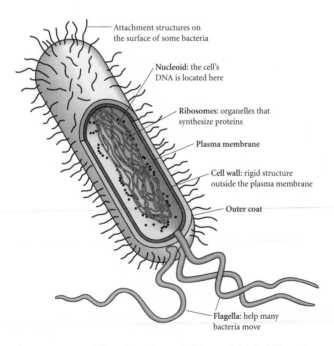

Attachment structures on the surface of some bacteria

Nucleoid: the cell's DNA is located here

Ribosomes: organelles that synthesize proteins

Plasma membrane

Cell wall: rigid structure outside the plasma membrane

Outer coat

Flagella: help many bacteria move

The structure and function of a typical bacterial cell. Although so tiny, such cells are capable of a huge array of chemical processes.

pools plumbed deep into the earth has the same effect in the wild. Except there is an added edge of danger, a sensation that the dragon might, after all, come out of its cave.

Pigments are one of the first products of metabolism and of life, and the pools and geysers of Yellowstone National Park are decked in primary colours. But not all of these bright splashes are the products of primitive organisms. When a yellow skin of elemental sulphur that has accumulated on the walls of a spring is seen through the prism of blue vent water, a lively green colour results—but somehow it is not the "natural" green of chloroplasts; it is too emphatically emerald. Precipitated iron compounds can paint patches of raw red, as happens within the Echinus Geyser in the Norris Geyser's Back Basin. However, much of the painting is the result of the metabolic activity of archaea and bacteria. The signature colour reflects the dominance of hordes of one or another species in slick masses as temperature, acidity or nutrients vary locally from one vent to another. Red, vermilion, carmine, pink, ochre, gamboge, purple, black and every hue and intensity of green imaginable can be found around this spring or that, often in some kind of concentric arrangement around the hottest part. Such a daubed landscape might offer a vision of what the world's surface looked like long before the greening of the land: a world of simple colours, crude face painting compared with the subtleties of moss and pine, but vivid and ever-changing like a lurid kaleidoscope. Land surfaces may have been decked out like this through the early part of the biological history of our earth, at a time when hydrothermal activity was more widespread, and the atmosphere richer in carbon dioxide. A visiting alien who needed oxygen to breathe would have shrugged in frustration at that time and passed on to a more hospitable planet. The early earth lacked the breath of life for animals. Oxygen would have been a poison to many of the founding microbes of life on earth, as it still is to those of them that survive. The Great Oxygenation Event would have brought death to them in many habitats; but since they are very, very small there are still always plenty of anaerobic places for them to hide and survive. A cavity the size of a pinhead is a world to a bacterium. Wherever something rots, using up oxygen, there is created a home for a host of anaerobes. And where a hot spring erupts, one lucky founder cell may proliferate into millions of progeny in short order, enjoying a brief resurgence of the glory days when the world was young and the air was foul.

Many of these primitive, surviving archaea and bacteria also relish heat. They need to live in hot water to survive and proliferate. This is not merely the kind of heat that makes you cry "Ouch!" This liquid can be virtually boiling. Such tiny cells are called thermophiles, and some are even hyperthermophiles. Imagine being adapted to sitting in nearly boiling acid. Much chemistry works faster at high temperatures, and life at all levels—and particularly at the single-cell level—is all about chemistry. It is helpful to think of these cells as tiny chemical factories wrapped in a membrane, in search of a source of enough energy to allow themselves to turn into *two* factories. The means of transporting that energy within the cell is the chemical Adenosine triphosphate (ATP), which also allows metabolism to happen in our own human cells. This is just one fact that makes it likely that all life descended from a common ancestor, but there are many more common threads that pull all animate existence together. Even genes with names like Elongation Factor feature in the common language of deep descent; every organism shares it. There is something rather wonderful about us having bits and pieces in common with organisms that thrive in nearly boiling water, and that have lurked somewhere on the earth for longer than we can conceive; but we all need to manufacture proteins in order to prosper, and that is where Elongation Factor vitally nudges into the lives of all beings from first to last. However, there remain such fundamental differences between the basic building blocks of life in archaea, bacteria and eukaryotes that Carl Woese of the University of Illinois and his colleagues have suggested that life should be divided into three "domains"—the largest conceivable unit of classification—to recognise the profound distinctions between the three groups. The two-fold division into prokaryotes and eukaryotes I introduced in the last chapter now begins to seem old fashioned: instead we need to talk about the domains archaea, bacteria and eukaryotes. Yellowstone offers proof that life was richer earlier than might have been thought.

Although this book is not about the chemistry of life, it is important to outline a few features of the domains because, as we have already seen in the case of photosynthesis, chemical systems are great survivors. Archaea have cell membranes containing lipids that are different in character from those in the other two domains.* Since the membrane is the fundamental barrier between the inside of the cell where life goes on and a (mostly) hostile external world, such differences really do lie at the heart of existence. What can or cannot enter and leave a cell is central to nutrition, and that in turn defines the possibility of growth. The next aspect of life's chemistry concerns ribosomes, contained within every living cell, and another essential component of every thing that might call itself alive. Ribosomes are the sites where proteins are created from the instructions of ribonucleic acid (RNA); they are like assembly yards where the architect's plans are converted into functioning structures. Although all three domains have ribosomes, there are major differences in the sequences in their RNA that highlight their deep separation. They all share a common ancestry, but they departed early on along different evolutionary paths. And if there really are three domains, then that opens up the possibility that one of them is more closely related to one of the other two, leaving the third domain in a more isolated position (somewhat on the principle of "two's company but three's a crowd"). Rather surprisingly, a considerable body of research points to the archaea sharing a more recent common ancestor with the eukaryotes than with the bacteria. I say "surprising" for no better reason than there might be a preconception that archaea were likely to be more archaic, if only because of the "antique" sound of its group name. But now the standard (but not universal) view among microbiologists is that archaea are more closely related to eukaryotes. The RNA of the ribosome is split into two parts known as the small and large subunits. Because the ribosome is such a fundamental building block of life, it is possible to use analyses of the amino acid sequences present in ribosomal RNA to make comparisons of relative similarity across almost the *whole* spectrum of living organisms. The way this works is

---

* In technical terms this means that archaea contain ether linkages between the glycerol backbone and the fatty acids, instead of ester linkages as in eukaryotes and bacteria. Archaea cell walls also do not contain peptidoglycan. These are such fundamental differences in life's chemistry that scientists regard them of importance in signalling early separation.

by the investigator constructing a tree diagram based on the relative similarity of sequences of gene coding for the RNA among a spread of primitive prokaryotes. In practice, the small subunit rRNA has been most widely employed in evolutionary studies because results can generally be obtained more quickly. Whereas sequencing genes used to be a time-consuming and expensive process, the technology arising from such massive endeavours as the Human Genome Project means that nowadays sequencing can be done very quickly. It is often just a question of booking time on the relevant machine to obtain a sequence of interest. This also means that trees tend to change in topology as more and more organisms are added as potential evolutionary "twigs." For more than a decade now, one curious and interesting feature of those diagrams portraying the relationships of primitive prokaryotic organisms has emerged: the lowest branches belong to organisms that live at high temperatures—even hyperthermophiles that can live in boiling water. Life, it could be concluded, started in hot water. The hot springs in Yellowstone are not merely curiosities; they are natural time machines. They take the visitor back to the early Archaean.*

The scientific names of these early and primitive organisms feature the Latin and Greek for heat and fire: *Thermoplasma volcanium*, *Pyrococcus furiosus*, *Thermodesulfobacterium hydrogeniphilum*, *Sulfolobus solfataricus* (Solfatara is a famous locality for vents near Naples in Italy) and so on. These names are blatantly descriptive. *Pyrococcus furiosus* does actually thrive in boiling water—it's a "furious fire sphere." The names have a drama that an individual archaean or bacterium does not usually fulfil in its morphology; one cannot help thinking that such creatures ought to look like little devils chortling gleefully in the heat of hellfire. However, most are tiny cylinders or spheres or threads a few microns across, the former sometimes equipped with a whip-like flagellum on one side, or even a small

---

* Just to make an important difference clear, the Archaean is the name given to the geological era further back in time than 2.5 billion years. Archaea are the superficially bacteria-like third domain of life. There were certainly archaea in the Archaean, but life in the Archaean was not confined to that group; bacteria were certainly there, and some even claim eukaryotes for such a deep time. The confusion of names is unfortunate.

bundle of them. They look far more impressive when they clot in millions to make bold splashes of colour on the ground. In nature, cells of one species frequently club together in strings to form wisps, or clots embedded in a gelatinous film. Mats and biofilms are usually collaborations of several species, with the metabolism of one species often interlinked with that of another in complex ways. What may look like a brownish scum on the ground can look like delicate tresses under a high-powered microscope. Individual species have tolerance ranges with regard to their optimum temperature for growth, so now there is an explanation for the concentric rings around a heat source. The most extremophile species will be in the middle, in the hottest part close to a vent, with less extreme extremophiles in a flanking position, and where the hot water spills out of a pool their tastes will be faithfully picked out in coloured rivulets draining away onto the surrounding flats, marked out as yellow swathes flanked by green or brown perhaps. The landscape of the extremophiles is perpetually shifting with the flux of water. If the water shuts off, the colour dries away to an undistinguished grey within days.

Now the painted hot springs can be seen in a new light. The geological roots of Yellowstone plumb into sources of heat and chemicals that feed a mass of very particular cells. Red and orange pigments in the hotter springs protect some of the photosynthetic bacteria from intense summer sunlight; the colour of the spring can even vary with the time of year as populations wax and wane. Bright paint in Yellowstone is often the signature of life. Yellow patches coloured by carotenoid pigments can be a feature of one of the first thermophiles to be discovered, the simply named *Thermus aquaticus** (which I take to mean something like "hot water"). Where there is energy to be derived, there will be a simple cell to utilise it, whether it is sulphur bathed in hot acid, or hydrogen, or iron. As a bacteriologist friend of mine put it: "if a bacterium can find an electron, it will steal it." The

* *Thermus aquaticus* was first isolated from Yellowstone but can be found very widely, including in domestic hot water systems. It was the source of *Taq* polymerase, an enzyme stable at high temperatures that permitted the development of the PCR techniques that "multiply" tiny amounts of DNA for further analysis. These are now vital to criminal investigations and have a hundred other applications—including the investigation of gene sequences relevant to the relationships of early life.

appropriate conditions would have been much more widespread in the very early earth. For example, *Hydrogenobaculum* derives energy from the simplest molecule of all, hydrogen, but only at temperatures above 65°C. In Nymph Creek, on the road from Norris to Mammoth, I saw strings of single cells of this organism joined together in yellow streamers, looking like fine blond hair waving beneath the acid water, a micro-organism made visible through sheer numbers. Under the microscope the streamers would have revealed an encrustation of sulphur. Because the chemical composition of water varies from one source to another, each spring or geyser will have some peculiarities all its own. Where iron is concentrated, bacteria prosper that can oxidise this common element to gain their energy, like *Galionella,* and they leave behind another kind of orange "rust." Cells are often surrounded by a sheath of mucus, so they really do approximate to that pernicious insult "pond slime." New kinds of archaea and bacteria are regularly discovered, but they don't often reveal their novelty at a glance; instead the secrets of the inner workings of the genome or the chemistry must be prised out in the laboratory. A whole group of hyperthermophile archaea called Korarchaeota was discovered in Yellowstone's Obsidian Pool. In 2008 they were claimed on molecular evidence to be most similar to the first living cells. Now we start to see the horseshoe crab as an afterthought, the velvet worm as a postscript, and even seaweed as a tardy arrival: we have taken a huge bound back to the earliest moments of life on earth.

Yellowstone encapsulates much of biological history within its vast caldera. It not only provides us with evidence of the earliest phases of living organisms on earth, but also reveals the later, oxygenated and "green" phases of evolution around the edges of the springs. On the flanks of Nymph Creek, where steam rises off the water and drifts off into the surrounding conifers, a shallow streambed is coated with a vivid green mat, unlike any other green in nature for its brilliant emerald intensity. It is made by the most acid and heat-tolerant alga known, an organism called *Cyanidium* that can tolerate 50°C. It is a thoroughgoing eukaryote for all its thermophilic tendencies. The bright green colour is down to a photosynthetic pigment called phycocyanin. Where the water drains further downstream into cooler pools, but retains its acidity, the mobile alga *Euglena* becomes important in its turn. It looks like a minute green sausage animated by a whip. In other less acid pools nearby, "blue-green" bacteria and diatoms (algae) weave intricate mats together, and recall the film of the surfaces of living stromatolites. These autotrophs are in their turn predated upon by single-celled animals of several kinds that engulf bacteria and algae within their protoplasm. Among these, amoebae and *Paramecium* are probably familiar from foundation biology classes, but there are many other less familiar species too. By now, of course, we have an ecosystem, a web of manufacturers and consumers, be it on ever so small a scale. Not all hot springs are acidic; it all depends on the local geology. If hot water moves at depth through limestone strata, then some of that hard rock dissolves. When the solution finally emerges at a surface spring, rather than depositing siliceous sinter, white travertine is precipitated, a material composed of calcium carbonate. The terraces that line the hillside at Mammoth Hot Springs are all travertine, and as dramatic as anything to be found in Yellowstone National Park. Many of the terraces are not currently being fed by "living" alkaline springs, and here the travertine ledges have turned as white as icing sugar. The cliff looks as if some giantess had been at work dribbling decoration on a mountain-sized cake ambitiously provided with many tiers. The springs appear and disappear with changes in the underground plumbing, but the total water extruded is thought to remain nearly constant *in toto.* I am surprised to see the famous Minerva Terraces are entirely dry, but nearby the Palette Terrace was in full flow, glistening and alive, with water spilling over the scalloped edges of staggered but delicate tiers. The conifers on the slopes are different species from the ubiquitous lodgepole pines around Yellowstone Lake, including twisted *Cupressus,* which resemble ancient *bonsai.* Where a new pool has overwhelmed them, as in Angel Terrace, bleached conifer skeletons attest to the lethal power of hot fluids over ordinary organisms. Bacteria and archaea thrive under exactly the same conditions, although alkalinity also ensures that many of the more choosy bacteria at Mammoth Springs are different species from elsewhere in the Park. But the same orange pigments are here again, colouring up pools, and in places dribbling over the edge of the pale travertine in lurid stains. Where the hot water flows away in channels, thermophilic microbial mats of many hues form gelatinous patches with bubbly or corrugated surfaces; some are vivid green, some olive, some brown, and all alive. Ignoring the curious glances of other visitors I kneel down on the ubiquitous duckboards for a closer look. A compulsion to look back into the Precambrian outweighs any embarrassment. I actually *see* bubbles of oxygen arising from one of these sticky mats, so I know that the rather scrofulous surface at which I am squinting must harbour cyanobacteria; maybe it is *Spirulina* with its tiny, spirally wound green threads, but more likely it is a club of several different kinds all woven together. Elsewhere, I notice that the microbes have made fine streamers waving in the hot currents: one of them is known as "angel's hair" as it is so pale and delicate. In Canary Spring at the top of the terraces, bacteria have conspired together to make structures starting as small pimples that then seem to amalgamate to form cauliflower-like heads, and all bathed in a kind of warm orange glow. I am reminded of one of Max Ernst's surrealist fantasies. Yet when the water supply is turned off, all life ceases immediately, and a stark, crispy grey or white wasteland is all that remains.

At the bottom of the travertine slope is the legacy of a long-dead

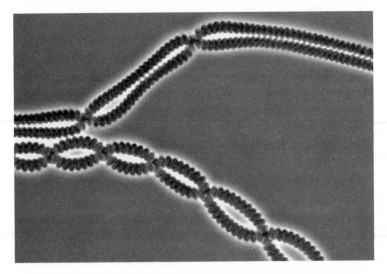

Threads of *Spirulina*—a cyanobacterium of distinctive corkscrew shape—are woven into microbial mats at Mammoth Hot Springs.

spring. This feature was named the Liberty Cap by the Hayden survey of 1871, whether from memories of the French Revolution or from more indigenous revolutionaries I cannot say. The Cap comprises the persistent travertine "core" of a spring whose terraces have eroded away leaving a steep-sided cone. It looks rather like an enormous tooth erupted from the soil. On the route out of town towards Gardiner the roadside cut exposes some "fossil" deposits of travertine, so evidently the hot springs must once have extended way beyond their present compass. On the broken rock surfaces, the successive layers of travertine deposited by the vanished spring are obvious. Some parts of the rock are slightly cavernous. Of colour there is no trace. Nor would I expect any jot or tittle to remain of the archaea and bacteria that would have thrived in their millions in the hot pools while the travertine was being deposited. Life does not always leave behind a signature that can be easily read by those who follow.

A similar problem applies to reading the very oldest fossil record. No doubt a world of sulphur, hydrogen and abundant hot springs would have suited our primitive and minute cells. But there is an obvious problem in looking for direct fossil evidence of creatures tinier than dust and lacking hard materials. Would it take some sort of miracle to preserve them? Around the acidic springs in Yellowstone I saw plenty of evidence of plants being overwhelmed by silica—on their way to being fossilised. Undoubtedly, Precambrian cyanobacteria can be very well preserved in silica, as the fossils associated with stromatolites have proved, so we can be confident in the evidence from the Proterozoic. But can we go back still further? In 1993 Bill Schopf described simple, thread-like fossils 3.5 billion years old from the Apex Chert from Western Australia. Martin Brasier later cast doubt on the authenticity of these fossils, showing that similar-looking organic tubes could form in other rocks that were not suspected of being organic. Schopf retaliated by demonstrating that the "walls" of the tube were carbon of the right type. Somewhat better preserved material from the Barberton Group (Hooggenoeg Formation) of southern Africa of nearly the same age has also been critically re-examined, and some sceptics are less convinced of the evidence than was once the case. We know that bacterial mats make subtle lumps and puckered surfaces, but can we be absolutely certain of the organic origin of very early examples in rocks that may have been heated and altered? I conclude that a measure of caution might be appropriate regarding some of the field evidence that has been claimed to reveal the very earliest days of life on earth.

Other lines of evidence of early life have become better established over the last decade. The information derived from gene sequences, and the discovery of new archaea with curious metabolisms appropriate to a primitive earth has already been mentioned. There is even evidence from the element carbon itself. It comes in two forms, the stable isotopes $C12$ and $C13$. Life activity favours the first, lighter isotope, for the simple reason that $C12$ is preferentially selected by the enzyme Rubisco—essential for the process of photosynthesis (pp. 82–83), and thence becomes incorporated into organic matter. If one compares the organic carbon present on earth with, for example, carbon from meteorites the earthly element is lighter, richer in $C12$. As far as we know, only life can perform this trick, and we can be fairly sure that light carbon tells us that life was present in quantity. Equipment to measure such delicate differences in molecular weight has become much more sophisticated, so very small quantities of carbon obtained from ancient rocks can now yield significant results. And, yes, certainly organic carbon was plentiful at 3.5 billion years: Round One to life. This helps to support the case for the existence of fossils. However, rocks surviving from such deep geological time are likely to have been baked and altered. For this reason, claims have been challenged about light carbon from even older rocks in Greenland; these are possibly the oldest on earth at 3.8 billion years. Our entire planet was traumatised about 3.9 billion years ago by meteorites during the so-called Late Heavy Bombardment. If archaea or their distant relatives were indeed present this early, they would certainly have lived in exciting times. But then there seem to be archaea and bacteria that can utilise almost any source of energy: there may have been extremophiles for extreme situations.

There are some places on earth which are also ideally suited for lovers of heat and sulphur, but they are not as readily visited as Yellowstone National Park. "Black smokers" have figured prominently in scenarios about early life and deserve a visit. They are hot water chimneys lying at an average depth of two kilometres in the sea along the mid-ocean ridges that mark out plate boundaries. The Mid-Atlantic Ridge is probably the best-known example of a mid-ocean ridge, and does indeed nearly bisect the great ocean in which it lies. The ridges mark lines along which magma is rising at depth, and are thus regions of increased heat flow and abundant mineralisation. Chimneys of iron sulphide build out into the sea where hot mineral- and gas-charged water vents into the cold ocean along the ridges. Scalding water belches out of the irregular conduits in total darkness, so nothing that needs light to survive can dwell down there. Much of the deep ocean is very sparsely populated with animals, but the area around the smokers is a kind of lightless biological oasis swarming with life, and all based upon metabolisms that do not require the mediation of sunlight. Archaea and bacteria that can use sulphur, hydrogen or iron for energy lie at the base of the food chain, and they are exploited in various ways by specialised crustaceans, clams and vestimentiferan worms—and predators in turn eat these consumers. The worms live in crowds of long tubes looking like clots of thick spaghetti, each strand with a red fleshy-looking tip. Remarkably, they have no gut; instead they rely entirely on chemosymbiotic bacteria for their nourishment. Nearby, long-legged crabs strut about the crusty chimneys like prickly steeplejacks. Fish that seem all mouth and stomach cruise around in the darkness, sensing prey. It is a strange and rather wonderful world, part Hieronymus Bosch and part Marvel Comics; not surprisingly, new species of organisms, both little and large, are quite regularly discovered in these dark depths.

It is not exactly routine to call in at black smokers: specialised vessels like the deep-sea submersible *Alvin* from the Woods Hole Oceanographic Institute are required to secure the information and specimens needed for science from such profoundly pressurised depths. Such research does not come cheap. The geological origin of the circulating heated waters powering the black smokers is not dissimilar to that described for Yellowstone springs, except that the pressure at depth means that temperatures can rise even higher without water turning to steam. The archaean *Pyrobolus fumarii* long held the record for survival in high temperatures (115°C), but an iron-reducing archaea dubbed "Strain 121" discovered five years ago

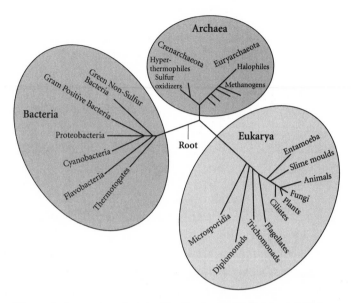

The evolutionary tree showing the fundamental division of life into its three great domains. Viewed this way, higher animals and plants are almost an afterthought at the top of the eukaryote tree.

in a black smoker may add 6°C to that (the poor "bug" went into suspended animation at 130°C, but could revive if placed into boiling water!). An organism like that is clearly a specialist even among archaea. Recently some scientists have begun to wonder whether microbial hyperthermophiles might *all* be secondarily adapted to very high temperatures. It seems that some of the special enzymes that help them cope with heat may be secondarily derived from a less tolerant variety. Maybe that special talent even helped them survive from deep geological time. It is now known that there are other kinds of vents elsewhere on the submarine floor (more are being discovered all the time), and these debouch at temperatures that are much less extreme than those along the mid-ocean ridges. It has been argued that a lukewarm or even cold seep might provide a better model for the cradle of life. These less dramatic vents are also rich in bacteria and other micro-organisms that metabolise in the dark. When more is known about the microbial life of cooler vents, a different picture of life on the early earth may yet emerge.

It is already clear that evolution proceeded according to different rules in those seminal times in the early Archaean, as indeed it still does among organisms close to the base of the tree of life. Some archaea and bacteria from Yellowstone have a lumpy or blistered appearance, and when they die, it is not in order to double up into daughter cells. Rather, these wonky-looking cells have been infected by viruses or bacteriophages, often simply termed *phages*. Viruses are even more minute than their microbial hosts, and essentially comprise packets of DNA or RNA surrounded by a protein shell. They are only capable of reproduction within a host cell, which they achieve by hijacking its cellular machinery to churn out more viruses. So when we spend a miserable day with a cold, our infected cells are experiencing a phenomenon nearly as old as life itself. Because of their apparent simplicity and mode of reproduction, the genetic structure of viruses is capable of changing very rapidly, as students of the AIDS virus have come to recognise all too grimly. Phages are also capable of transferring genes of one bacterial host to another (by a process known as transduction). Indeed, viral DNA may itself eventually become incorporated into the DNA of its host cell. Hence using viruses as vectors, evolution can proceed sideways: genetic material can be exchanged across what might be conventionally thought of as species boundaries.* In fact, such exchange proves to be commonplace at the bacterial level, where it is known as lateral, or horizontal, gene transfer. The label highlights the difference from our own father/mother to daughter/son transfer of genetic information. We might think of this as "upward" descent, perhaps, which moves vertically following the inescapable onward arrow of time and the rules of sexual reproduction. The microbes have different ways of achieving modification of their generations.

The most radical way of achieving horizontal gene transfer in simple cells is by means of plasmids. These are additional DNA components of archaean and bacterial cells that are separate from the

* Genetic engineering and gene therapy is making full use of these discoveries. Viruses and other vectors can be used to "deliver" a biological function, like resistance to a disease, to where it is needed on the genome. The insertion of a gene from a jellyfish that makes for green fluorescence now has myriads of different uses in research.

DNA on the chromosome. They are capable of independent replication, and so they present a different take on evolution. A plasmid that develops an advantageous mutation will prosper. They have been the focus of particular attention recently because they are implicated in allowing some bacteria to become resistant to antibiotics. Plasmids are circular, and can be as large as two microns across, which is big enough to be photographed with an electron microscope. During the process called conjugation between two bacterial cells, a kind of bridge is constructed through which a plasmid can pass into a new host cell. Many scientists regard this process as akin to the endosymbiosis I have described (see p. 81) at the origin of the eukaryote cell, when one prokaryote was tucked usefully inside a larger cell. Conjugation, too, is a way of generating more complex and multipurpose cells. Indeed, plasmids may have once enjoyed independent existence in the same way that photosynthesising chloroplasts were once free living. A new plasmid may confer advantages on the recipient; this is what happened in the case of antibiotic resistance. MRSA and other resistant strains are becoming more common as the chemical tools to control them lose their effect. Many of the interactions between simple organisms resemble a kind of chemical biological warfare; even at this small size a creature seeks an "edge," and getting the right kind of plasmid on board might be just the trick. Some simple cells function like a kind of *poste restante* where plasmids can successively be delivered. The main structure of the "post box" remains the same while many of the other contents shift. The ability to fix atmospheric nitrogen, for example, is an extremely useful talent in certain circumstances, as farmers know very well, and the right plasmid might be capable of delivering that capacity. Once a useful function has been delivered to a cell the lucky individual will prosper and reproduce with great rapidity. The normal rules of natural selection will apply, thank goodness; the cell most fit for purpose will out-compete its brethren. For those of us raised in the traditional view of descent, thinking evolution sideways-on requires new gymnastics from the old muscles of received ideas. Sometimes it still hurts a little.

This field of science is itself evolving fast—in fact, with almost bacterially reproductive haste. As with all science, and particularly in the case of evolution, simple ideas rapidly become more complicated.

THIS AREA IS CLOSED TO ALL HUMAN TRAVEL

This area provides critical habitat for grizzly and black bears emerging from winter hibernation. Bears depend upon this food-rich area in the spring due to the concentration of winter-killed bison and elk, large numbers of bison calves, and the availability of early snow-free vegetation.

Research has shown that the presence of people displaces bears, reducing the bears' ability to use the area. Also, the presence of people may be viewed by the bears as a threat to their food. Therefore, for your safety and that of the bears, this area is closed.

Path closed to Grand Prismatic Spring. Bears take precedence on one of Yellowstone Park's most spectacular trails.

Endosymbiosis is an attractive concept, and easy to grasp, but it now looks as though endosymbiosis happened many times over in complex ways. More and more new microbes are being recognised, and with a variety of internal chemistry that belies simple external appearances. Plain parcels may wrap up all kinds of exciting secrets that can only be revealed by patient work in the laboratory. My journey to Yellowstone becomes not just a journey into the past, but also an excursion towards the edge of discovery.

I really want to see Grand Prismatic Spring before I leave Yellowstone National Park. It is the mighty apotheosis of hydrothermal features, and must have been photographed about as often as Mount Rushmore. I have admired aerial photographs of it for years, with its blue heart surrounded by a swathe of brilliant green, a thin lip of sulphur yellow, and orange spillways all around like a circle of flames stretching across a brownish barren plain. But there is a notice: THIS AREA IS CLOSED TO ALL HUMAN TRAVEL. Nobody could have access to Grand Prismatic because grizzly bears are patrolling the area after their long hibernation, searching for carcasses of winter-killed elk or bison. This is one drawback of visiting so early in the season. "The presence of people might be viewed by the bears as a threat to their food," the notice adds, helpfully. It does not mention that the people might actually be seen as a dietary item. I must be content with watching hot water whoosh towards me in periodic waves from the invisible spring. A mile or so up the road I see that the warning is correct. A huge bear tears up an elk carcass to satisfy its hunger; it must be ravenous after its long sleep. A ranger politely but firmly makes sure that onlookers keep at a respectful distance from the carnivore so it can munch on, in peace. Long-lens cameras are unpacked. I feel slightly embarrassed, as if I were one of a pack of paparazzi snooping on a private feast. It is intrusive, somehow, to be spying on such unbridled gorging. Wolves have been reintroduced into the Park, and are doing well enough, apparently, although the natural caution of these clever canines makes them much more elusive than black bears and grizzlies. Perhaps they don't like being spied on. Yellowstone National Park is not just about bears and wolves. They may be where the glamour lies, but the real wonders require an electron microscope to get to know at first hand.

This book has arrived as close to the base of the tree of life as it is going to get. I emphasise again that the ancient bacterial world lives on everywhere. I like the thought that if everything except its bacteria were magically made to vanish a fuzzy outline of a grizzly bear would still remain, painted in the air by millions of prokaryotes. But as with velvet worms or horseshoe crabs, few of these much more minute organisms have stayed exactly the same. The world has been transformed in the billions of years since life began, and its smallest living inhabitants have moved with it. I have hardly nodded towards bacterial diversity. If I had delved into some of Yellowstone's lakes beneath the sediment surface, where mud is thick and black, I would have disturbed a range of methanogens producing bubbles of marsh gas, methane. If I had found some corner of a streambed where oxygen was low but where light still penetrated, there would be a thriving population of purple sulphur bacteria* using light for a special kind of photosynthesis that does not produce oxygen, and which would have predated the "blue greens" in the Archaean; a whiff of rotten eggs (hydrogen sulphide) might well betray their presence. Tiny autotrophs are everywhere, prospering in secret.

During the long adventure of evolution, microbes have moved into the guts of animals that provide the main attractions in Yellowstone National Park. In the spring the bears need to gorge upon the corpses of herbivores like elk or bison to get their weight up after the winter's fast; later, they can move on to berries or anything else that takes an omnivore's fancy. The herbivores in their turn rely absolutely on a range of bacteria to process much of the vegetable matter they consume. Cellulose cannot be digested without the help of enzymes produced by symbiotic microbes housed in their billions in the rumen of these animals. This chamber houses a brew of archaea, bacteria, fungi and single-celled animals—all interlinked in a kind of churning feeding frenzy. Solid flesh is the product, bear food, in its turn: the by-product is methane, now known to be an important greenhouse gas. Some of the microbes are second cousins to those in bacterial mats; they have been around far longer than mere grazers and cud-chewers. Where there is a niche that resembles some Precambrian crevice or spring, there they will revive prolifically. Even if the bear's tough stomach is so acid that only a few extremophiles can survive there, his great intestine is another microbial haven where nutrients slowly percolate into the beast's metabolism. As for the colon, it is a paradise of sorts.

Yellowstone National Park may be the best place there is to reflect on the fact that we are all pond slime. Every cell in our body acknowledges a deep history, a time when organelles floated free in a world we would have found insupportable. Our close mammalian relatives,

---

* I have mentioned a relative of these bacteria briefly in the last chapter as a possible cause of the "boring billion." The journey from non oxygen-producing photosynthesis to an oxygen-producing photosynthetic eukaryote is a long and arduous one, and the stops on the way are far from resolved, but certainly involve more than one endosymbiotic event.

the bears and the bison, all of us little twigs on the branches of the evolutionary tree, could not survive without the simple cells that are near to the base of the trunk. We could do without the horseshoe crab or the velvet worm, but we would be lost without bacteria. The lodgepole pines that line the lakeside in Yellowstone could not grow without help from symbiotic fungi coating their roots and scavenging phosphates from the soil, molecule by molecule. The story of life is not only a narrative of the triumphs of one organism over another organism; rather, it is as much about communal shelter and mutual benefit. It is a web or it is a crossword puzzle with no final answer. It is certainly not simple. When you touch a brown mat, blistered and glistening beneath a veneer of slime, you are communing with the deepest part of your own history: deeper than mind, deeper than sex. Now you know where you came from.

5

An Inveterate Bunch

We are in search of worms and shells, specifically a famous old timer, but we need help to find it. Dr. Paul Shin from the City University of Hong Kong is just the man to know if you want to find anything in the seas around the former British possession. For decades he has been scouring his home patch for interesting species. Even so, he has taken the precaution of sending out one of his students a couple of days in advance into the New Territories to make sure we will not be disappointed. He also supplies us with wellington boots and a minibus. We are leaving Hong Kong behind, that vigorous "city state" in the Renaissance sense, an *entrepôt* as it has always been, but now imbued with its own special energy. We are heading away from the signature buildings designed by I. M. Pei and Richard Rogers, which now dwarf the last colonnaded, but still pompously dignified, remnants of British imperial architecture. Away from the undistinguished tower blocks and factories that have now spread so far beyond the showcase centres of Victoria Island and Kowloon, with their Armani and Bulgari. In this part of China granite makes steep-sided mountains, and humans are filling up all the spaces in between. Richly diverse jungle still hangs on in the highlands, and occasionally I can hear whistling birds, while even in the city large, brilliant butterflies swoop out lazily over the highways. Fortunately, a few areas have been preserved as national parks, allowing glimpses of the natural countryside. I notice that mountain slopes are steep enough to provoke landslips, which expose little scars of underlying granite, although profuse vegetation soon covers them up again. When the coast is reached, it is dramatically indented with bays and promontories, some of which still have no roads reaching them; so a few small peninsulas might still be described as remote even though they are within sight of brand new skyscrapers.

Sai Kung East Country Park is only twenty kilometres or so northeast of Kowloon but it is a different world, surrounding a big bay encircled by wooded hills, where the trees come right down to the shore. The minibus winds through groves of acacia trees, which were introduced during the Second World War to stabilise the slopes. Hong Kong's national flower, *Bauhinia blakeana,* is in blossom everywhere; small trees brightening the roadsides with extravagant white or dark red blooms. Crested bulbuls sit on the tops of trees crying: "stick no bills." We park by a small white house mysteriously called "Lok Ness," pull on our boots, and Paul leads the party along a track following a stream towards the sea. There's a light breeze and just a few clouds. Where can the old survivor be lurking? The stream enters a narrow bay flanked by boulders where at low tide a series of muddy and sandy flats extend along the estuary. Adjacent to the shore the forest gives way to a fringe of small mangroves; their aerial roots pop up from the surrounding sand looking like so many chopsticks. Where small boulders lie around, they are completely encrusted with rock oysters. We walk further out on the shore. I can see that the area must be rich in nutrients, for the muddy sand flats are covered with millions of little snails, making tiny curled spires. There are many small crabs fastidiously grooming themselves with the air of a nineteenth-century gentleman checking his whiskers. I assume the little stream brings in extra nutrients from the surrounding forest to add to the daily refreshment provided by the tides. Nearer the sea a few Filipino women dig for shellfish buried in the sands, including very large clams, and a venerid species looking much like the European cockle: they wear conical straw hats resembling lampshades, and red neckerchiefs, as protection from the sun. It's a timeless trade, as old as mankind.

Quite suddenly, Paul and his colleague start attacking the flats with their mattocks and spades. We are about mid–tide level. I realise that the object of our search must be hiding in the sand and mud. Lumps of sediment are prised out, all sticky and black. The clods on the spade have creatures within—I can see signs of things wiggling, but not my animal. After several failures, I start to wonder whether we have actually come to the right place. But then one sodden sample falls in two on the spade, and something unusual is revealed. Here lies a glistening greenish shell as big as your toenail; and hanging down from it a gelatinous or rubbery "stalk" or pedicle, which is almost transparent when freshly lifted. The animal separates easily from the black sediment that encloses the pedicle, and then we wash it in a pool for a better look. Now I can see that it actually possesses two tongue-shaped shells, or valves, pressed together like hands in prayer. The pedicle dangles from a more pointed end; along the junction of the two valves some little whiskers protrude. The odd little living object is called *Lingula anatina,* and animals very like it have been around for nearly 500 million years. It is one of the great survivors, and I am thrilled to see it move from the pages of a familiar textbook into real life. Its Chinese name means "green sprout," and this animal's pedicle is indeed on the menu—but then almost everything is in these parts. My Chinese friends like to quote the old proverb: "in China, we eat everything with legs except the table." The clam pickers tell us that the little shells are not as common as they once were, and neither are they so popular as food. Certainly, they do not seem to offer much more than a chewy mouthful.

Lingula lives with its pedicle buried vertically in the sediment, and it secretes a gluey mucus at its tip; as the stalk contracts, it allows the valves to be retracted safely into a burrow when the tide is out. When the tide comes in, the two valves open upwards and gape sufficiently to take seawater inside the animal, where tiny edible particles are removed from suspension by a kind of ribbon carrying cilia called a lophophore. The little hairs around the valve edges prevent large unwanted particles entering the feeding chamber. Any edible material is eventually passed into a simple digestive system. This uneventful mode of life is enough to allow for *Lingula*'s growth and reproduction, the latter by means of tiny larvae that disperse as part of the plankton and are carried to a suitable surface on which to settle. Much of the valve material is made of calcium phosphate rather than calcium carbonate, or calcite, that is the common material for other "shellfish." Biologically speaking, *Lingula* has very little to do with the clams that the Filipino ladies were gathering, for it is no mollusc. It is a living example of a brachiopod. The most important unit in the classification of animals is the phylum, and *Lingula* is a member of the Phylum Brachiopoda—a group that is hardly related to the Phylum Mollusca. *Lingula* is one of the animals most commonly termed a "living fossil," and indeed if this description is true of anything, it is true of *Lingula*. When I was studying rocks of Ordovician age in Wales, I broke out a fossil *Lingula* from a shale quarry with exactly the same tongue-like shape as the one from Hong Kong; its shell glinted blackly at me in the sunlight. The same rocks contained fossils of trilobites that had been extinct for 250 million years along with colonial organisms called graptolites that had disappeared forever nearly 400 million years ago. *Lingula* had apparently travelled through the aeons, defying time and change.

Lingula is an invertebrate. This means no more than it lacks a backbone. Whereas vertebrates are a compact group descended from a common ancestor all fitting into the Phylum Chordata, invertebrates are an evolutionarily heterogeneous lot, and are classified into more than thirty phyla. *Lingula* brings us back again to about the same level in evolutionary time in the great tree of life as the velvet worm, another invertebrate, so our excursion back to the beginning is now over. Many phyla have their first fossil record in Cambrian strata alongside the trilobites, so they have more than 500 million years of history behind them. A friend of mine with a rather approximate knowledge of biology once referred to this group of animals as "inveterates." After snorting rather pompously at his gaffe, I thought of *Lingula*, and my Oxford English Dictionary definition of inveterate as "deep rooted, obstinate" and came around to the idea that the word was, after all, rather suitable. This chapter is about inveterate invertebrates whose lineages have survived from a time before the colonisation of land by plants more than 400 million years ago. I have chosen a bunch of species from different animal phyla that tell us much about life's early days. I have been able to visit many of them.

Lingula and all other brachiopods are filter feeders, sweeping a living by harvesting the minute organic flotsam and jetsam of the ocean. The muddy sands on the foreshore in Sai Kung East Park house other kinds of organisms that thrive in the same habitat. What is required to prosper here is the ability to make a feeding chamber from within a burrow or buried beneath the sediment surface. Peering closely over the superficial muddy film, I see small fishes and shrimps apparently doing the same job, inconspicuous animals that have to work hard to keep their tunnels open. Then I spot the tell-tale paired holes that prove the presence of siphons coming from clams buried deeper beneath the sediment surface. Another ancient habit. Next, our spades turn up a somewhat disgusting-looking, limp, pallid object about the size of my forefinger. Greyish white and worm-like, it has a fat "body" end with a gut, and thin "head" end which resembles a proboscis. There is indeed a small mouth at its open end, surrounded by about twenty tentacles. The head end can be pulled back into the body thanks to strong muscles in the body wall. This animal is another survivor, and one with a lineage as distinguished as that of *Lingula*. It is a sipunculid worm (and a member of another phylum called Sipunculida), otherwise known as a "peanut worm," because when retracted it does look like a large peanut. Paul Shin tells me that these "worms" are collected, made into cubes with gelatine, frozen and then eaten. Catholic though I am in my culinary tastes, I cannot say that this idea is particularly attractive. However, were I to sample either peanut worms or *Lingula* pedicles, then I would be able to take my taste buds on a journey back to the Cambrian. For sipunculid fossils are known from the early Cambrian Chengjiang faunas of China (see p. 45). The names *Archaeogolfingia* and *Cambrosipuncula* say all that needs to be said, when you learn that the living genera are called *Golfingia* and *Sipuncula*, respectively: they really have not changed all that much over 500 million years.*

---

* Although I didn't find an example, penis worms (Phylum Priapulida) deserve a mention here. These not-really-worms are still widespread on the sea floor in today's oceans, where they eat sluggish organisms like worms (real worms, such as polychaetes). They have a proboscis equipped with hooks, which can be extended,

The cry of a seabird breaks my quiet contemplation of *Lingula* and its companions. It is an extraordinary thought that on the mud flats where my wellington boots now disturb the peace of native invertebrate life, countless numbers of individuals live together belonging to species that root into very different levels on the great tree of life. Tiny towered snails (*Cerithidea*) and rounded winkles; fishes like gobies and some shrimps, are of relatively recent evolutionary origin. I recognise some of the clams like ark shells and *Anomia* have a much longer history, certainly extending into the Palaeozoic. Then there are the longest timeservers of all, *Lingulas* and peanut worms, still thriving in the muddy sands 500 million years later after all that geological time could throw at them. The point is obvious: there is nothing intrinsically inferior in adaptive terms about organisms with older origins. In the world of the mud flat an old timer can successfully live alongside a latecomer. Filter feeding is a very particular style of earning a livelihood. So it may be that in a site with food aplenty, space is at more of a premium, rather than direct competition between feeding strategies. Once a patch is successfully occupied, maybe the resources can look after themselves. Occupation is all. It could be that some of the older organisms are less tasty to predators, for example, or in the case of our two Cambrian survivors, only the human species is foolish enough to feast on a piece of tough jelly. But whatever the reason, it is certainly true that in this environment the passage of time alone does not eliminate geologically ancient organisms nor ensure the hegemony of the new. On the other hand, there is always room for an appropriately adapted species; everyone else just has to shift along a little.

I ought momentarily to expand on the subject of worms and worminess. If we take the common, or garden, earthworm as our

suggesting a comparison that lends the group its common name. Simon Conway Morris of Cambridge University demonstrated that in the Cambrian Burgess Shale they were more varied than they are now. Recent discoveries in China emphasise the same point. It has been supposed that some priapulids may have been important predators in those early days. Penis worms, like velvet worms and arthropods, are animals that have to moult in order to grow; modern classifications often group them together within the "supergroup" Ecdysozoa, and assume they shared a common ancestor.

model, then plenty of creatures commonly called worms, including the peanut worm, are not really worms at all. The earthworm is a member of the Phylum Annelida, so one might reasonably say that only organisms belonging to the same phylum have a claim to be called worms in the true sense. This group would then also have to include the ragworms that are commonly dug up on sandy seashores as fishing bait. These belong to a marine branch of the Annelida called polychaetes (bristle worms, generally). This important group also roots back to the Cambrian, as proved by a well-known species from the Burgess Shale (*Canadia spinosa*) in British Columbia, Canada. These have probably changed rather more through time than *Sipuncula*. When applied to other phyla, "worm" is merely a description of something vaguely long and wriggly, and no statement of biological relationship. So our peanut worms are a different phylum (Sipunculida) and not worms at all, nor are roundworms, flatworms, horsehair worms or penis worms truly worms.* There is a famous cartoon by Linley Sambourne published in *Punch* in 1881 bearing the legend "Man is but a worm," showing an evolutionary succession beginning from an earthworm and progressing via an amphibious thing to a passable ape and eventually to Charles Darwin himself. Most biologists are rather fond of this caricature of evolution, which is emphatically wrong in every detail. Whatever "worm" may have lain at the foot of the evolutionary tree which eventually led to humankind, it certainly was *not* an annelid. So whatever he may be, man is certainly not a worm.

Long ago, brachiopods thronged almost everywhere in the sea. In America, near Cincinnati, I have scooped handfuls of Ordovician fossil brachiopods from a streambed; these had washed out of a limestone formation nearby. The sea floor must originally have been nothing but brachiopods as far as the Palaeozoic eye could see. The

---

* In fact, they belong to different phyla. They should be classified in the phyla Nematoda, Platyhelmithes, Nematomorpha and Priapulida, respectively. But marine beard worms (formerly Pogonophora) and spoon worms (Echiura) probably are true worms, since recent molecular analysis shows that they are really extremely specialised annelid worms. They were formerly treated as separate phyla. So categories can change, though this is a rare discovery.

great brachiopod *Treatise on Invertebrate Paleontology** runs to five big volumes, and just to leaf through them is to understand something of past diversity. For here are illustrated spiny species, huge oyster-like forms, encrusters, radially ribbed species and others smooth as pebbles. Their only common feature is that they were all filter feeders. *Lingula* is the longest-lived survivor among them. You could argue about dating the heyday of the group: for sheer abundance it would be hard to beat Silurian and Devonian strata, when the insides of the valves of many species developed miraculously wound lophophores hung on calcareous loops and spirals to increase their feeding capacity. Certainly, there was a huge extinction of brachiopods at the end of the Permian, but they prospered again in the Mesozoic. Through it all, *Lingula* pottered onwards. Textbooks like to describe brachiopods as a group "in decline," but my colleagues who study them would point out that they are still numerous in some parts of the world, such as New Zealand. The majority of living species belong to a group of brachiopods with calcareous rather than phosphate shells, and they do not burrow in the sea floor like *Lingula*, but live attached to hard surfaces by a shorter tough pedicle. Since they are found in tens of fathoms of water, they are not encountered on the usual beach walk. This may have exaggerated an impression of their rarity.

Nobody could describe the molluscs as in decline. There are probably as many species alive today as there have ever been, and every gardener despairing over slug damage knows that these particular invertebrates have successfully invaded land, and can be almost indestructible. The octopus has been portrayed as one of the peaks of invertebrate evolution, and with some justice. It has a large brain, and eyes that mimic many of the features of those of the higher vertebrates. It can be a master of camouflage and disguise, and can pass its boneless body through small holes. Its relative the giant squid is

---

* I should point out that in the latest version of the *Treatise*, *Lingula* and its fellows have been split into more genera than was once the case. Some of the early relatives of *Lingula* would now be called *Lingulella*. I have found fossils in burrows that show that *Lingulella* already had the habits of its living relative.

the largest invertebrate on earth. Visitors who stare at a specimen preserved in alcohol in the Natural History Museum in London are hardly able to believe that muscle alone could support such a monster. Even the word "tentacle" has alien connotations. If an alien invader fails to be modelled on an insect, then it will almost certainly be half octopus. Clams and snails may seem rather unspectacular by comparison, but they have been around for far longer, and still prosper. The Phylum Mollusca as a whole boasts a long list of survivors from deep in the evolutionary tree, and they are important to my story. Not all of them are easy to visit. A simple cap-shaped shell called *Neopilina* was discovered alive and well in the deep sea in 1952. The group to which it belongs (Class Monoplacophora) had previously been described as fossils from rocks more than 400 million years old. I have found them in Cambrian rocks in Oman; at that time they seemed to have been able to live in the shallows. A handful of living species of monoplacophorans are now known, all of them in the deep sea. They show an interesting series of repeated muscles and other internal organs that used to be thought of as indicating a primitive kind of segmentation. Recent evolutionary studies have decided that these little cap shells might rather represent a side branch in molluscan history. I am rather relieved not to have to go down in a submersible to find one.

I did discover one molluscan survivor by chance. Turatao Island lies on the southern border of Thailand, where that country abuts Malaysia on the western side of the Malay Peninsula. Golden sands and coconut palms speak of a tropical paradise, an impression marred immediately by drifts of plastic bottles piled up on the beach, swept in by storms to lie unrotted for decades. Whitish coral rock along the shore is a legacy of past reefs; it is a mass of dead coral and algae welded together. Where the sea crashes over, the rock is soon fretted into sharp crests and puckered ridges, which mercilessly gash unprotected legs. Nobody with any sense would attempt to walk over the abrasive surfaces without shoes. I am ambling cautiously over the dangerous bluffs, peering at the variety of marine life, just pottering. Tucked into a hollow in the old coral is an odd-looking, pale-greenish organism pressed hard against the surface as if pretending to be part of the rock itself. It is the size of the palm of my hand, and divided cross-wise into a series of eight plates, surrounded by a frill of something vaguely resembling chain mail. It looks vaguely like a huge woodlouse, but unlike that animal, it has no legs. I know it well; it was on my "shopping list." It is a chiton, another old timer, which thrives in exposed sites much like its distant relative the limpet in similar habitats in northerly climes. I make an attempt to remove it from the rock by plucking at it with my fingers, but it sticks—well— like a limpet. I eventually do succeed in dislodging another of the animals but only by using the sharp tip of my geological hammer. Turning it over I see its massive muscular foot, rubbery and slightly glistening, that provides such a secure anchor against the worst that waves can do. The presence of such a foot is a uniting feature of all molluscs. The plates are articulated, allowing the animal both protection and freedom of movement. After death, the chiton falls into separate plates, and these are sometimes preserved as fossils; its scientific name Polyplacophora is in recognition of the shell being made of several calcareous plates. Like many molluscs, chitons have a toothed radula to help with feeding, a kind of organic chain saw. When the seas wash over them, chitons graze algae off the coral limestone; when the waters retreat, they hunker down. It is not a complex life, but it is

Chiton *Ischnoradsia australis,* a species that lives under rocks and stones on the coasts of southern and eastern Australia.

an ancient one. I am convinced of this by chiton fossils of Ordovician age—the same age as the rocks that I was investigating for trilobites nearby on Turatao Island. There is something particularly eloquent about such persistence. The trilobites have long gone extinct along with many of the organisms that lived alongside them. The chiton lived on. Other molluscs, like ammonites, rose to prolific abundance and variety after that extinction, only to become extinct in their turn. The chitons hung on to their rocks and they seem to be doing just fine today. Some authorities contend that they go right back to the days of the Cambrian, more than 520 million years ago; although there is a measure of disagreement about the true identity of the earliest fossil plates. No matter how the oldest fossils are interpreted, these molluscs are undoubtedly serious marathon performers.

A journey to North Stradbroke Island in Moreton Bay, Queensland, Australia, is a more planned quest for a seriously unusual mollusc. Moreton Bay is a famous site for marine biologists, rich in all kinds of invertebrates. John Hooper of the Queensland Museum tells me that it is an exciting place in which to dive to make a collection, because of the exceptional abundance of sharks. "We come up in pairs, back to back, with knives in our hands," he says. I think I might be counted out of this particular survey. "Straddie," as the locals call it, is about sixty kilometres long, and reached by a leisurely car ferry across the Bay. Arriving this way at an island is somehow always exotic. But it is paradoxically homely for me because for more than thirty years I have been holidaying in Suffolk in eastern England, where the local aristocrat was Lord Stradbroke. When the boat docks at Dunwich on Stradbroke Island, the illusion is compounded, because the small village of Dunwich is a mile or two from Stradbroke's seat of Henham Park. It seems that a little bit of familiar Suffolk had been snipped out and moved to the other side of the world, but the commonsense explanation is that a branch of the Stradbroke family moved to Australia many years ago and adopted familiar names to make themselves at home. The landscape could not be more different from eastern England. The island comprises a wooded dune system, the sand originally blew up from the south during Australia's great arid events. There is just one road around it. In the centre there is a big mine. The sands include commercially important concentrations of small crystals of titanium minerals. The deposits have been mined in a discreet way, rather unusual in the mineral industry. The rest of the country is prime eucalypt bush, watered by plenty of rain. Some parts are swampy, full of paperbark *Eucalyptus* with scruffily peeling trunks and ferns sprouting beneath. Coral reefs lie off the seaward side of the island beyond Moreton Bay. Our destination lies on the other side of the island near Amity Point. This is a sandy coast, with flats exposed at low tide. Thousands of tiny, blue-tinted soldier crabs (*Mictyris*) live and feed on the sand flats, searching the sediment for scraps and tiny organisms—the whole beach is covered with the pea-sized balls of sand they have processed. They are pretty cowardly for soldiers, as they flee in waves before intruders like herds of minute sheep being chased by invisible dogs. In the distance, where the tide has exposed low flats, everything is somehow darker. A few large spoonbills and a heron are fastidiously picking their way over the surface, prospecting for food. As I move in closer, I see that the dark colour is a dense mat of a sea grass called *Zostera*. It really is a grass—one that has returned to live in the sea—and it produces regular meadows, which harbour all kinds of life to attract hungry birds at low tide. At high tide the meadows are covered by a metre or two of water.

Once more it is back to digging, this time in the *Zostera* beds, with John Hooper as our guide. Under the matted roots that cover the surface, the sediment is both muddier and soggier than I expected. It is also quite black. A hermit crab runs away in its borrowed apparel: a battered snail shell. Then I notice something coming from the sediment; a foul smell, perhaps a hint of rotten eggs. In life, *Zostera* trap fine sediment in their leaves and this is added to the sea floor. Then the dead leaves of the plants themselves, together with any unconsumed organic material, become incorporated as the grasses continue to grow. Decay of this material uses up nearly all the oxygen in the sediment, which soon becomes black and foul. Under these anoxic conditions, certain kinds of bacteria flourish, while organisms that breathe oxygen cannot survive. The smell is hydrogen sulphide gas, which is released as part of the metabolic processes active in low oxygen quarters. Evidently, just a few centimetres below the *Zostera* field a special world is hidden away. Then, a small clam turns up in the dark sandy mud; this is the object of my search. Dark coloured and not much bigger than a large bean, its two valves gape at one end. The clam is called *Solemya*. The genus has been known for a long time. *Solemya* was named in 1818 by the great French biologist, Jean-Baptiste Lamarck. But novel species like the one on "Straddie" are still being discovered.

Although it looks anything but spectacular, *Solemya* is a very odd and interesting animal. To begin with, it lacks any real gut—so how does it feed? Clams are generally equipped with a rather obvious digestive system. Next, *Solemya* evidently lives in the low oxygen region beneath the sediment surface, a special place by any measure; but it is proving to be quite common where this habitat is searched. For many years *Solemya*'s lifestyle remained mysterious. Now it is known that the little clam has turned adversity to its advantage: it is designed to live where little else can survive. The mollusc actually cultivates special bacteria that can thrive in the anoxic mud, and these colourless sulphur bacteria are grown on the gills of the animal, where the nutrients they produce are absorbed directly into the body—a gut is redundant. *Solemya* is a special kind of underground gardener, living where its microscopic partners can be happiest. The bacteria, one called *Thiomicrospira* particularly, use energy derived from sulphide to convert carbon dioxide into nutrients, in a kind of lightless equivalent to the photosynthesis I have described previously. Photographs taken under the electron microscope show bacteria festooned on the interior of the clam like thousands of minute beads. The bacteria are technically described as "chemoautotrophic symbionts," which is the kind of technical mouthful it is always useful to have at one's fingertips in case of a lapse in conversation at parties. Using its flexible foot, the clam maintains sufficient contact with the surface to allow it to continue to absorb oxygen. But it is living life "on the edge."

Fossils prove that *Solemya* is another of the great survivors. Not far from where my trilobites are collected in south Wales, my colleague John Cope hammered out very early fossil clams from a quarry in strata of early Ordovician age. A lightly wooded, hilly area near the ancient town of Carmarthen is a particularly pleasant place to do fieldwork in springtime, with primroses on the banks and the nettles not yet knee high, so breaking rock is not the burdensome labour it might sound. Nonetheless, bashing stones for hours on end requires patience and inextinguishable optimism. One of the molluscs John Cope recovered proved to be very similar to the living *Solemya*. The clam seems to be another *Lingula* or, like the chitons, *Solemya* just chugged on through time. Enduring from the days of the trilobites and the heyday of the brachiopods, eventually one particular *Solemya* species came to lie snug beneath sea grasses that only appeared 450 million years later. Through all that time there must always have been a billet for the clam somewhere. It seems that if a niche is specialised enough, it may provide safe passage through some of life's great crises.

A marine snail called *Pleurotomaria* is a much more beautiful object than *Solemya* but has equal claims on antiquity. It displays a lovely, elevated spire, something like a Buddhist *stupa,* and is recognisable by having a distinctive slit in its aperture. Some specimens are as large as spinning tops. Snails of this kind were once more diverse than they are now, but they are still widely distributed. They were not on my list for a visit because they live only at considerable depths today. I have instead touched specimens reverentially in a museum drawer. Many other living molluscs root back into the Mesozoic. Their fossil relatives are readily identifiable to experts with knowledge of living sea creatures: I should mention relatives of scallops and oysters. Some molluscs that were abundant worldwide in the sea during the Age of Dinosaurs have since become much restricted in their distribution. I have on my mantel shelf a curious object that I picked up from the old Jurassic quarries at Portland on the south coast of England, source of the finest building stone in the country. The quarrymen called these fossils " 'osses 'eads" because of a passing resemblance to horse's faces, but actually they belong to clams called trigoniids, which are abundant in many localities in Jurassic strata across Europe and much of the rest of the world. They all have a very distinctive rubbly ornament confined to one side of the shell. In Australia, I handled a living species called *Neotrigonia margaritacea*, which can readily be dredged from the sea floor near Adelaide. It instantly took me back to the Jurassic. Five living trigoniid species are

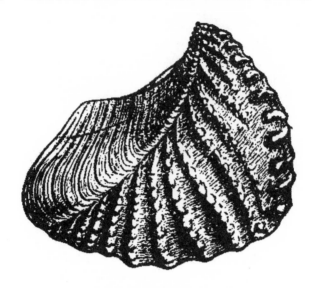

Scaphotrigonia, a Jurassic fossil from Europe that is closely related
to *Neotrigonia*, which still lives in Australian waters.

now only found around the coasts of Australia and Tasmania. They
contracted in scope, but carried on.

Shales of the same Ordovician age that yield both trilobites and
the clam *Solemya* contain further mollusc fossils. One of these is a
simple-looking tapering tube, divided on the inside by a series of
walls that partition much of the body into chambers. It's a fossil nautiloid, and among the earliest of a great group of cephalopods that
include those most complex of living invertebrates, the octopuses and
squids. Today, cephalopods are important predators among the molluscs, with their tentacles, jet propulsion mode of swimming, highly
developed nervous system and complex eyes. Some modern cephalopod species use indigo ink as a way of confusing their enemies and
making their escape. They live at every depth in the sea, down to the
darkest and least known regions of all; they are adaptable enough to
survive there. Who can resist the name *Vampyroteuthis infernalis* for
one of the deepest? It looks like a weird animated umbrella equipped
with its own lights: infernal by any measure. But there is a nautiloid surviving from the very early days of the cephalopods, called,

unsurprisingly perhaps, *Nautilus,* the same name as Captain Nemo's vessel in Jules Verne's *20,000 Leagues Under the Sea* and taken from the Greek for "sailor." *Nautilus* has outlived many of its descendants.

Nautilus lives below the reach of a shallow diver, but I was lucky to see film at the Queensland Museum of the living animal taken in 2009 from a remotely operated undersea vehicle (ROV). Andy Dunstan guided me through the footage: he is writing a doctoral thesis on the ancient creature. *Nautilus pompilius,* the pearly nautilus, has an elegant, tightly curled spiral shell attractively patterned over one half with broad, pinkish-coloured bands. The inside of the empty shell shines brightly with mother of pearl, nacre. Sectioned shells are pretty enough to be popular tourist trinkets—much too popular for the good of the animal. Over fishing is a major problem, particularly off New Caledonia in the Coral Sea. Sections show well how the interior of the shell is divided into compartments by gently curving walls, along the centre of which passes a duct, the siphuncle, connecting the outside chamber with the others further back down the spiral. The living animal occupies the large chamber at the open end, and as it grows, progressively walls off chambers behind. The creature itself looks like an untidy mass of tentacles—about ninety—stuffed into the end of the shell, with large eyes to either side, and a prominent hose-like tube that provides the jet propulsion the animal needs to get around. *Nautilus* does not have sophisticated suckers on its tentacles like its more advanced relatives; instead it has rough ridges that help the hunter grasp its prey. Through the eyes of the roving undersea vehicle, I followed the nautiluses living on the Queensland Plateau around Osprey Reef, offshore from Cairns.

The Queensland Plateau lies further out to sea from the Great Barrier Reef. It is a piece of foundered continental crust, and has proved to have a fascinating fauna, about which little was known until recently. The animals living there are different from those of the famous reef despite its proximity, but then it lies deeper in clear water, before a drop-off to 2,000 metres or so. Australian biologists tell me with a hint of desperation that despite the riches of their nation, they are still desperately short of the special equipment they need to explore this newly recognised marine world. They have been able to spy on *Nautilus,* which lives at about 200 metres depth during the day, but can live happily at 700 metres if needs be. Off New Caledonia some individuals even come almost to the surface at night. The ancient mollusc can adjust its buoyancy and depth by using the siphuncle to move gas and fluid into the chambers no longer occupied by the body. The shell bobs around purposefully when it swims. *Nautilus* makes short work of any bait fish, using a tough beak in the centre of the head. Andy tells me that the beak can readily tackle chicken wire. Pearly nautiluses do not need to eat very often, perhaps because they bumble along using so little energy. Cells sensitive to dissolved chemicals on their arms can pick up the "scent" of carrion. Some of the bacteria that consume organic remains at this depth are luminescent, so maybe *Nautilus* can seek out this kind of food. They are certainly not fussy eaters. *Nautilus* can live for twenty years or more, a fact that had to be laboriously established by tagging individuals and then recapturing them; by this age they may have thirty or so internal chambers. This is a long life for a cephalopod. They do not reach reproductive maturity until eight years old, which implies that when they are exploited too vigorously their natural populations recover only slowly. They lay a few rather large eggs almost constantly, but, mysteriously, juveniles are not caught in the wild. They are thought to hide away discreetly at depth. New research shows that the pearly nautilus can move ten kilometres a day, and can negotiate currents if it has to. But its eyes are weak—something like fifty times less efficient than fish eyes: they function as "pinhole cameras." It must have been an eye of this kind that was honed and polished into the sophisticated octopus organ. Nautiluses are themselves consumed by sharks and octopus; in the latter case, it is a bit like being eaten by your second cousin.

I have on my desk a fossil of Jurassic age. I bought it in the flea market in Beijing, which is famous for having more reproductions on sale than anywhere else in the world. If it can be imitated, then ingenious Chinese will reproduce it, faithfully. I bought the fossil from a stand that was selling jewellery (fake jade). The items for sale were draped over my object, and the proprietor was slightly discombobulated when I pushed aside the jewels and asked how much for the fossil. I got it for a good price and I am convinced it is genuine. I may be the first person to get a real bargain from the Beijing flea market. The fossil is a pleasing pale brown colour, and has been sectioned down the middle to display its insides. The curved walls that divide the interior into chambers are obvious: a few nearer the exterior end have fractured and collapsed; well, that happens with fossils. The little cavities so produced have allowed crystals to grow around the walls. Nobody could fake that. Along the middle of the whorls a calcified tube shows the passage of the siphuncle. It is obvious that this fossil is a nautiloid very similar to the living pearly nautilus. If anything, its siphuncle is more heavily calcified so it clearly links all interior chambers together. But nothing much has happened to this animal in about 150 million years. The lineage of *Nautilus* can be traced all the way back to the top of the Cambrian, 500 million years ago. The Ordovician fossils I discovered in Wales did differ from the living animals in that they were mostly straight, tapering cones. At some stage in its history the tube evidently "curled up" into a spiral, and this shape endured. Like several of the animals described in this

One that did *not* survive—the ammonite mollusc *Dactylioceras,* widespread during the Jurassic Period, about 180 million years ago.

book, nautiloids were much more diverse and varied early in their history. Some of the Ordovician species were monsters with shells as long as a man, and since they must have been predators, they must also have been among the top echelon of the marine food chain at that time. A number of early forms curled up into spirals, but they also assumed other shapes: gently curved like a pronghorn deer's antler or dumpy like a small barrel. It is likely that in their early days nautiloids pursued more than one way of earning a living. From this variety, one design persisted through all the following mass extinctions, moving continents and greening of the land, a simple shape which now rests next to my word processor. But during the Devonian Period, about 400 million years ago, a branch of the nautiloids began to develop much more complicated internal walls. These became progressively folded and crimped. At the same time, the exterior of the shell often became more elaborately sculpted with ribs or knobs. These ammonoids radiated into one of the most diverse groups of marine organisms in the fossil record, with thousands of species displaying every conceivable size and variation on the spiral shape over more than 300 million years. Ammonites must have been as numerous in the Mesozoic seas as squid are today, and they lived in swarms or schools at many different depths. Their history was punctuated by several extinction events, notably at the end of the Permian, from which they recovered to move on to new glories. At the end of the Cretaceous, ammonites passed from the world forever, a marine casualty to match the dinosaurs on land. Only *Nautilus* bobbed onwards, the last representative of a great tribe.*

The Queensland Plateau is an unusual habitat. Surveys 150 metres below present sea level show that its platforms were eroded at the time of the last glacial maximum, when mean sea level was much lower as a result of the water locked up in huge ice sheets. *Nautilus* was free at that time to move down to still greater depths to survive what might otherwise have been a lethal crisis. When sea levels recovered, the platforms were drowned, and the ancient mollusc could move up again. Seamounts made by subsequent coral growth

---

* There is also a second, little known living species *Allonautilus scrobiculatus* living off New Caledonia.

now separate individual populations of living *Nautilus*. The plateau became a favoured habitat for colonisation by surviving "Mesozoic" animals. Alongside *Nautilus*, stalked sea lilies (crinoids) thrive, waving elegant, feathery arms in the passing currents to harvest drifting plankton. They belong to an animal group with a total history as long as that of *Nautilus* itself; living crinoids also have close relatives among Jurassic fossils. Elsewhere in the ocean many living species of stalked crinoids are found in the perpetual blackness of the abyss. They are old timers from the Phylum Echinodermata, another great natural group in the roll call of life, which includes living sea urchins, starfish and sea cucumbers—the "spiny skinned animals." Sea lilies recognisably like those still living on the Queensland Plateau waved their arms while trilobites scuttled in the mud nearby.

I think of survivors as the last descendants of some noble family, whose portraits hang crowded on the walls of a stately home that has perhaps seen better days. The portraits remain the best way to study the features of the forebears. I spent some years studying fossils called graptolites, which are among the most abundant fossils in marine rocks of Ordovician and Silurian age. Although they often look like no more than little white lines drawn on the rock, they were animals of marvellous complexity. They were colonial, each graptolite comprising a mass of tiny feeding individuals that were housed in tubes usually no more than a few millimetres long; but the colonies were highly organised. Graptolites floated freely in ancient oceans harvesting minute plankton; 400 million years ago the sea must have thronged with them. Their colony shapes were designed efficiently to reap a thin crop: they evolved spirals, tuning forks and all manner of branching shapes. Graptolites disappeared from the earth 350 million years ago, for reasons still unknown. But they have an inconspicuous living relative called *Rhabdopleura*, which encrusts sea-shells and pebbles. Together with a small handful of other living creatures it is placed in yet another phylum, called Hemichordata, which molecular studies relate distantly to echinoderms. *Rhabdopleura* is little more than a series of tiny interconnected, upright organic-walled tubes joined together by a kind of miniature rope that includes a nerve strand. With some difficulty, I examined a specimen attached to the underside of a shell dredged in shallow seas off Scotland. The tubes are very fragile, and the pair of feathery-looking, filtering "arms" belonging to the little animal that inhabits these frail houses can be destroyed by an ill-considered pin. This animal is a true survivor. A fossil very similar to the living *Rhabdopleura* was recently discovered in Cambrian rocks, so this colony has a 500-million-year history. It both preceded and outlasted its extravagantly evolved graptolite relatives. This is starting to seem like a familiar story.

SPONGES AND JELLIES

John Hooper, peerless sponge expert at the Queensland Museum, points out to me that the *real* survivors have to be his favourite organisms. Hooper's narrow office is lined with jars containing sponges in alcohol and dried specimens in packets according to species. It is the kind of place small boys might dream of, with microscopes and books full of brightly coloured marine creatures. Sponges, once collected, fade. The back rooms of the Museum are stacked with previous collections, making a permanent archive for one of the most diverse areas for sponges in the world. Hooper is as devoted to sponges as he is to conservation. Like many people who know a lot, he knows how little we know.

Sponges, Phylum Porifera, slot into our narrative below the other invertebrates. All the animals discussed so far in this chapter are symmetrical about a line running along the length of the body. This great assemblage of organisms is known as Bilateria because the left side is a mirror of the right side. You and I are bilaterians, and we share ultimate common ancestry with brachiopods, horseshoe crabs, velvet worms and slugs, who are also bilaterians. Admittedly, things do get twisted about a bit in snails, but there is no doubt that the primitive shape for the molluscs was also bilaterally symmetrical, just as in *Nautilus* and the chiton. No such constraints apply to sponges. Many sponges have simple shapes resembling vases and goblets, but they can assume almost any form, from cauliflowers to candelabras to cushions or crusts. Some can grow as big as a man. Individual species often vary in shape and size according to local conditions on the sea floor, which can make for problems in their identification. Sponges may lack a nervous system or a stomach, but they do reproduce sexually, and they assuredly do not lack organisation. Little cells closely resembling the single-celled animals known as choanoflagellates cooperate in lining the sponge's body to produce currents, their tiny flagellae beating in sympathy to help harvest bacteria and other tiny organisms from the seawater. Since free-living choanoflagellates use their tiny "tails" for swimming, one can imagine a scenario in which a cluster of individuals moved together and found cooperation an advantage. An apparently simple event of this kind may have laid the lowest layer of the foundations of complex animal life deep in the evolutionary tree. Choanoflagellates were joined by more than half a dozen other kinds of cells before a simple sponge could be constructed—but one of them is an amoeba-like cell that seems to be capable of transforming itself into many of the others. And chains of cells will soon require struts to support them if they are going to achieve a workable shape for harvesting food. Sponges have developed supporting "skeletons" of extraordinary beauty composed of little elements termed spicules. The bath sponge is not such a familiar object as it once was (over harvesting yet again), but most of us will recall its remarkable combination of toughness and flexibility. It is made of a material called, rather unsurprisingly, spongin—a variety of collagen—that makes wonderfully interlinked frameworks at a microscopic scale. Other sponges are less flexible. Some employ calcium carbonate to construct their spicules, the same material as employed by many invertebrates. Others use glass to do the same job. It is not exactly glass, but it is the same material chemically: silica, silicon dioxide, commonly known as rock quartz. Some of the spongin sponges also add silica spicules to their construction, but they are different from those of the glass sponges themselves. The "skeletons" of glass sponges are among the most wondrous structures in nature; they look as if spun by a magical geometer. One has to remind oneself that the beautifully regular shapes are just a matter of circulating water efficiently to harvest the meagre nutrients available at the depths at which these sponges live. Some of them are centuries old, like trees.

John Hooper is more than an ambassador for sponges. He has discovered dozens of new species. Many new forms can be found even in the shallow seas around Australia; there are simply not enough people to study them. Just because sponges have a comparatively simple body plan does not mean that they are evolutionarily sluggish. Quite the reverse: on the Great Barrier Reef alone there may be as many as 2,500 species. In regions between the reefs there might be another thousand species, and among these even the commonest often lack scientific names. Some ancient sponges have been hiding undiscovered deep in the oceans. As recently as 2007, glass sponges forming whole reefs were discovered off the coast of Washington State in western America. Composed of clustered cups and cones built on the skeletons of their predecessors, they make up a miniature ecosystem, one which is now under intensive study. Fossils of glass sponges are familiar from Cambrian rocks in many parts of the world, and at that time they were capable of living happily in relatively shallow sites on the continental shelf. They seem to have retreated to the deeps today.

John Hooper handed me another object, a small humpy thing. This sponge was recovered from Osprey Reef in 2009. He told me that it was called *Vaceletia,* a coralline sponge, very heavily mineralised. It is a familiar fossil in Jurassic and Cretaceous rocks, so it is yet another survivor, a true living fossil since it was actually known as a fossil first. David Jablonski of the University of Chicago dubbed such organisms "Lazarus taxa" because they apparently come back from the dead. Some specimens are known to grow only a fraction of a millimetre a year, which means a coin-sized stony sponge might be centuries old. It is practically a fossil while still alive. Such slow growth has obvious implications for its regeneration if its habitat were ever damaged. Finding a relic of this kind was an unexpected gift to molecular scientists, who were amazed to discover that its DNA allied *Vaceletia* to the bath sponges. It had been thought that *Vaceletia* was related to an ancient group called Sphinctozoa with fossils going all the way back to the Ordovician Period. It now seems more likely that heavily calcified sponges had more than one evolutionary origin. With relatively simple organisms it is always necessary to be careful about classification; nature may pull off almost the same trick more than once. The living *Vaceletia* is practically indistinguishable from species living more than 100 million years ago, and genuinely related to them, but this may not be true for similar-looking sponges four times the age. There is no doubt that sponges were already numerous and diverse in the Cambrian; the famous Chinese Chengjiang fauna includes a dozen species, including *Choia xiaolantianensis.* There is nothing like being a palaeontologist to challenge the vocal cords.

Most scientists accept that sponges root back below bilaterally symmetrical animals. Nobody disputes that glass sponges are true marathon runners among organisms. The basic body anatomy of all sponges, comprising a mere two layers of cells separated by a jelly-like layer, is surely primitive, to which one can add the capacity of some of their cell types to change into other types. Problems come about when biologists try to decide which of the major sponge groups is closest to more advanced animals. A number of researchers favour the idea that sponges are not a single entity at all, and that the three major skeletal types (spongin, calcareous and "glass," in short) are three separate phyla, of which the first named is closest to "higher" animals. Silica spicules of the first group of sponges have been recovered from Ediacaran rocks 580 million years old. So it is probable that sponges were around at the same time as the enigmatic animals we met at Mistaken Point. In August 2010, Adam Maloof of Princeton University claimed sponge-like fossils from South Australia 650 million years old, which, if true, would take them back even into the late Proterozoic. This should be of vital importance in constraining the time range during which the animal phyla split off the trunk of the tree of life. However, recent scientific papers partly based on molecular evidence have revived the idea that *all* the sponges belong together as a group on their own, pushing them to one side in the story of animal evolution. It is not uncommon for such deep roots in evolution to offer contradictory results; even molecular signals are sometimes hidden in "noise" generated by hundreds of millions of years of later modification. I would not like to predict how this particular issue will be resolved in the next decade. It would be disappointing if the sponges were displaced from their seminal position if only because it is so tempting to subvert that well-known Darwin cartoon legend of 1881 to read: "Man is but a sponge." So much more appropriate than a worm.

John Hooper has an unusual source of funding for his Porifera research. He receives a grant from a giant pharmaceutical company. There is a good biological reason for this rare confluence of interests. As they have been around for such an inordinately long time, sponges have permuted an astonishing array of chemical defences. Even today, animals such as sea slugs and snails find them nutritious. Like every other living thing on the earth, sponges have been infected by viruses and bacteria. Over hundreds of millions of years they have developed resistance to these attacks. If they had not, they would not still be thriving in the seas to tell us their story. They are goblets filled to the brim with interesting organic chemicals. Such compounds are described as "biologically active" if they have inhibitory effects on actively growing bacteria colonies, to select just one example. Now that our current antibiotics are becoming increasingly ineffective against bacteria such as MRSA, the hunt is on to find other ways of staying one step ahead of the versatile and protean prokaryotes. New chemicals that will inhibit bacterial growth or reproduction are top of the list. So far, sponges have yielded more than a hundred compounds of interest in medical research; sponges are not so much goblets as cornucopias of chemicals. None of them, however, is going to provide a "quick fix," as it takes years of research to determine molecular structure, safety, medical utility and side effects, before a new name can be added to the pharmacopoeia. Nonetheless, "Humans saved by sponge" would be a headline worth waiting for.

Other animals refuse to be bilaterally symmetrical, instead having radial symmetry, like a sunflower. Jellyfish are familiar to anyone who has encountered a stranding on the beach. No child can resist poking the curious, motionless blobs with a stick. Jellyfish seem lumpish on the shore, inert, drying and dying in the sunshine. Yet in the open sea they are elegance incarnate, pulsing, contracting, almost flying through the water, apparently dancing to some music that our ears cannot sample. Below the bell, tentacles dangle. Some are twisted round and round like ringlets; they might belong to an eighteenth-century femme fatale. Others dangle distantly, or wiggle nervously. The orange colours displayed by some species look threatening to our eyes (perhaps with justice), but the whitish transparency of others conforms to the palette of ghosts. Yes, this is the nub of it: jellyfish are spookily beautiful, spun out of something that has little substance yet can engender complex form. They seem to drift without purpose, yet they are constructed with a precision that insists on function.

One does not have to embark on a journey into a remote corner of the world to find jellyfish; they come to us, driven on currents. Their dangling tentacles earn them a living in the sea's economy, for they are decked in stinging cells that paralyse their prey. These tentacles are frequently several metres long, and trail wispy skirts. Some species are powerfully poisonous, most notoriously the box jellyfish (*Chironex*) of Australia, which occasionally takes a human life. Others feed on plankton, and their "stings" are no more than a curious tickle. Many species harbour photosynthesising micro-organisms and live a peaceful, symbiotic existence. On a summer voyage across the Arctic Ocean many years ago, I was astonished by the profusion of life. Waters pumped full of nutrients from melting ice grew a throng of jellyfish and crustaceans. The sea was like one of those virtual universes that are used to suggest motion through infinite distance, where stars and galaxies whiz by in sprinkled abundance. In this cold sea every astral body was an organism drifting endlessly onwards. Some of the jellyfish were of the common kind known as medusae, with bells and tentacles, but others looked rather different—more like barrage balloons, divided into segments lengthways. I was looking at another kind of jellyfish, belonging to the comb jellies, or sea gooseberries (Phylum Ctenophora). The lines running along the sides of the animal were clustered cilia beating synchronously to drive the jelly onwards through the sea, making these animated balloons the largest ciliated organisms on earth. This separates them from their pulsating, cruising medusan neighbours in the sea; modern molecular studies have led to the separation of the comb jellies from the rest of the jellyfish on deeply different evolutionary pathways. It is truly astonishing that there are fossils demonstrating the presence of such confections of jelly and tendrils in ancient oceans; yet undoubtedly Cnidaria and sea gooseberries are now known from Cambrian strata. While I was in the old Chinese capital of Nanjing, I was thrilled to handle a specimen collected from the Chengjiang fauna (Lower Cambrian, Yunnan Province) that was instantly recognisable as a fossil of one of the barrage balloons. It even showed the terminal mouth and "combs" typical of the Ctenophora; it rejoiced in the name *Maotianoascus octonarius*. It was preserved as a delicate film, almost like a watercolour that had been impressed upon the shale that had enclosed the fossils for more than 500 million years. Such fine preservation is almost a miracle (see image 37).

The medusae that drifted and pulsated through the Arctic seas dangling their delicate tentacles were all species belonging to the much larger Phylum Cnidaria. These kinds of jellyfish have another way of arranging their anatomy if they are attached to the bottom of the sea (or it could be a lake). This is the polyp stage, perhaps most familiar in the guise of the sea anemone, anchored to a rock. It is as if the floating animal had been turned upside down. A circle of tentacles surrounds a simple mouth, but their goal is still to ensnare passing prey. A simple nerve net warns the tentacles to retract if danger threatens. Some cnidarian species alternate a polyp stage with a medusa (the latter being the reproductive phase). By now it will not be surprising to learn that a Cambrian fossil cnidarian of the polyp variety, *Xianguangia,* is preserved, tentacles and all, in the Chengjiang fauna. A separate and large group of cnidarians are even better adapted to life on the sea floor: corals. Many corals build hard skeletons out of calcium carbonate over which their polyps are draped, so that in feeding mode they resemble a field of voracious flowers spread out to catch prey rather than sunshine. As their skeletons are accreted layer by layer, they grow upwards and outwards, competing for space. Many reef corals compound the analogy with flowers by using light for sustenance: anemones *sub aqua,* indeed. They house symbiotic algae (dinoflagellates called zooxanthellae) in their tissues that supply most of their nutritional needs. Since this association flourishes best in the warm, clear seas of the tropics, most coral reefs are to be found there; as fringing reefs around atolls, or barrier reefs, or scattered patches. Dozens of coral species jostle for light and space, almost as if they were forest trees struggling for space in the canopy. In due turn they make dozens of specialised habitats for fish, urchin, prawn or worm, which is why reefs dazzle with their biological diversity. I should add that deep-water corals that are capable of building quite respectable reefs have been collected in recent years, so corals can thrive even in the absence of their photosynthesising lodgers. Many new invertebrate species must await discovery in the dark and inaccessible crevices made by these hidden cnidarians.

Corals and jellyfish are still evolving: nothing ever stays completely unchanged in the oceans. Yet medusae and comb jellies are also messengers from deep geological time. They have moved with the ages, but only along their distinctive and fundamental tracks. They insert on the tree of life above sponges, but below the bilaterally symmetrical animals that have been the subject of much of this chapter. That they are only one step up from the sponges is proved by their internal organisation; unlike their goblet-like precursors, they are blessed with a stomach and a rudimentary nervous system. Some of my human acquaintances could hardly claim more.

There are no corals in the same Cambrian rocks. The trick of building limy skeletons was not acquired for another 50 million years. During the Ordovician Period the first reefs grew, constructed by the new marine builders. The potential for a productive habitat is something that nature always discovers very quickly. Trilobites scuttled then among the branched corals; brachiopods were tucked into cavities inside the reef front; early relatives of *Nautilus* cruised the surface, looking for prey. We would all recognise a reef teeming with life. Yet the truth is that these early reefs are *not* related to the similar-looking structures that now contribute to our Western ideal of a tropical island. The corals that built the first coral reefs belonged to a different group of cnidarian monumental masons. They have internal structures quite distinct from those of living corals. The early corals were not survivors—they disappeared along with so many other marine organisms in the mass extinction at the end of the Permian Period. Corals that construct reefs today were almost certainly a separate group of cnidarians that learned the same constructional skill all over again. If there is an opportunity, life will find a way, even for something so specific as building a coral. As I have said, just because a group roots low in the tree of life does not imply evolutionary idleness. Jellyfish can be as inventive as mammals, in their own spineless and quivering way.

What one *can* predict from their position on the tree of life is that jellyfish (of both major types) should have appeared before all the bilaterian animals.* They have some, but not all, of the characteristics of animals higher up the tree, and life moves ever onwards in discrete steps. The fossil record of many bilaterally symmetrical animals extends to the early Cambrian Period, perhaps 520 million years ago, or even further. The invertebrates described in this chapter still survive in today's oceans, living evidence of such deep ancestry. So it is necessary to look in rocks that are still older than Cambrian to read the earlier pages of history; evidence of the earliest jellies ought to be found in rocks laid down during the Ediacaran Period. Are they there among the fossils of those curious, soft-bodied creatures that flourished before animals got hard, mineralised shells? Ten years ago, a number of fossils of this age were indeed identified with medusae, but now that the form of many Ediacaran animals has become better known, several disc-shaped fossils have been reinterpreted as the "holdfasts" of animals like *Charniodiscus*. Others have been rejected entirely, as being of inorganic origin. But in the latest review of the genuine animals by Mikhail Fedonkin and his colleagues, I am happy to say that there are still Cnidaria listed among the Ediacaran fossils, such as *Medusinites*. So it seems that jellyfish must have drifted onwards for hundreds of millions of years, outliving all the dramas that beset the rest of life. When seas became poisoned in the Ordovician or Permian, there was a safe corner of an ocean for jellyfish; when meteorites struck, jellyfish drifted onwards on the other side of the world; when ice covered the continents, there were open seas enough to ensure their survival. The surface of the sea was a good place to be if oxygen became scarce in deeper waters. Cnidaria are tolerant of different temperatures, so phases of global warming held no fears for them. Even today, when fish stocks are over exploited, jellyfish become more abundant; they "bloom" in the hard times. While

---

* Some authorities prefer to regard the comb jellies as closer to the bilaterally symmetrical animals than the Cnidaria. Others maintain that the two kinds of jellies are "sister groups." We have seen previously how placement of early groups is often controversial. Creationists have taken this as a sign that science is "wrong." In fact, it is only to be expected that there will be difficulties in resolving early history, precisely because there are ambiguities introduced by a richness of intermediate forms. Or, to put it another way, because evolution happened.

plants and animals colonised land, leading to the next great adventure in life, the jellyfish continued their dignified procession through the ocean as they had always done. Bells and dangling tentacles will see us impertinent humans disappear into oblivion. Compared with human history, the seas are eternal, and the medusae pulse on and on, like an unstoppable heartbeat.

6

Greenery

To create a little flower is the labour of ages.
—WILLIAM BLAKE, *The Marriage of Heaven and Hell*

Now I am after a small herb with a distinguished history. I have come to Norway to look for the early history of land plants. Norway is also probably the most fortunate country in the world, although most Norwegians are far too modest to boast about it. Perhaps they recall that it was not very long ago that they were rather a poor nation dependent on codfish and potatoes, and glued together by an unsparing Lutheran church. Now they have oil revenues and a government prudent enough not to squander the proceeds too quickly. Many Norwegians living in towns still have family cabins out in the country, so they enjoy well-ordered urban lives during the week, but can sail on the fjord or wander through spruce and birch woods in their free time. Many Norwegians stop work in the early afternoon during the summer months, the better to enjoy these opportunities. They have the good sense not to treat long hours as an innate measure of virtue. Many of the same people are also tall and good looking. As I say, they are a fortunate nation.

My old friends (tall, good looking) take me to their cabin in the hills north of Oslo on a warm September day that reveals the first colours of autumn. The more fertile, undulating open land has been cultivated for centuries, and when we arrive the cereals had already been taken into the barns. The hillier areas are still largely wooded, thickly covered with conifers. Off the major roads, tiny byways take us into the Palaeozoic hills. It turns out that virtually all the cabins in this particular neck of the woods are the property of Norwegian geologists. Alongside vigorous unpolluted streams that bounce through the wooded slopes aspen and birch trees make charming glades all mossy underfoot. For a while I am distracted from my search. It is a treat for me to go to the cabin while the wild mushroom season is at its peak, because the ground seemed to heave with the things, showing off every hue and fungal form. I recognise familiar genera: brown *Boletus*, yellow or scarlet *Russula*, orange *Cortinarius*, pinkish *Lactarius* . . . the list goes on. Many of the species that I knew only as rarities in the woods at home are common everywhere: cryptically coloured hedgehog fungi lifting up from every bank, chanterelles as yellow as egg yolks by the path sides, unusual pale relatives of the deadly death cap (*Amanita phalloides*) springing from mossy carpets. Not for the first time I wonder what possible biological function the brilliant (or come to that the pallid or dark) colours of fungi could perform, since the fruit bodies are basically machines for distributing spores, where pink is as good as purple, or so one might think. Then my eye is caught by clumps of a dark green plant that looks at first glance like an unusually tall moss. It makes patches rather more than hand height in the open glades, a series of narrow leafy columns with a slightly bristly appearance. It is immediately recognisable as a lycopodiophyte; a herb my friend Anne Bruton narrowed down to the genus *Huperzia*. It is one of the survivors I had been seeking.

Compared with some organisms rooting far back into the tree of life, *Huperzia* is not uncommon. It is present in many northerly localities in damp places, though I doubt it would be as pretty anywhere else as it is in the Norwegian woods. This lowly herb is a descendant of a great plant group that included massive trees that flourished in the Carboniferous Period about 300 million years ago. Tree trunks from these lycopod giants supplied one of the major components of coal seams, and so helped power the industrial revolution. One of the first fossils I found in my childhood was an impression of the bark of the Carboniferous tree *Lepidodendron* that had arrived on the back of a piece of shale as part of a delivery of household coal. It had the same surface texture as the crocodile skin handbag my mother used for smart outings, a raised pattern of rhombuses and lozenges, but all as black as a bowler hat. I remember feeling it gently, much as a blind man strokes a sheet of Braille. I was later to learn that the fossil tree was related to clubmosses such as *Huperzia* and *Lycopodium*, but blown up in size a hundred fold. The links on the tree of life connecting my fossil with the living herbs stretched back through two mass extinctions, one at the end of the Permian Period and one at the end of the Cretaceous. *Huperzia* is a botanical equivalent of the horseshoe crab.

Huperzia takes the story of life back even further than the Carboniferous, as far as the first days of the greening of the land. During the Silurian, about 420 million years ago, plants were well under way with the most important scene change in the history of our planet since the adoption of oxygenating photosynthesis billions of years earlier. Plants had moved out of the protective and supportive bath of the waters into thin air. It was a move with unprecedented consequences. After that transformation, animals could follow, and the boundless possibilities of terrestrial life could begin to be exploited by organisms small and large. It started with a few species of herbs around muddy flats. One of them was *Baragwanathia*, a fossil genus first known from Silurian strata in the state of Victoria in Australia. This ancient herb does bear close comparison with *Huperzia*; it has the same prickly columnar build, the same little leaves, and it too bears its spore-carrying structures in the axils of the leaves. Botanists would insist that the "leaves" of both plants have a single conductive strand in the centre, and should strictly be called "microphylls" to distinguish them from the more complex veined leaves of higher plants. Most important of all, both *Huperzia* and *Baragwanathia* are *vascular* plants, which means that they have stiffened water-conducting tubes (tracheids) within their stems. This is the most important piece of mechanical engineering required to achieve terrestrial status; without it, plant life on land would have been a flop, in the most literal sense. Think of the hopelessly flaccid condition of seaweeds when the tide goes out. Woody, or lignified, tracheids put the backbone (not in the literal sense, this time) into the invasion of the land: they were supply pipes and scaffolding in one.

Squinting closely at *Huperzia,* I have to imagine the tree concealed within the little herb: after all, once established on land any plant that captures the most light by growing up and away will steal a march on its fellows. Plants have been competing for light ever since the early days. It would be helpful for an ambitious plant to divide into roots for absorbing nutrients and water, and into stems and leaves for reproduction and photosynthesis. But other, subtler changes were necessary before upward growth could happen. Water loss into this new thin stuff—air—was going to be a problem in the drying sunshine: a thin waxy coat (cuticle) over the plant's surface would fix it. However, this created another problem: how to perform the gaseous exchanges like absorbing carbon dioxide, or releasing oxygen and water vapour, that are fundamental to the processes of photosynthesis and respiration? This problem was solved by locating these functions in special sunken cells inside the leaf. These important cells were protected by "guard cells" that close up to save evaporation during hot weather—a kind of cellular air conditioning and gaseous treatment centre rolled into one. "Stomata" is the name given to these structures. Fossils need to be well preserved to reveal them, but careful preparation of fossil cuticles has shown that stomata were indeed present in the early plants as they groped upwards into the air. In one locality in Scotland known as the Rhynie Chert, 400-million-year-old fossils of early plants were preserved in the deposits formed around a siliceous hot spring very like some of those in Yellowstone Park. Every plant cell is retained in this fossilised sinter with almost miraculous precision. One could say that the anatomy of these fossils is known as intimately as that of their very distant relatives growing in the Norwegian pine forests.

But even they were not the first. Some pioneering little green pad crept over a moist muddy riverbank before there was any ability to lift up into the air. It probably had little in the way of roots and shoots, but it certainly spread by means of spores that could be carried on the slightest breeze to another suitable habitat. We do not have to go to Norway to find such a plant, nor to the Sahara Desert, nor yet to Amazonia. Any damp dell will do, any neglected bridge over a swampy drain, any dank bank on a forgotten canal. I have several suitable sites within a short walk of my home. The plant is called a liverwort. The name indicates a very general resemblance to the shape of liver,* and the old botanical name of hepatics reflects the same etymology. Somehow the modern name Marchantiophyta—though doubtless correct—does not have the same appealing antiquarian ring to it. I'm off to collect hepatics!

The ones nearest to where I live grow beside streams and in shady corners by the River Thames. Dark green and crimped, somewhat like the leaves of the more recherché varieties of lettuce, the thalli of liverworts cover low banks by the water's edge. They grow up onto the damp boles of trees where the water splashes up; they will happily adhere to vertical surfaces. The habitat has a particular smell: a kind of watery and mossy perfume, invigorating like freshly poured beer. Small umbrella-like structures that arise from the pads are responsible for producing and shedding their spores. They do not have stomata, not even roots in the conventional sense. They are just about the simplest photosynthesising laminae one could imagine. Despite the passage of 400 million years or more, liverworts still have an ecological niche, even if hardly a showy one. There is evidence that liverworts, or something rather like them, came out of the water ahead of any relatives of *Huperzia,* or even the still simpler *Psilophyton* plants that are among the earliest plants in the fossil record. Spores with morphology consistent with dispersal in air are found right back into strata of Ordovician age, perhaps as old as 475 million years. A spore sac with a wall structure rather like that of modern liverworts was discovered by my colleague Charles Wellman in rocks of this age a few years ago. Unfortunately, most of the fossils that might represent these earliest pioneers are just small black blobs of carbon on the surface of shale that do not preserve any details, and a sceptic could insist with some justice that they cannot be nailed down as a liverwort with any degree of confidence. It remains likely that these green

---

* There are several charming old plant names with "wort" attached (it is the Anglo-Saxon word for plant). "Quinsywort," "rupturewort" and "woundwort" indicate something of their herbal use. One species of liverwort, however, was reputedly used as a protection from the bite of a mad dog; but apparently not for the treatment of the liver.

pads pressed against damp soil do indeed furnish a vision of how life weaned itself away from the necessary embrace of water. Mosses are hardly more complex, making soft green cushions wherever moisture and nutrients allow, and they also probably appeared very early in the history of colonisation of land. They still thrive on walls and on forest floors, and tucked away among grasses.

The colonisation of the land by plants had one consequence that justifies the somewhat overused expression "feedback loop": decaying organic matter interacted with rock to produce soils, which in turn supported better-nourished vegetation. Plants made soil made plants. Bacteria and fungi must have been part of this process, even though their fossil record is so rarely visible. Remember that the fungi branched off from the tree of life far down in the Precambrian, so they must have been silent partners in the subsequent story. Their skill in breaking down cellulose and helping generate soils was as essential in those early days as it is now. The enduring presence of such quiet survivors was as important as the newest breakthrough in evolution. The association of fungi and root systems may have been present from the first. Fungal threads became enclosed in the earliest roots to make what we know today as mycorrhiza. In this association fungi contribute nitrogen and phosphate to the growing plant, and receive sugars in return, in the most intimate of symbioses. Humid soil then provided a sheltered habitat for a "degrader community" of small animals like mites and nematodes; diminutive predators in turn preyed on this secretive soil fauna. They still do. A remarkable fossil occurrence from the latest Silurian strata of England (423 million years old) shows that some animals in this community had already established themselves from the earliest days of terrestrial life. Pseudoscorpions (and they really do look like minute scorpions) preyed on mites as well as tiny flightless insects, the springtails. It is more than likely that soft-bodied "worms" of various kinds were there, too, although they left no fossils. Tucked away from sight, this community endured while the world above ground went through transformations and extinctions, surviving the great biological crises at the end of the Palaeozoic and Mesozoic Eras. They just quietly got on with the job of turning litter into soil, a process that Charles Darwin called "the formation of vegetable mould." In their own way, the vegetable "degraders" comprise as remarkable a survivor as the stromatolite community, for all their ubiquity. Common as dirt, as my mother would have said. Zoologically speaking, of course, they belong in a different part of this book, but I am going to tuck them in here, where they have belonged for so long, under the plants.

I have met with another relative of *Huperzia* all around the world, but only in warm climes. This little plant grows on field edges in Thailand, creeping inconspicuously where water buffalo trudge through muddy paddies. It grows up in the cloud forests of Ecuador along paths snaking through trees dripping with epiphytes. It abounds in the wetter part of New Zealand, forming charming moss-like patches in the podocarp forests. I have even seen it as a "weed" in tropical houses in botanical gardens, under the stages, keeping out of sight and out of trouble. The plants are as delicate as lace, and like lace, often make frills and dainty folds. *Selaginella* is another lycopod, and when it "flowers" it throws up little spore-bearing cones (*strobili*) on stalks; *Huperzia* had matching structures tucked among its "leaves." Similar organs can be found in fossils from strata hundreds of millions of years old. *Selaginella* has been called "spikemoss" because of its characteristic appearance in its reproductive finery. Its erect strobili bear two kinds of reproductive bodies in separate receptacles: they become large female megaspores, or tiny male microspores, respectively. The microspores can be readily dispersed by the wind to fertilise other plants elsewhere (they germinate inside the spore wall and in due course release sperm). The megaspores stay put, and undergo internal changes until they are ready to be fertilised by the sperm, whereupon they remain as a semi-parasite on the parent plant; finally, they fall to the ground to begin an independent life, and repeat the cycle all over again. It may sound a little complicated as life cycles go, but it provides a hint of what is to come: a larger female awaits the arrival of a tiny male travelling from another plant. This is a precursor of the seed habit, which dominates the green world today.

Ferns are the least fussy of plants and have been around for millions of years. They maintain a certain measure of exuberance even in the shadow of trees where nothing else will flourish. Some resemble fountains of fronds, delicately divided, leaves curling upwards and outwards. They unwind like those honking party tooters blown by irritating guests. The young fronds are called fiddleheads in north-eastern Canada, and they can be eaten, in moderation. One fern has supplied food to aboriginal Australians for centuries, the nardoo (*Marsilea drummondii*). But its edible bean-shaped spore-bearing body has to be prepared very carefully for food. The transcontinental explorers, Burke and Wills, died slow deaths by failing to follow a necessary dilution of the nutritious soup as practised by the native peoples. Despite their appearance of luxuriance, few ferns provide nibbles for animals. They have been around for far too long and developed too many chemical defences over millions of years to make comfortable forage. The tough creeping fern called bracken seems to be almost indestructible; there are a handful of specialist insects that can make a meal of it. After disaster sweeps over a landscape, ferns are always among the first colonisers. They spread by means of spores to find a foothold in damp crannies, greening the land all over again in an accelerated reprise of that first painting of the land in photosynthetic drag that happened in Devonian times. When Mount St. Helens erupted in 1989, ferns sprouted up from under the ash itself, regenerating from buried rhizomes. After the crisis that led to the extinction of the dinosaurs and many other creatures at the end of the Cretaceous, there was a brief triumph of the ferns in many parts of the world. Those who study fossil pollen and spores recognise a "fern spike" typified by enormous quantities of fern spores in strata laid down immediately following the disaster. Landscapes blasted by wildfire or destroyed by some mass mortality soon provide an opportunity for ferns. Ferns do not need much to thrive. I have seen mighty examples dangling as epiphytes from tropical trees looking for all the world like sets of mounted antlers hijacked from some English stately home—except for being almost lurid green. I have seen them appear as if by magic in an abandoned well or at the entrances of caves. In New Zealand and Hawai'i tree ferns are so common in some places that the visitor might wonder whether he had wandered out of time and into a Jurassic landscape. There is nothing apologetic about the survival of ferns. I do not doubt they will survive our own species. Even in one of those dystopian landscapes imagined by writers like

J. G. Ballard where the whole world has become an endless building lot, there would surely still be ferns hanging on in the cracks in the drainpipes. So I do not have to make an expedition to find them. I can see a hartstongue fern from where I write these words, happily growing out of a wall (nor did I invite it, it just arrived).

True ferns have a fossil record going back to the Carboniferous Period, perhaps 345 million years ago. However, as Robbin Moran points out in his delightful book *A Natural History of Ferns,* the majority of living types of ferns probably evolved alongside the flowering plants. The group as a whole may have roots in deep time, but has continued to evolve new forms throughout its long history. Nonetheless, there are a few genuine survivors among them. In the second decade of the twentieth century a species of *Dipteris* was discovered buried in the forests of New Guinea, a fern that belongs to a family that was diverse at the time of the dinosaurs. When working in the Malay Peninsula, I saw another relic from the same time period called *Matonia.* Both plants look somewhat like miniature palm trees or perhaps outstretched hands with too many digits, rising from rhizomes creeping along path sides and in clearings. Their perched fronds just seem to pop up from nowhere in particular. By contrast, filmy ferns are as delicate as tissue paper, with fronds so thin as to be almost transparent, and as complex as filigree. They grow only in conditions of high humidity. I have seen them on tree trunks on the wet side of the South Island of New Zealand and in damp cave mouths in southern Europe. It is surprising that something so flimsy could be fossilised, but there is a Triassic example that proves that where spray drifts from a waterfall, or where damp moss decorates moist undergrowth, these specialists have lasted for nearly a quarter of a billion years.

When I was a child there was a little glasshouse tucked away at the edge of Kew Gardens in London that housed the collection of filmy ferns. It was a secret place, and esoteric compared with the orchid house with its blowsy showiness. It attracted few visitors. I can still recall its damp and slightly astringent smell. To go there was to understand the inexhaustible subtleties of shades of green, and of crimping and cutting of fronds. Under the slatted stage a moist haven was provided for growing baby ferns. Spores could germinate there to produce tiny green pads, where the sex organs developed on prothalli, both male and female. Coddled in the dampness male sperm swam to the egg, fertilised it, and the resulting cell (zygote) produced the embryonic fern, which was then free to grow. In most ferns, the brown dust of spores is released from little packets called sori on the underside of the fronds. The lightness of these minute reproductive packages allows the rapid spread of ferns to colonise fresh lava flows, or even the wall outside my study window. For serious survival, it sometimes pays to blow with the winds. Fern groups did try out the alternative seed habit, investing more energy in a large, female "propagule" which is well supplied with food after germination, thereby ensuring a greater individual chance of survival. Seed ferns, which are now known to be a very heterogeneous collection of plants, are actually rather abundant fossils in the Palaeozoic, but they did not survive to the present day. The dusty spore strategy won out in the longer term.

The same technique worked just as well for the horsetails. Gardeners I know would ban these particular survivors if they could. Spreading from underground rhizomes, they seem to be impossible to exterminate from the vegetable patch. Lacking leaves, their curious growth is comprised of a number of joints that pull apart rather readily into similar-looking units. They feel somehow like do-it-yourself assembly plants, as if they could be supplied in kits. They could be described as resembling delicate-looking, miniature Christmas trees, but they can also make thickets if conditions are just right. Their spore-bearing cones poke up from the top of the plants. Around freshwater springs horsetails may be the dominant vegetation. In South America I have walked underneath giant examples in clearings in cloud forest. I have seen specialised horsetails growing as scrawny shrubs in deserts. In the long race of life, horsetails are among the marathon runners. More than 300 million years ago, rather leafier giant horsetails lined boggy creeks, and they make common fossils in strata of that age. Horsetails survived into the era of ruling reptiles 150 million years later with their vigour hardly diminished. Laid out on a rock, their fossil stems might be a little reminiscent of bamboos, to which they are utterly unrelated. Today, horsetails creep along wood edges, discreetly colonise streamsides or pop up among the potato plantings. They are not coy survivors from a previous era slinking down some sheltered redoubt. Horsetails are well able to cope with conditions in the twenty-first century for all their long pedigree. One could imagine an early amphibian crawling between their feathered forests, or maybe an early dragonfly launching off into the sky from one of their jointed branches. Not showy, just a herb doing the job.

I have mentioned trees several times in this book. These have mostly been evolutionary trees, man-made constructions that portray the relationships between organisms. Such trees are in a philosophical sense, theories, continually subject to updating and revision as more information from fossils or molecules is gathered; not everything is permanent about them. It is therefore something of a pleasure to be able to turn to a *real* tree, another survivor from deep geological time, *Ginkgo*. Although this attractive tree is now extensively planted, I need to see where it came from. Why should it survive for more than 100 million years? Its "wild" habitat is confined to just a couple of sites in a mountainous forest region of southeast China. I require help to find the surviving survivors. An old friend from Nanjing, Zhou Zhiyi, offered to take me there. Since everything in China is done on a personal basis, Zhiyi brings along a botanist friend as well to take us to the celebrated trees, and a driver. It begins to be quite a cosy party. My Chinese friends all chat among themselves about where it is best to have lunch. With the possible exception of Italians, the Chinese people I know take food more seriously than anyone else.

The ancient city of Hangzhou on the Yangtse plain is surrounded by a forest, but it is a forest of new developments with tower block endlessly following upon tower block. To one who had visited the old China, the relentless new buildings come as a shock. Many of the new towers retain a vestige of the traditional roof with turned-up corners, but elevated implausibly atop thirty floors or so of identical apartments. I am reminded of the peculiar, kitsch towers that provide cheap accommodation in Las Vegas, where a "Roman themed" roof might be perched above a nondescript tower block, supported by a few token columns beneath. When I look at the restrained good taste of the traditional pagodas, I wonder about these new towers as the price paid for China's current affluence. But their vigour cannot be

Bas relief of the leaves and fruit of *Ginkgo biloba*, Eastern Parkway
entrance of Brooklyn Botanic Garden, New York.

denied. It came as a relief to see mountains in the distance looming
in misty ranks just as they do in the classical Chinese paintings. Little
by little, traffic jams are left behind as the road climbs upwards into
gentle hills. The lower slopes are clad in ranks of elegant bamboo
canes, which are likely destined to become scaffolding for the new
buildings. Old-fashioned farmhouses with wooden balconies survive among the bamboo groves, each only a couple of storeys high
and topped by roofs made of dark pottery tiles. The road continues
upwards, and the bamboos eventually give way to plantations of the
conifer *Cryptomeria*. Above that, again the road loops this way and
that among small chestnut and pine trees: real countryside at last.
As is the tradition in China, we get lost, which involves much cross-
questioning of local peasantry, who gesture vaguely at the mountains
beyond. We reverse down one little road and go up another one, and
this time we are on the right track. In the Tianmushan Forest Reserve,
Zhejiang Province, the little road terminates after dozens of hairpin
bends at about 1,800 metres above sea level in a small, neat square of

traditional low buildings made of dark wood and fronted by veran-
dahs. As everywhere, there are hopeful souvenir sellers trying to
attract the Chinese tourists to buy their trinkets; there are hardly any
Western faces in these parts. "Here Buddhist and Taoist relics coexist
harmoniously" a notice informs us. From the square onwards, the
way is on foot along a well-made stone path which weaves dramatically along a precipitous forest edge through groves of what must
be largely native trees perched on the slopes. So much of China is
under human bidding that it is rather wonderful to discover an area
of apparent wilderness. The view outwards from the path takes in a
series of sharp, inaccessible peaks clad with trees and disappearing in
the distance progressively into the mist. There are even a few whistling birds, which have all but vanished from the plains. This is the
China of my imagination.

The path starts rather higher up than the appropriate habitat for
ginkgo. Many of the plants here have a familiar look, and the genera—if not the species—are well known from gardens and hedgerows
in the United Kingdom. Roses (*Rosa henryi*), dogwood (*Cornus*),
walnut (*Juglans*), hornbeam (*Carpinus*), oak (*Quercus*), *Viburnum*
and ash (*Fraxinus*) are all old friends to an Englishman (or a North
American). Even the more exotic trees like *Liriodendron* (tulip tree)
and *Lithocarpos* have a place in old landscaped gardens around the
stately homes of England. It is hard to overstate the contribution
of Chinese species to European horticulture. The high mountains
in eastern China allow for a tongue of this cool climate "Palaearctic
flora" to extend further south than the rest of its global distribution
running across wide swathes of north temperate and Arctic latitudes;
such trees are not typical of much of the warmer far east, but they
can readily transplant to the milder gardens in suburbs and parks
in Western Europe. The same is true of the adaptable ginkgo. The
path now takes a dizzying dive downwards along a sheer rock face
and through a cleft in massive limestone cliffs. In the old days priests
would have taken the same path into the hills without the benefit of
handrails. Twisted conifers grow from the tiniest crevices. They make
natural bonsai, with the elbows, doglegs and crooked turns that are
so appreciated by specialist gardeners. One of them is stated to be
eight hundred years old. From Four Sides Peak, a label tells us, "one

can see rolling mountains rising and falling like 10,000 horses galloping . . . one cannot help wondering if they are in fairyland." No doubt it is less purple in Chinese, but I know what they mean.

Ancient shrines appear as niches beside the path as the steps plunge further down. An attractive conifer called *Cunninghamia* with bristly, dark pointed leaves like old-fashioned pipe cleaners briefly attracts my attention. But then I see the oldest ginkgo tree sitting at the top of the steepest scarp slope. It is still early April, so its leaves have not yet fully appeared. When the path finally winds upwards again, the tree can be seen to be an aged natural coppice. The thoughtful Reserve staff have placed a wall around it to stop ginkgo lovers plunging fifty metres over the vertical crag. The most ancient trunk is a gargoyle of a thing, with a twisted shape and covered with warty lumps, and with zig-zag twigs adding further decoration, looking almost like forked lightning outlined against the sky. But a dozen or more younger trees sprout from around its base showing straighter trunks of various sizes. This must explain why the tree is known locally as "the family of five generations." Rather than dying, the tree has sprouted young clones to keep the line alive. Buddhist fragments retrieved from nearby shrines prove that the area was once holy ground. This fact has led to a protracted debate about whether the ginkgo trees here are truly wild, or whether they were planted in the mountains as a tree revered by the priests. Could it have been religion that saved the species from extinction? There is evidence for the special place held by the ginkgo in this part of China as a kind of substitute for the *bodhi* tree under which the Buddha spent so much time in meditation. The climate is too cold hereabouts to allow growth of the original *bodhi* tree, which is the spreading evergreen fig (*Ficus*). So the ginkgo with its slightly leathery, triangular leaves split into two lobes may have seemed a suitable replacement. Priests also found another use for the ancient tree: they sucked the seeds to suppress the desire to urinate while they were meditating. The giant conifer *Cryptomeria,* a Chinese version of the giant redwood, grows along the same mountain path, and is thought to have been a native of Japan introduced in ancient times. One of these mighty trees is claimed to be more than a thousand years old, and has the same kind of majestic, soaring dignity as their ecological equivalent on the western coast of America. A millennium is not so vast a time span within the leisurely compass of Chinese culture. There is still a small Buddhist temple functioning in the hills, and it is not difficult to imagine the centuries drifting by in daily ritual as little happened except for the ginkgo putting up another shoot from its ancient rootstock. However, it does seem difficult to imagine anyone deliberately planting such a remote tree as "the family of five generations," and this is certainly a habitat in which the ginkgo seems at home. There is at least one other site in the inaccessible hills in Zhejiang Province where the ginkgo survives in similar fashion. Molecular work has suggested that the two populations may have been separated for several thousand years, taking us back to a time before Buddhism had spread to China, but this evidence has not gone unchallenged. The true "wildness" of the ginkgo might be an historical question without a definitive solution. There is circumstantial evidence in the endurance of other survivors in the same area, which I shall come to later, that suggests the whole region might provide a refuge from extinctions elsewhere. The answer to the ginkgo question partly depends on how this part of China fared during the last ice age. During the high point of the glacial period twenty thousand years ago, ice caps on the mountains would have driven the ginkgo into warm glades lower down the slopes: in this case, it has been claimed that the two "wild" populations may have been the last survivors. Perhaps the enigmatic smile of a small Buddha on a pedestal recovered from near the ginkgo site is a comment on our pretensions to know everything. If he knows the answer, he's not telling.

Far from its native redoubt, *Ginkgo biloba* is now one of the most widely planted trees in the world. It is tolerant of high levels of pollution, so that the attractive tree can be found on city streets, apparently thriving where little else will grow. Beijing and New York both have avenues lined with ginkgo trees. They have been made into bonsai, and trained neatly up walls. The two sexes are on different trees (*dioecious*), and the male tree is generally the one under cultivation outside China. This is partly because the fleshy seeds produced prolifically by the female smell so bad. They are soft (often wrongly called "fruit" for that reason) and pale yellow in colour, looking something like small grapes. They are widely used in Chinese cookery and medicine:

it is claimed that they are good for the memory. They taste slightly creamy—the Chinese name actually means "white fruit." "No more than ten seeds a day," my Chinese hosts inform me. Presumably, eating ginkgo seeds would help one remember how many ginkgo seeds one had already eaten. It is probably not so surprising that *Ginkgo biloba* has an impressive biochemical array, for its kind has been around for a very long time, and survived the attention of browsers among the dinosaurs, and the mammals that followed them. After the atomic bomb was dropped on Hiroshima at the end of the Second World War, ginkgos were the trees that regenerated from scarred trunks one year after the total destruction. Their fan-shaped leaves are not like those of any other plants, having veins radiating into them from their petioles, like spokes. They are therefore easily recognised in the fossil state, and fossils tell us that once they were widely distributed. At present, the earliest fossil member of the Ginkgoales is early Permian in age, about 280 million years old. This means that the ginkgo line must have survived several mass extinctions. Atomic bombs and even the Beijing atmosphere are probably small potatoes in comparison.

The range of the ginkgo tree extended early from China, accompanying the spread of Buddhism. It was cultivated in both Japan and Korea. The doyen of ginkgo studies in China, Zhiyan Zhou, told me that it had already been portrayed in paintings from the eastern Jin dynasty (312–420). Elsewhere in China, and a long way from Tianmu, some vast and ancient trees survive that have undoubtedly been placed in sacred sites. In Shandong Province there is some evidence of trees planted as early as the Shang Dynasty (1600–1100 BC). The long history of the cultivated tree evidently meshes with the expanded Chinese view of time. Further afield, the ginkgo was one of the first trees to be planted in Kew Gardens in west London in the reign of George III. We know that Goethe had such a tree in his garden in 1815, for he wrote a love poem that year using the cleft leaf as a metaphor:

> *Does it represent One living creature*
> *Which has divided itself?*
> *Or are these Two, which have decided,*
> *That they should be as One?*

The original draft of the poem survives, with a pair of *Ginkgo biloba* leaves charmingly crossed at the bottom of the page, but, sadly, Goethe's original tree does not.

The ginkgo tree is also at a crossroads in the evolutionary sense. It is certainly a primitive plant, and inserts low down in the tree of life, but the question remains where and what are its closest relatives. Different lines of evidence suggest several candidates. The fertilisation of the ginkgo female ovule by swimming male sperm was demonstrated in 1896 by Sakugoro Hirase in the Tokyo Botanical Garden, and could be said to mark the beginning of scientific botany in Japan. The discovery that the sperm of ginkgo swam with a flagellated "tail" is recorded in a plaque in Tokyo even today, although when I was there this memorial was largely ignored by thousands in search of pretty cherry blossoms. Such a mode of fertilisation is evidently primitive, and the flagellated sperm cells are similar to those of another group of plants surviving from the Age of Dinosaurs known as cycads. These are tough plants that are frequently cultivated as pot plants, for they are long lived, tough, and happily tolerate city life. Both cycads and ginkgos were much more diverse 100 million years ago, and it might seem appropriate at first to assume that they were closely related. Together with conifers and an interesting plant called *Gnetum,* cycads and ginkgo comprise the gymnosperms, that is, plants in which the female "seeds" are naked and exposed in the unfertilised state. However, when Zhiyan Zhou reviewed the state of the science in 2009, he favoured another theory in which ginkgo is partnered with early conifers and an even older, Carboniferous tree called *Cordaites,* taking us back almost to the dawn of forests themselves. The shape of the answer depends how the investigator interprets and tots up the characteristics of wood, fertile shoots, pollen and embryology in all these plants, including fossils, for which data may be incomplete. It is rare for different specialists to agree on all such characters, and different interpretations result in different evolutionary trees. At this point in the story new evidence from the molecular sequences often comes to the rescue. However, even this evidence is not unequivocal. One Japanese researcher in 1997 apparently found much evidence to reaffirm the ginkgo–cycad connection, but other molecular studies joined the old survivor with primitive conifers. In a way, these very ambiguities point up the singular importance of ginkgo as an intermediate form—why should evolution make it so easy to read the signposts as if at a simple fork in the road? This odd survivor might have borrowed features from more than one evolutionary line, while retaining a whole lot of primitive characters just to further muddy the waters. Trying to parse the language of events that happened more than 280 million years ago was never going to be easy. Unresolved mysteries are all part of the allure of the ancient tree.

Cycads, too, are successful survivors from the Mesozoic. I have come across them in rain forests in Malaysia and in Australia, evidently thriving among the undergrowth. They do not require a special expedition to discover them, although many of the African species are rare in the wild. Most of the ones I have seen are about my height or smaller, but some species make respectable trees, for they live for a long time. They are found through much of the tropics, and many species are tolerant of drought. Usually growing from a single stem, at first glance they look like some kind of palm tree, which they are not, although their leaves sprout from a terminal growth point in a palm-like way, and they do have handsome dark evergreen leaves several feet long, divided into leaflets. These are tough to the touch, and rather sharp and unwelcoming, so any competing resemblance to a tree fern is soon disqualified. Like the ginkgo, cycads have males and females on different plants, and they reproduce by seeds. Reproductive cones on cycads appear at the centre of the plant, in striking structures obviously different from the leaves—if anything, they superficially resemble pineapples. A minimum of one male and one female is therefore required to perpetuate any species. There is actually one species living today, *Encephalartos woodii*, known from males alone, and it is doomed; this is a great pity as it is a handsome plant. It no longer exists in the wild in Natal, South Africa, and existing specimens in collections are all clones of one plant. Although they were once thought to be wind pollinated, it seems that insects may be involved, too, since cycads carrying weevils seem to set more seed. The fact that cycads are very poisonous seems to imply that they needed protection from grazers in the past. They carry a neurological toxin so potent that it can result in death. One cannot readily imagine anything wanting to sneak a cycad snack, although a surprising number of pets are poisoned this way. It comes as another surprise to learn that cycads *can* be eaten, if the seeds are beaten and leached in a special way to extract the toxins. A type of sago produced by this process is bland and more or less pure starch, but there is something rather satisfying about the idea of feeding on something that first evolved in the late Palaeozoic.

In reconstructions of "the Jurassic world" in books and movies, cycads often figure prominently in the background while dinosaurs steal the glamour centre stage; ginkgos and appropriate conifers may appear decorously elsewhere, presumably as fodder. A further group of plants that need to be added at this point are the extinct bennettites, which superficially looked very similar to cycads, although some botanists now believe that their reproductive structures ally them more closely to flowering plants. They possess something resembling a dramatic spike of flowers, although the structure of the individual flowers themselves is not very like that of living flowering plants, and is currently believed to be an example of parallel evolution. All these plants together presented nothing but hard work if they were to provide food for the large herbivores among the dinosaurs: eating the leaves must have been rather like chewing through old copies of the weekend magazines. These animals must have had a way to digest the cellulose with the help of appropriate bacteria, and presumably also a means of countering the hefty doses of alkaloids they ingested at the same time. There was nothing like grass around in the Mesozoic, nor were the shrubby titbits appreciated by browsing animals today available until the late Cretaceous. It was a time of tough, leathery skins and tough, long-lived plants. There have been suggestions that the evolution of large seeds like those of the ginkgo was an adaptation to passing through the guts of large animals, as a dispersal strategy. A similar trick is found in tropical rain forests today: big seeds need a way to travel to new opportunities. Even now, the thin "skin" that covers the cycad seed does indeed prevent germination until it is ruptured. It is intriguing that this feature may have evolved to cope with a sojourn in the stomach of a dinosaur. The evolutionary memories of a vanished world may be impossible to erase.

As for the conifers themselves, a remote valley in Hubei Province, central China, hides another example of a living fossil in the most

literal sense. This tree was described first from fossil Japanese material by Professor Shigero Miki of Kyoto University in 1941, and then
found in the wild, still very much alive, two years later.* *Metasequoia
glyptostroboides,* known by the common name "dawn redwood,"
is a charming and graceful deciduous conifer, which is commonly
planted in damp sites in botanical gardens in Europe and America. In
spring, the fresh foliage makes the tree look as if it has sprouted pale
green feathers all along its branches. Fossil examples of *Metasequoia*
are widespread in sedimentary deposits, the oldest dating back to
about 100 million years ago. This conifer was once distributed widely
across the northern hemisphere. It was virtually unaffected by the
events that exterminated the dinosaurs and so many other organisms
at the end of the Cretaceous Period. In the unusually warm climates
that followed the trauma, *Metasequoia* was even found as far north
as Greenland. Since it would not have been able to grow there in the
dark of the winter months, this may well have been where its deciduous habit arose. In Japan, fossils as "young" as about 1.6 million years
old have been recognised. The tree's range has evidently been shrinking for about the last 10 million years. Now, only about 4,000 trees
remain in the wild in its last redoubt in central China. One outlying
specimen from Hunan Province known as "the Paomu tree" is all on
its own. It is a splendid example, forty-four metres high, and it may
be 300–400 years old. The microscopic details of this particular tree
are said to resemble those of the fossils most closely. At the moment
there is some debate about whether the Chinese population is truly
a last pocket of wild trees, or whether they recolonised the mainland
from Japan after the last ice age. The family to which *Metasequoia*
belongs—which includes cypresses and several related conifers—
has been analysed using molecular sequences in the year 2000. One
interesting result was that the most primitive member of the whole
group proved to be a tree that I passed rather hurriedly on the path
as I searched out the wild ginkgo: the dark green, bristly *Cunninghamia.* I wished I had examined it more closely, for it seems that

---

* The story of its discovery and introduction as a cultivated tree is a complicated one that has been documented in detail by Jinshuang Ma in *Harvard Papers
in Botany* 8 (2003).

Fossils (left) and living leaf (right) of the deciduous conifer
Metasequoia. A small number of trees remain in the wild in China.

this particular conifer reaches down even deeper into geological time
than *Metasequoia.* The mountains of China seem to be something of
a time capsule.

I always associate another Mesozoic survivor, the monkey puzzle
tree *Araucaria,* with cemeteries because a tall and melancholy example dominated a Victorian graveyard near my childhood home. I only
seemed to notice it on wet afternoons. Something about the pen-
dulous branches or its sombre colours seemed appropriate to such
a habitat, and I have struggled to rid myself of its deathly connotations ever since: *Araucaria,* the gloomy gymnosperm. This is not
only unjust, but fails to acknowledge the variety of form in a large
genus of trees, some of which are nothing less than exuberant. The
Norfolk Island pine, *A. heterophylla,* is a tiered architectural essay of
a tree. I love to see these not-really-pines* ranged like guards along

* Although customarily referred to as "pines," these trees are only pines to the
extent that they are gymnosperms, and are distantly related to true pines of the
genus *Pinus.*

the promenade at Manly Beach in Sydney, Australia. The Moreton Bay pine (*A. cunninghamii*) reminds me instead of a firework display, with little bursts of branchlets at its distal tips swishing up into the sky. All are covered with the scale-like leaves typical of their family. Old trunks become pocked, wrinkled and lobed like elephant legs. Even the dolorous monkey puzzle tree (*A. araucana*) is magnificent in its natural setting on the slopes of the Andes in Chile, where it makes groves that look naturally Mesozoic. There are nineteen different species of *Araucaria* scattered around the southern half of the world. Like *Ginkgo biloba,* most monkey puzzle trees have the sexes on separate trees. The characteristic big spherical female cones, looking like green pineapples, are usually perched on branches high up in the tree, while the more slender, and often leafier, male cones are carried throughout the male plant. The seeds were important as food for native peoples in South America. The "hot spot" for diversity in *Araucaria* is on the island of New Caledonia, where a baker's dozen of endemic species occur. Elsewhere, different species tend to be found in separate areas or islands, each one presumably having evolved in adaptation to its own environment. Some species grow happily in rain forests, so they seem to be able to compete with the flowering plants (angiosperms) that are often thought to have displaced gymnosperms in that habitat. The South America–Australia distribution will by now strike the reader in a familiar way: it's that ancient Gondwana map still written on the ground today. The implication is that these trees originally date back to the time when today's dispersed southern hemisphere continents were united as one (recall also the example of the podocarp forest in Chapter 2). In this case the story is a little more complicated, because back in the Mesozoic, *Araucaria* or its close relatives were found in the *northern* hemisphere as well. The monkey puzzle family (Araucariacea) as a whole goes back to the Triassic, more than 200 million years ago. I collected an *Araucaria* fossil or two in the Jurassic rock of Yorkshire in northern England when I was an aspiring young palaeontologist. Unlike *Metasequoia,* monkey puzzle trees were greatly affected by the mass extinction 65 million years ago, and the northern hemisphere varieties did not survive. Recent molecular studies have suggested that even the many species in the southern hemisphere may have evolved within the last

60 million years, while still retaining their ancient Gondwana signature. The most astonishing coda to the story of the monkey puzzle trees is the discovery of the Wollemi pine by the explorer David Noble in 1994. A small group of trees growing in a remote valley in New South Wales, Australia, lined a ravine so well protected that abseiling down from the forest above provided the only means of reaching it. The tree proved to be a completely new genus of plants called (unsurprisingly) *Wollemia*. Although many survivors from deep time are of relatively modest size, the largest of the Wollemi trees was dubbed King Billy and has been estimated to be forty metres tall. It seems almost inconceivable that such giants could remain undiscovered until close to the end of the twentieth century. The discovery naturally caused a sensation, and immediately spawned headlines of the "dinosaur tree" variety. At first glance the spiky foliage looks like that of yet another member of the monkey puzzle family, the kauri (*Agathis*), similarly a conifer with a Gondwana heritage, which I had seen in New Zealand in my hunt for the velvet worm. However, the bark of the mature Wollemi pine is like that of no other tree. It is usually described as "bubbly," but I prefer an account that compares it to being completely covered with a swarm of fat bees. The spherical seed cone is borne at the tip of the branch, while the slim male pollen cone dangles downwards. I would dearly have loved to travel to the remote Wollemi locality as one of my field trips in preparation for this book. But the site is rigorously protected, not to mention the logistical problems that have to be considered in getting there. The trees are probably best left as undisturbed as possible in their secret redoubt. Even though I did not get there, it must earn a mention. Nowadays, it is possible to buy small seedlings in certain garden centres in the United Kingdom. How soon the recherché becomes commonplace.

Further research on *Wollemia* revealed that the tree was once much more widespread. The male cones produce very distinctive pollen. Surprising as it may seem, tiny pollen grains make enduring fossils once they are trapped in sediments. With the proper chemical treatment, the grains can be readily recovered and studied in the finest detail under the electron microscope. In 1965 a palynologist (pollen expert) called Harris had named as *Dilwynites* a new kind of pollen covered in little granules. The plant that produced *Dilwynites* could not be identified; in fact, this is not unusual in palaeobotany. Pollen travels far and wide, and often well away from the plant from which it originates. Quite frequently a name has been given to the pollen in the absence of knowledge of the parent plant. Fossil *Dilwynites* is now known from New Zealand as well as several localities in Australia. Its distribution was conforming to that old Gondwana signature yet again. When *Wollemia* was at last discovered and its life history investigated, scientists were astonished to realise that its pollen was one and the same as *Dilwynites*. The tree must have once grown widely in the southern continents. Since the fossil record of *Dilwynites* goes back at least 90 million years, here we were again, back among the dinosaurs. Another piece of the puzzle was supplied when, within the last few years, molecular sequence analyses were applied to the whole of the plant family to which *Wollemia, Araucaria* and *Agathis* belong. A 2005 study showed that the most primitive of all the living araucariacean plants is *Wollemia,* so trees like King Billy could well have predated the monkey puzzle tree and its allies. The Australian survivor begins to look even more ancient. The molecules also revealed another extraordinary fact: the genomes of all the *Wollemia* trees investigated seemed to be so similar that they were effectively clones. This means that the population of *Wollemia* had at one point shrunk close to the point of extinction. A few lucky individuals, at most, survived. It is known that Australia experienced climatic crisis while the Pleistocene ice ages came and went at higher latitudes. Other organisms, from the koala bear to the Tasmanian devil, still bear its imprint in their genomes. What a miracle that hidden gorge now seems, where an ancient tree could ride out the hard times with just enough moisture to survive; and what a message its belated discovery conveys to humankind, at a time when our species is single-handedly forcing species as grand as tigers into genetic cul-de-sacs by relentless exploitation and habitat destruction. From time to time I have to justify the kind of scientific work palaeontologists pursue in comparison with a "hard" science like physics. Our science can be accused of "just telling stories" by the more uppity of these contemporaries. Well, I cannot think of a better story that marries geology, botany, palaeontology and molecular biology than the tale of *Wollemia*. It may not require atom smashers costing millions to reveal the truth, but who could deny the intrinsic fascination of this messenger from the past, or its relevance to current questions about maintenance of a biologically diverse planet?

Moving across the southern hemisphere to Namibia, the relentlessly dry desert there is home to a plant that can live for more than a thousand years with only two enormous leaves. It is called *Welwitschia mirabilis*, "mirabilis" meaning "marvellous," which indeed it is. I should add that Welwitsch was the German botanist who discovered it in the nineteenth century. The plants are not rare in their special habitat, although *Welwitschia* is found nowhere else in the world. Since it is the only plant in its genus, family and order, no other organism on earth can lay such a claim to being "one of a kind." I have to come clean now: I didn't visit it myself. But I did grill Sir Ghillean Prance about it after he had been on a special pilgrimage to see it in 2009. As a former Director of Kew Gardens, he is surely the most reliable of all surrogates for my own pair of eyes. During its long lifetime, *Welwitschia* builds up a short woody trunk, and develops a considerable taproot for water storage. Its two leathery leaves originate from the top of the trunk and issue forth like dark green, never ending conveyor belts. They often split into several lobes, so old plants look like a weird species of giant starfish stranded on the sands. When they get to a couple of metres long, the ends of the leaves simply wear out and curl up and fray into grey whiskery threads, so it is only their continual renewal from the centre that allows the plant to carry on growing and photosynthesising. During the night in the desert, the temperature falls dramatically, and dew droplets condense on the leaves. This meagre supply is enough to keep the plants alive. If any organism truly deserves the soubriquet "survivor," it is surely *Welwitschia*. I have seen a photograph of a grizzled *Welwitschia* plant growing alongside the trunk of a fossil tree. Which is truly the more ancient—the fossil or the living survivor? There is a kind of poetic symbolism in their close juxtaposition today.

Welwitschia has separate sexes on different plants (we are used to that idea by now). But when the female produces its seed-bearing structures, they poke up all around the wide centre of the plant like so many greenish knobbly fingers. The cones (or strobili, to use the correct term) could be mistaken for flowers at first glance. There is something almost flower-like about the male organs, too. *Welwitschia* is classified together with its more conventional-looking, distant relative *Gnetum* in an order Gnetophyta.* I have already mentioned the extinct, superficially cycad-like bennettites (formally called Bennettitales); these contemporaries of the dinosaurs are also usually placed in the same group. Such seed-bearing plants have been regarded in the past as a kind of halfway house between gymnosperms and flowering plants (angiosperms), but their interpretation has always been controversial. Some species of *Gnetum,* for example, have broad leaves that would readily remind most observers of those of a typical flowering plant, such as a tropical vine. Its leaves are certainly most unlike the needles or scales of a typical conifer. There are details in its water-conducting strands and other fine structures that are also reminiscent of those of flowering plants. Although the theory that flowering plants actually evolved from something like *Welwitschia* or *Gnetum* is not widely believed, there are several botanists who would like to think that Gnetophyta and angiosperms descended from a common ancestor. This would imply that they belong together in the same natural group. On the other hand, there is a large (and now probably larger) school that would prefer to place Gnetophyta within the gymnosperms—alongside the conifers, but in a separate group. This would imply that any resemblance between these curious plants and true flowering plants was a matter of evolution along separate pathways leading, eventually, to similar morphology. This is a common enough phenomenon in evolutionary history. My colleague Conrad Labandeira at the Smithsonian Institution in Washington believes from fossil evidence that the convergent evolution even extended so far as to have different insects pollinating the seed plants in the Mesozoic—a group called scorpion flies took the place of today's bees and beetles. This implies a separate evolution of insect

---

* There is a third living genus in the Gnetales called *Ephedra*. It is a shrubby mass of green thin photosynthesising sticks, lacking respectable leaves, and I often stumbled across its bundles when working in the semi-deserts of western America, where it is sometimes known as Mormon tea. It is the most widespread genus of the Gnetales, and like *Welwitschia* has become a specialist for dry habitats. Most DNA studies suggest that it is also the most primitive member of the group.

pollination from that which followed in the flowering plants. It is an interesting fact that insects are alleged to help to fertilise *Welwitschia* today, apparently attracted by sweet chemicals secreted by its reproductive structures. At this point in the scientific story, data derived from sequence analysis of DNA might be expected to step in to provide an independent arbitration on the question of the relationships of *Welwitschia* (and friends) to flowering plants and conifers (and other gymnosperms). Most of that evidence rather *weakens* the connection to flowering plants, supporting the notion that Gnetophyta are gymnosperms that have developed some features resembling those of angiosperms. However, even the molecular evidence is not without ambiguities. A bewildered reader might begin to wonder whether the answer matters one way or the other. But since flowering plants dominate the botanical world today, far outnumbering all the other groups, the question of their origins is of great significance in unscrambling the events that led to our modern biological world. However the relationships of *Welwitschia* and its relatives are eventually resolved, nobody questions that this plant and its allies must root far back into geological time, to a period at least 200 million years ago when the main groups of seed plants were establishing their biological personalities. The ancestors of the most curious of all survivors must have colonised the desert habitat, carving out a niche from which they have never been displaced. Their leathery leaves survived drought and fires and the events that saw the extinctions at the end of the Cretaceous Period. When things got tough, *Welwitschia* could hunker down into the sands and wait for the dew.

Which brings us towards the crown perched atop the evolutionary tree, to the flowering plants (angiosperms). Flowers are fragile: "the grass withereth, the flower fadeth" (Isaiah 40:7). It is surprising to find that flowers have any fossil record at all, particularly from their early days while they were still relatively rare. Sedimentary rocks can occasionally preserve even the most evanescent of biological membranes if they are entombed quickly enough, before decay sets in. Fossil flowers show up as pressed flowers might upon brown paper, delicate as organza. Such preserved flowers are now known back to the early Cretaceous. Water lily flowers about 120 million years old have been discovered in Portugal. Although they have a similar structure to living species, these blooms were much smaller. The almost blowsy, showy blossoms that grace so many ornamental ponds today developed subsequently through millions of years. An even earlier flower from China (*Archaefructus*) was originally claimed as Jurassic, but is now known also to be early Cretaceous in age. This particular fossil has excited a lot of discussion regarding its closest living relations, but the most sober assessment is probably that it, too, is related to the water lilies (Nymphaeales). The fact that early flowers are already so widespread points to a still earlier history as yet unknown. It is perhaps not so surprising that the commonest fossils found belong to an aquatic group with an appropriate chance of preservation in rocks laid down under fresh water. Who knows what botanical treasures Cretaceous mountains may have held that have left no record in the rocks at all? Even in the absence of preserved flowers, angiosperms betrayed their presence in their characteristic pollen grains, in the same way as I described for the unrelated *Dilwynites*. I was still a student when Norman Hughes and his students in Cambridge University identified such telltale signatures of Cretaceous fossil flowers; thirty years ago the microscopic evidence was still unmatched by preserved leaves or blooms. The variety of pollen increases through the Cretaceous Period, so it is clear that the early diversification of flowers happened during an epoch when dinosaurs were still the largest land animals. As more strata are searched, more and more fossils (mostly leaves) are discovered that add to the list of early angiosperms. Even the split into its two major component groups is believed to have happened deep in the Cretaceous. The "dicots" have a pair of seed leaves (dicotyledons) which appear above ground as those tiny green ovals so different in shape from mature leaves. They include many of our familiar garden flowers and virtually all our fruits and trees. The "monocots" have only one seed leaf, and embrace lilies, orchids and sedges, and (a later innovation) that most ecologically important of all plants for the welfare of animals, the grasses. It seems that a serious radiation of flowering plants happened well before the extinction of the "ruling reptiles," and that the event at the end of the Cretaceous Period was only a relatively minor setback for the plants that are today more varied and diverse than all the rest of the flora put together.

Nowhere has the molecular sequencing revolution had a more profound impact than on our understanding of the relationships between the different families of flowering plants. If "revolution" is an overused word, it is nothing less than appropriate for the rational ordering of the huge variety of angiosperms into a single tree demonstrating their evolutionary history, and suggesting how they can be classified. This is the modern expression of the system with which the Reverend Gilbert White sought to organise the natural world in *The Natural History of Selborne* (1789). It is the culmination of the efforts at plant classification begun by Linnaeus in Sweden earlier in the eighteenth century, based largely on the sexual organisation of the flowers—so much so that it was deemed unsuitable for the ladies by some of his prudish contemporaries. It is also a perpetually unfinished project, forever becoming more refined. If one had to put one's finger on the first key work, it might be a molecular-based classification tree of angiosperms by Mark W. Chase (now at Kew Gardens) and more than forty of his colleagues published in 1993. It is the best example I know of science by committee actually working: the huge task of putting in data from numerous different plants was shared between many workers, and the appropriate computer program worked out the tree solution that best fitted the information. The cooperative model continues into the new millennium, using sequences from a variety of bits of DNA, together with other biological molecules. This is not the place to develop the details of this work, a task admirably fulfilled by the Angiosperm phylogeny website.* But for my purposes, what can be taken from the broad picture of plant evolution is the identity of the most basal living flowering plant, a survivor from low in the tree of life. It is called *Amborella trichopoda,* and it is a native of the island of New Caledonia, which, as I have mentioned, is also where Araucarias are most diverse. It is an evergreen shrub, and, to be frank, not very startling in appearance. Somehow, one expects a messenger from the Cretaceous to be more flamboyant, after the fashion of *Welwitschia.* In the wild, this herald of flowers is under threat from deforestation and mining, but since its discovery it has been adopted by many botanical gardens (I caught up with it

* http://www.mobot.org/mobot/research/apweb/.

in Hawai'i). *Amborella* bears longish leaves with serrated edges alternating along the branches, no different from the foliage of fifty other tropical and subtropical plants. Its tiny yellowish flowers are born a few together in clusters, with males and females developing on different plants. It is impossible to distinguish petals and sepals on the flowers, as we can on so many of our garden plants. The flower itself is distinguished only by tiny, spirally arranged coloured tiny leaf-like lobes (so-called tepals). A small, red berry fruit less than a centimetre long contains a single seed. Unlike almost all other shrubby flowering plants, *Amborella* lacks the water-conducting ducts known as vessels, which allow a supply of moisture that permits tree canopies to spread in the sunshine without wilting. Once you get your eye in, *Amborella* does indeed look somewhat primitive. It has been assigned to its own plant order, but it is scarcely as bizarre as that other strange survivor *Welwitschia*. No fossils of *Amborella* are known. It is encouraging that water lilies, which do have fossils, appear near *Amborella* low in the angiosperm classification. Of course, they are not true lilies, which are "monocots." Nor is *Magnolia* far away. This attractive tree was once considered among the most primitive of plants, and like *Amborella* its stamens are spirally arranged. *Magnolia* is known among fossil assemblages of pioneer angiosperms. I saw a *Magnolia sieboldiana* blooming extravagantly white with the merest flush of pink in the foothills in China, its flowers like a flock of white pigeons among the branches. It is pollinated by beetles, which are attracted by a sweet exudate produced by the flower, a system predating the evolution of bees and butterflies. Ginkgo trees grew higher up the slope. Plants have much of their history mapped out by survivors. During the course of a short walk through the damp forest, we might pass greenery that would chart floral progress from the Devonian to the Cretaceous. History has left many green mementos.

Amborella has a lot to answer for. The diversification of flowering plants set off one of the profoundest enrichments in the life of our planet. We talk about the "flowering" of a civilisation or culture to imply some creative climax, and the metaphor is not misplaced. If a flower is fundamentally a simple mechanism that helps in the transfer of genetic information in pollen to insure cross fertilisation, then the variations upon that simple theme are the source of endless wonder. We have seen that insects may have been used as vectors of pollen even before flowers evolved, but when they did appear, a mutual orgy of invention was triggered, when flowers competed to be more attractive to their increasingly dedicated pollinators. No colour was too sumptuous, no perfume too seductive. Plants that wished to attract flies developed a scent of rotting meat. Others smelled of heaven to attract fairies, which are always portrayed with butterfly wings. More prosaically, the fossil record tells us that the great burst in evolution of butterflies and bees followed on from the extinction at the end of the Cretaceous Period. This was a time when angiosperms largely displaced conifers and other gymnosperms from warmer latitudes; although anyone who has travelled across Canada will know that conifers still rule where winters are fiercest. Flowers became "designed" specifically to be pollinated by one particular species or another. Tube-like flowers with long corollas evolved for appropriately long tongues; nocturnal lures of pallid hue and compelling odour attracted moths and bats, while hummingbirds evolved to mimic bees or hoverflies when the sun shone. Charles Darwin wrote one of his most famous books on the exotic deceits used by orchids to favour their chosen pollinators, which evolved twisted and mottled waxy blooms more fantastical than anything mere animals could devise. Plants living on poor ground or suspended in trees devised flytraps to supplement their meagre diet. The rather tepid phrase "co-evolution" hardly seems to do justice to such wild inventiveness. Richard Dawkins would probably remind us at this point that all this was driven by natural selection, which is rather like saying a symphony is composed of notes. Undoubtedly this is true, but the thrill is in the interlocking multitude. But then, some flowering plants turned their backs on show, and became self- or wind-pollinated. From the Miocene onwards, grasses covered the plains, and made life possible for millions of cloven-hoofed grazers and ruminants by growing perpetually renewed from their bases rather than their cropped crowns. Yet other grasses returned to the sea and provided food for dugongs. Life cycles ranged from a few days of floriferous glory in deserts after rain to a dogged millennium in an oak forest, where craggy and burred veterans saw human revolutions come and go. Seeds spread in a hundred different ways, buoyed by breezes, shot forcibly from pods or sequestered within delicious fruits. Cacti colonised American semi-deserts by storing water in fat stems. *Euphorbia* in Africa independently evolved almost identical forms in response to similar stringencies. Both reveal their angiosperm credentials, and their separate ancestry, in the brief splendour of their flowers.

The Victorians used blooms as a form of sentimental signalling, a "language of flowers," of which the red rose for love is one lingering remnant. The Bible employed flowers for moral commentary: "consider the lilies of the field . . . even Solomon in all his glory was not arrayed as one of these" (Matthew 6:28). The elegant lily is in turn emblematic of the Art Nouveau movement; a stylised chrysanthemum provides the Imperial seal of Japan. It seems that angiosperms can be turned to almost any human purpose. Charles Darwin himself returned to flowers in a famous passage nearly at the end of the *Origin of Species* (1859), where he considered "an entangled bank" as a marvellous summary of the intertwined lives of so many organisms, governed by the interplay of evolution and natural selection from their common origin. It is indeed a marvel that every plant we know could arise from something rather like a liverwort. It is another wonder that species recording many of the steps along the way are still living their green lives, some modestly and some flamboyantly. It may be true that "All flesh is grass, and all the goodliness thereof is as the flower of the field" (Isaiah 40:6). But the brevity of the individual life is nothing. If we can only understand it, the language of flowers tells of enduring lineages that link together all green things.

7

Of Fishes and Hellbenders

The road running northwards from Brisbane commands open views of the Glasshouse Mountains, isolated peaks that are the remnants of extinct volcanoes. Conical protuberances stick up above the rain forest, their bareness contrasting with lushness everywhere else: they are probably "plugs" composed of hard rock that filled vents at the end of the eruptions. Seawards along Queensland's Sunshine Coast discreet developments are tucked away, designed for a newly affluent class that can work remotely. Further inland, Maleny is a small, neat town of low, painted wooden buildings with a half-concealed hippy past, lying on a ridge in the Blackall Mountains. It offers a view both ways, towards the sea and down into a wooded valley below. We choose the latter, for the lungfish lie that way. A valley floor is reached by minor, paved roads, and supports a series of dairy farms that would not seem out of place in some of the richer parts of England, such as the Welsh Borderland. The difference becomes apparent when I spot a distinctive species of monkey puzzle tree, and notice that these cows are sheltering in the shade of a huge, dark-leaved fig. Peter Kind is showing us the upper reaches of the Mary River, where the lungfish hide. We are in search of the great survivor among the vertebrates, the animals with backbones. This is a creature that traces our own evolutionary line back to the days when our ancestors broke free from having to live under water.

Queensland is the only place in the world where this particular species of lungfish (*Neoceratodus forsteri*) survives. Among a handful of living lungfishes, it is regarded as the most primitive living form. Much of the research on its biology has been conducted in the last decade. Before that, the fish pottered along for hundreds of millions of years minding its own business. In the right habitat it is not rare, the Burnett and Mary catchments being its two main redoubts. At one time 20,000 fish were tagged on the Burnett River catchment, and a low rate of recapture was a good indication of a large wild population. That does not mean that the fishes are safe from human interference. Growth in the rural population of the Brisbane hinterland in Queensland has meant a greater demand for water, and the Mary River seemed to provide a ready source. The Traveston Crossing Dam project of 2006 sought to solve future water shortages by damming the Mary River. No doubt somebody in the planning office thought that more water might do the lungfish nothing but good. But it was absolutely necessary to discover more about the needs of lungfishes before the dam could go ahead. Peter Kind was part of a research team tracking individual fish to learn something about the way they patrolled the river, and how and when they reached their spawning grounds. Because some of the sites are very hard to reach, and because the fish are peripatetic, radio tags were attached to them that could be tracked from the air. There were also other fish species that attracted concern: the Mary River cod is rarer than the lungfish and provides better eating. Maybe both required conservation. So we were off to see a typical lungfish habitat on the Mary River near Conondale. We leant over the bridge, looking at the stream below. Patches of a waterweed called *Vallisneria* make the shallows look quite green with stretches of simple ribbon leaves covering muddy areas. This is where *Neoceratodus* likes to lay its eggs, a few at a time, attached to the weedy stems in fully oxygenated water at about ten centimetres depth. The eggs have several layers, of which the outermost has a gluey surface to help their adhesion. Males vie with one another to fertilise the freshly laid eggs with their milt. A dam would have destroyed their favoured habitat very effectively. Peter discovered that lungfish are generally attached to a particular stretch of river, patrolling it from a home base; a safe place under a "drowned" log might suit very well to rest up during the day, for they are often crepuscular in habit. Since the big patches of weed that make the best spawning grounds move around from year to year, fishes are also obliged to seek them out and may then move far from their home turf. They cannot simply be ponded up like a goldfish. As to diet, lungfish are catholic in their tastes. When they are fully mature, they chomp up large masses of waterweed, even mud, extracting freshwater snails and "shrimp" which they process with teeth designed for crushing. Anything they do not eat is excreted. It is not an exciting lifestyle, but it is a living. Downstream, we peep out along a stretch of river lined with feathery she-oaks with black trunks, trees as wispy as those in a Corot painting. We were hoping to catch a sight of our quarry, but the water is too high and a personal encounter would have to wait. There had been a lot of rain in recent weeks. I see nothing particularly unusual about the Mary River to qualify it to harbour such a piscine Methuselah, but animals have more exquisite tastes than we might think.

A dammed lake was clearly not a good idea for the welfare of one of the most important living links with our remote past. Fortunately for the lungfish, it had a champion in the person of Jean Joss, doyenne of the scientific study of *Neoceratodus*. Jean lobbied everyone from the Premier of Queensland to ecologically minded pop stars. She drummed up support from professors of ichthyology all around the world. Professor Per Ahlberg of the University of Uppsala, an authority on fishes close to their emergence onto land, declared that *Neoceratodus* was an incomparable treasure for understanding this crucial phase in the evolutionary history. Politicians did their usual bluff and bluster to try and swing decisions their way in favour of the dam. It was a close thing: the government had already bought up several farms ready to immerse them in water. Local people from the Mary River valley joined in the protests, discovering a warm affinity for a cold-blooded fish that they had previously taken for granted. Finally, in 2009, the Traveston Dam was abandoned. The government offered to sell back the farms at cost to their original owners. Even if there were no lungfishes lurking in Yabba Creek off the Mary River, a fine piece of countryside has been saved for the future.

In the course of this labyrinthine political process, much has been learned about the biology of *Neoceratodus*. These fish turn out to be as remarkable as any of the survivors in this book. In the first place, they can live for a long time, certainly to fifty years in the wild. One fish in captivity is eighty years old at the time of writing. A really big fish weighs up to 30 kilograms; when landed, they drape as a dead weight between their captor's arms. A typical lungfish is one metre long at twenty years old, weighing in at about 12 kilograms, and the females are bigger than the males. The tail is a kind of blade surrounding the rear of the body, while the two pairs of fins low down are strangely reminiscent of legs; the forward pair has a fleshy lobe at its centre. Lungfish do indeed have lungs, and this may be key to their survival. During the heat of summer, the Mary River becomes a different place, often low in oxygen, and then a fish on the move will take gulps of air. Peter Kind tells me that they make an audible gasp. This capacity may well give these ancient fishes an edge over more evolutionarily recent competitors that are wholly dependent on dissolved oxygen. *Neoceratodus* fishes can get through hard times thanks to their emergency procedure. They also breathe air when they need extra puff during busy periods like egg laying, when they will come to the surface every few minutes. Their eggs are large compared with those of most other fishy species, fully a centimetre across. Because the eggs are also yolky, the hatchlings do not have to feed for a few days. When they begin to hunt for edible morsels, they have little needle-like teeth, the better to catch their tiny prey. This immature dentition transforms gradually into the crushing teeth of the adult. Large overlapping scales are a striking feature of the mature animal, and one shared with many other ancient fish. Scales do grow incrementally at their edges as the fish gets bigger, but the "growth rings" are so minute that they cannot be used as a reliable indicator of age. Their bones are not hardened, so the fish feels weirdly flexible to the touch. Lungfish do not have particularly acute eyes, although they do have colour vision. In the relatively murky world they inhabit, they are helped by special organs in the head capable of detecting movement in the mud below them. These have been compared with the "ampullae of Lorenzini" in sharks. As for internal organs, there is no true stomach, but they have a long and convoluted spiral valve in which the food is retained for a long time. Everything about the lungfish seems to happen at a leisurely pace; as an American might say, they take life real slow.

Jean Joss has studied lungfish at Macquarie University in Sydney for several decades. Although retired now, mere years are irrelevant when she enthuses about her favourite animals, punctuating her conversation with those splendid uninhibited laughs that seem to be special to Australians. She has her lungfish on the roof. She leads me upwards to inspect large circular tanks above the biology department which house a number of fine mature fishes, olive black in colour. At last, I am allowed to embrace one of her charges. It feels remarkably slimy and dense, but at the same time curiously amorphous, like holding a sack stuffed with small beans. The fish struggles a little, but it lacks bones to inflict any pain. I have never handled such a curious living object before, and I am not altogether sorry to be handing it back. When not fighting City Hall, Jean has done profound work on the structure and development of this marathon champion among the fishes. *Neoceratodus* lies as close as is possible in the living fauna to the point where fishes left the water for life on land. As if to demonstrate the point, she holds up a fish to show the paired nostrils in the mouth "ready to turn into a nose," as Jean says, with a soupçon of irony to indicate that it certainly was not as simple as that. *Ceratodus* was a familiar fossil from strata about 220 million years old when the Australian lungfish was discovered. The "neo" in the name of the living fish just means "new," which is a fairly measured response to such a zoological Lazarus. Jean has run a breeding programme since 1993 using two large ponds in another part of the Macquarie campus, although now she is worried about the long-term future of this research resource. Her fishes may need to return to the wild. Lungfish do not become sexually mature until they are fifteen years old, in which regard they are just like us. This implies that if they experienced a mass death in the wild from whatever cause, dams or otherwise, it would take them a long time to recover their numbers. Jean thinks that part of the reason for the survival of the species in Queensland is that in the Mary River they live in a special ecological "window," south of where crocodiles are found, but where temperatures in summer are hot enough to induce temporary fouling of the water, a time when lungs are critical to survival.

The development of the embryo of the Australian lungfish offers clues to the evolutionary history of the vertebrates. Lungfish are classified along with the coelacanths in a group (class) of fishes called Sarcopterygia, which share the curious, lobe-like fins I had noticed while holding my slippery prize. The related class of bony fishes Actinopterygii, with rayed fins like bass and herrings, includes 25,000 species or so, and comprises the bulk of vertebrate diversity on earth. It is well established that four-footed land animals (tetrapods) were descended from a member of the former group, much the less conspicuous nowadays. Fins became legs, and walked on land. So studying how the lungfish develops is likely to cast light on details of our own development and relationships. Jean and her colleagues have shown how groups of cells in the neural crest, a region of the developing embryo within the egg of *Neoceratodus*, are destined to develop into different parts of the adult anatomy. "Fate maps" are built into them. They have predictable ways of developing into cell and muscle types that are similar (or "conserved") in amphibians and many other land-dwelling vertebrates. The unfolding of these patterns is at the bidding of regulatory genes, which act rather like site managers directing the construction of an architect's plan. If something goes wrong, then a pathological adult results—if it lives that long. It will be clear that such research links in with current work both on gene malfunction and stem cell research. We may well have reason to be grateful that the big, black breathing fish survived to tell its story.

I have chosen to dwell on *Neoceratodus* at some length, and now I should explain why I am not going to take you to the home of the coelacanth, *Latimeria chalumnae*. This large and (apparently) cumbersome fish is often considered the very example of a living fossil. When Ogden Nash wrote, "Consider now the coelacanth / our only living fossil," he perpetuated a popular misconception.

Its significance is neatly encapsulated by two of the books written about it: *Old Fourlegs* (1956) and *A Fish Caught in Time* (1997). No other living fossil has been so fully put down on paper. The first book stressed its importance as an "intermediate," between regular fish and land-dwelling tetrapod; the latter its surprising survival from the dinosaur age. However it is interpreted, *Latimeria* is certainly something of a rarity, and lives at some depth in the sea, so one cannot guarantee a sighting. It is possibly commonest around the Comoro Islands, which were fomenting a ferment of fighting at the time of writing. In any case, without special diving equipment the fish is impossible to find. Instead, I decided to take the easy way out. I went to visit the coelacanth in a glass case at the Natural History Museum in London, which is much safer. Like all preserved specimens behind glass, it exudes a certain air of gloom. It is a chunky, meaty-looking, coarse-scaled blue fish with scattered white spots. It has a thick tail with a curious extension of the fin at the back, and along the body it seems to have rather too many fins compared with your average fish. It certainly has several more than the Australian lungfish, including some along its back. Like *Neoceratodus,* the fins have fleshy bases (hence "fourlegs"), a feature that is reflected internally in the structure of the bones. Unlike many other organisms I have considered in this book, the coelacanth seems to tickle the anthropocentric fancy (it must be those "legs"). This may be why its appearance engendered such a sensation.

The discovery of the coelacanth happened in 1938 thanks to Margaret Courtney-Latimer, a young curator at the East London Museum in South Africa. When a curious-looking fish came her way from a local fisherman, she knew at once it was something special, and managed to preserve it long enough, wrapped in a formalin-soaked double bed sheet, to secure the attention of Dr. J. L. B. Smith, the ichthyologist who would confirm its identity. The fish made Smith's reputation, but it also guaranteed Margaret immortality in its generic name, *Latimeria,* a Latin tag secure forever in the litany of biodiversity. Smith recognised that its nearest relative had been discovered as a fossil in strata of Cretaceous age: old fourlegs' ancestor was a contemporary of the dinosaurs and the ammonites, and its descendants had lingered on unseen in the ocean while the world was being utterly transformed. If anything was a living fossil, this was it. Even the bony plates on the head and the thick scales on the body seemed to hail from a vanished epoch. The type specimen still resides in the East London Museum, preserved by the taxidermist's art. I particularly like the headline that greeted the discovery in *The Illustrated London News*: "Best fish story for 50,000,000 years." We might add another 20 million years with modern knowledge of geological time. Fortunately, Latimer's fish was only the first in a line of similar piscine prizes. The fish is now known to be quite widespread along the eastern coast of Africa and its outlying islands. It has also been fished up off Indonesia, and there is currently a debate about whether or not specimens from this new locality represent a second species of *Latimeria*. It is clearly good news that the type "living fossil" is not quite as rare as was once thought, though still hard to find.

Film of the living fishes taken off the Comoro Islands has proved that they do not live at abyssal depths as had once been thought. They are far more graceful under water than stiff-looking specimens in glass cases might suggest: the leggy fins (or finny legs, if you prefer) seem to wave back and forth in efficient ways to orient the creature. A surprise is that the fishes seem to want to align themselves in life vertically, head down. They have organs in their heads that may be sensitive to slight vibrations from the sediment below, wherein worms reside, probably the same batch of cells as in *Neoceratodus*. The curious posture is very likely a hunting technique—a specialisation that has evidently served the old fish well. Coelacanths lay comparatively few, large eggs. Among the native peoples that turn them up, they are not a popular fish for culinary use; maybe even this has helped the species survive for so long. One is obliged to observe at this point that the eggs of another old survivor, the sturgeon,* yield the most luxurious food of them all: Beluga caviar. The authority on fossil coelacanths and their relatives is a former colleague, and near namesake, of mine at the Natural History Museum in London, Peter Forey. (I have been benefiting from the confusion of our names for years when Peter's scholarly and beautifully illustrated works have been mistakenly attributed to me.) He has the distinction of having the oldest fossil coelacanth named after him, a 410-million-year-old specimen from Australia called *Eoactinistia foreyi*, described as recently as 2006. This takes us back to just before the time when backboned animals were moving out of water and onto land, colonising the new habitats made available by the advance guard of the plants. The case of the coelacanth invites comparison with that of the horseshoe crab

* The sturgeon (*Acipenser* and relatives) is one of the most primitive of surviving bony fish, with several species growing slowly to a great size. It has a largely cartilaginous skeleton, and is covered with pointed, bony scutes, rather than scales, as in its more advanced relatives. It is not an easy fish to visit in the wild, although it is widely distributed. If this book ever runs into a second edition, I may be able to catch up with it.

with which we began our story. It has not stayed still; it has continued to evolve while retaining its intriguing primitive features. So the end of Ogden Nash's short poem is also wide of the mark as nothing in nature ever remains completely unaltered:

> *Old Coelacanth, so unrevised*
> *It doesn't know it's obsolete*

Nor is the "old fourlegs" story still quite what it once was. However appealing it might be, few scientists now believe that a relative of this fish crawled onto land on its stumpy pseudo-legs, as a stepping-stone towards becoming the first terrestrial tetrapod. Modern analyses favour the lungfish group as more closely related to living land vertebrates, which is why this chapter began where it did. The lungfish idea had been mooted on evidence from detailed analysis of skeletal structures by my palaeoichthyological colleagues, who had shown that more evolutionarily advanced characteristics were shared between fossil lungfish and tetrapods than with the coelacanth and its allies. This conclusion was eventually supported by molecular evidence, but it took a decade for that evidence to become available. It seems that the definitive living fossil was not quite at the Central Station of evolution as had once been believed. Instead, one of the ancient fishes that belonged near the base of the *lungfish* family tree took the crucial step from fresh water onto that muddy riverside. In Singapore, I saw small fishes called mudskippers hopping athletically on their fins on damp mud under mangroves. Film footage of these animals has often been shown on television to suggest how the first steps towards land were made. However, since all the fossils of earliest tetrapods and their fish-shaped relatives seem to have been relatively large animals, probably comparable with an adult lungfish, one imagines the whole business to have been much clunkier; more scraping and waddling than skipping. No matter; once those small steps were made, it was a giant step for all subsequent life. There was plenty of time to become more elegant, even if today one could still harbour doubts as to the elegance of a newt. Once on land, there were endless new opportunities for animals with backbones: one of the strongest branches on the tree of life had developed its first twigs.

As to when that happened, a few years ago I would have confidently asserted that it was in the Devonian Period, about 380 million years ago. This is when the appropriate "ancestral" fishes subsequently gave rise to the first undoubted and relatively well-known terrestrial tetrapods, as proved by fossils from several localities in Greenland and Arctic Canada. In 2010, a report in *Nature* described fossil tracks from the Devonian strata of Poland that apparently showed a terrestrial tetrapod lumbering across an ancient sand flat at least 10 million years before the previous estimation. If it proves to be true, this is an important modification to the timetable. We have found elsewhere in the tree of life that the first steps are often the least known, and that transitions may have happened earlier, more obscurely; the first appearance of terrestrial plants is another example. Maybe the origin of tetrapods will prove to be a similar case. One of the best things about palaeontology is that fresh and unexpected discoveries are still being made. The day people stop looking is the day nothing new will be found.

Having started this chapter at one of the most pivotal biological thresholds in the history of the biosphere, there is a choice of moving either up the tree of life, or down to see the beginnings of the vertebrates. Since it is important to know where we ultimately came from, let us go downwards first. Once again, there are living animals that carry the imprint of a still earlier history, even before lungfish, before "old fourlegs." All the animals with backbones mentioned so far share one feature, one so obvious that it may be hard at first to think of it as an innovation: they have jaws. In technical language, they are gnathostomes. Jaws have stayed the evolutionary course, grabbing, biting, chewing, filtering, billing, and cooing. But there was a time before jaws. When one thinks about it, jaws with their hinges and their coordinated movements, upper and lower, are complicated bits of machinery, even in a fish. It takes evolutionary work to make a jaw. There are fishes that root back to a time before that work had been carried out: obviously, it is necessary to find a lamprey.

Sometimes it is just not possible to be in the right place at the right time. I had heard that in the Baltic States the lamprey was still to be found on the menu. I knew that the Lithuanian language was itself a survivor, and somehow the coincidence of two relics in the one place seemed tailor-made for my purposes. There were similarities between the history of language and the history of life that seemed worth investigating.

The old part of Vilnius is curious and rather lovely, a walled city stuffed with ornate Roman Catholic churches. It is quite unlike the severe, if inspiring, buildings that are the norm around the Baltic; for Lithuania is a Roman island within a Protestant sea. The country was affiliated for centuries with Poland, which is also largely Catholic, and the former aristocracy held lands in both countries. Many of the streets in Vilnius are flanked by courtyard tenements that must be entered through deep archways that go back to the early days of the city. They create a charming sense of surprises waiting around the next corner. The present citizens recall the oppressive Soviet days with some distaste. Many ancient areas became run-down, including a delightful artists' quarter reached by a bridge over the River Neris that recently declared itself an independent republic. The chief legacy of the communist period is the concrete Stalinist residential blocks outside the city walls; identical buildings blight other cities across the former Soviet Union. Lithuanians rightly regard the recent sprucing up of their capital city with a measure of pride. To English ears, their language is curiously without location. The names of Lithuanians I know, Petrus and Gediminas, seem to have an almost Roman flavour. Overheard on the street, however, the first impression perhaps recalls a Slavic tongue, but then the language sounds vaguely Mediterranean, Corsican maybe? Or could that be Welsh? There is a good reason for this ambiguity, for Lithuanian originates from deeper stock than all of these languages. A professor of linguistics told me: "If you want to know how Sanskrit sounded, listen to Lithuanian!" That may be something of an exaggeration, but Lithuanian is a survivor of sorts from the roots of the Indo-European family of languages, the group that includes the classical tongues, the modern Romance languages (Italian, Spanish, French, etc.), as well as Celtic, German, Slavic and many minor languages living and extinct. Together, they comprise a huge part of human culture. From an origin on the Indian subcontinent, the "proto language" and its peoples spread westwards, splitting into regional "species," now the modern languages. Perhaps those tribes that pushed northwards to the Baltic remained sufficiently isolated to change least from the original form. They remained within the thick woodlands, living the simple life. Nothing remains completely unchanged, of course, and language usually changes more quickly through time than anything else, so there is no question of Lithuanian actually *being* Sanskrit. But there is an interesting parallel in these surviving cultural features with the kind of morphological survival already described in horseshoe crabs and velvet worms. There is no such thing as no change at all, but survival from deep in the tree of evolution is always informative. Languages have evolutionary trees just like organisms. The lamprey seemed appropriately to belong in this land of strangely enduring phonemes.

I learned the Lithuanian word for lamprey, *nege*, and trotted around the cobbled, sloping streets seeking restaurants in Vilnius to ask if they had the primitive fish on the menu. Almost everywhere, I was met with puzzled looks. I suspect it was as if a Lithuanian had gone around London asking if eagle was on the bill of fare. The experience was, I confess, rather dispiriting. Finally, I met a restaurateur with good enough English to explain. "In my grandfather's day," he said with a broad shrug, "*nege* was common . . . Now nobody eats it." It appeared that my marriage of early fish and linguistics was not to be. I was a generation too late. Apparently, they still eat *nege* in neighbouring Latvia, nearer the sea.

I did eventually run down a lamprey much closer to home. In fact, at one time it *was* my home, a little village called Boxford in Berkshire, England, through which runs a stream called the Lambourn. When I lived there as a child in an old thatched cottage, I had no idea that brook lampreys (*Lampetra planeri*) could be found in the little river at the end of my garden. Five decades later not much has changed, although the village is better groomed, the thatch neat as an Eton crop. The Lambourn is still as pure a chalk stream as it ever was. Waterweeds wave like rich green tresses in the fast flowing, clear, shallow water. I remember a gamekeeper telling me it was the coldest water in the South of England: it wells up from springs plumbed deep into the chalk, a pure white limestone of Cretaceous age. My father fly-fished the Lambourn for trout, and they were the best I have ever eaten. I notice one dart away as soon as I hang over the bridge to get a better look. The Environment Agency is sponsoring a survey of river lampreys, and one of their sites is at Boxford, a few hundred metres away from where I used to live. I could take it as an omen. The survey is using electric fishing, which is as clunky as fly-fishing is elegant. Electrodes are placed in the water, and fishes cannot resist swimming to the positive electrode; it is, of course, an ideal way of conducting a census. Four of the local wildlife officers in waders handle the equipment. Within twenty minutes a small tubular fish, perhaps twelve centimetres long, swims to one of the electrodes. I notice it has small, round, beady eyes. It looks as if its front end has been bent downwards and there is no sign of a jaw. It is a lamprey.

Lampreys were among the favourite foodstuffs in the medieval kitchen. Their oily and powerful taste evidently appealed to the rich baron who measured his worth by the variety of his table. King

A Fatal Case of Lampreycitis. A.D. 1135.
Henry I died at St. Denis, in Normandy, after an illness of
seven days, brought on by eating an excess of lampreys.

Henry I of England, fourth son of William the Conqueror, died in Normandy on 1 December 1135, from what is always described as "a surfeit of lampreys." Quite how many of the fishes constitutes a surfeit is not recorded.

The city of Gloucester was famous for its lampreys, and is enjoined to provide the monarch with lamprey pies. One such was sent to the present Queen of England to celebrate her silver jubilee in 1977. Whether Her Majesty also regarded the twenty-pound pie as a surfeit is also not recorded. The popularity of lampreys in the kitchen continued long after her namesake, the first Elizabeth's reign. Of a famous poet's end in 1744, Dr. Samuel Johnson wrote in *Lives of the Poets:* "The death of Pope was imputed by some of his friends to a silver saucepan, in which it was his delight to heat potted lampreys." Whether Alexander Pope was interested in another alleged property of lampreys is difficult to judge. They were supposed to heighten sexual desire. As with so many such associations, a simple Freudian explanation comes to mind, given the size and shape of the marine lamprey. John Gay exploited its reputation in his poem *To a Young Lady, with some Lampreys* to comic effect:

> *Why then send Lampreys? fye, for shame!*
> *'Twill set a virgin's blood on flame*
> *This to fifteen a proper gift!*
> *It might lend sixty five a lift.*

Lampreys are jawless fish (Agnatha), but they don't lack teeth. Their teeth line a circular, rasping mouth that attaches mercilessly on to the fishes (ones with jaws) that are their prey. It is the revenge of an organism near the bottom of the vertebrate tree on those lying further towards the crown. Lampreys can drain a salmon dry. If they become numerous, they also deplete fish stocks. Recently, they have grown into something of a plague in the Great Lakes in North America. In general shape the larger species look superficially like eels, which are good bony fish (or teleost), and no close relation. There are several species, the largest being sea lampreys (*Petromyzon*), the smaller river lampreys being distinguished as "lamperns" (*Lampetra*) in the west of England. The little lamprey I fished out of the River Lambourn has actually turned its back on parasitism, and the adult fish is solely concerned with reproduction. For the rest of them, there is something instinctively revolting about seeing the dark parasites dangling from the flanks of a handsome game fish; they look indecently large, like perched devils in a painting by Hieronymus Bosch. Along the back of the lamprey runs the main conduit for the nervous system, the notochord. This is the structure that gives the name to the Phylum Chordata to which we all belong: lamprey, lungfish, lion, you and me (the "higher" vertebrates still retain the notochord in the embryo). What we eat is the long muscular body, although one chef in Latvia is reputed to serve notochords en masse as a separate delicacy. The head region is most remarkable for our understanding of vertebrate history, for here are seven paired openings along the side of the animal which admit water to irrigate the gills—they are called "gill slits" though they look like dark round punctures in a freshly caught fish. One popular name for the lamprey is "nine eyes," referring to the seven openings plus eye plus nostril on each side, anatomically not very respectable, of course, but one knows what is meant. From "nine eyes" to four legs to two legs is the shortest possible cartoon history of the chordates, leading to mankind. The life cycle of the sea lamprey is a little like that of the salmon. The fishes spend the first part of their lives in the rivers where they hatch from the egg, mature in the sea and then return to the rivers once again to spawn and die. The larva (ammocoete) is a little pale wisp of a thing that lives in river mud for about four years and filter feeds.

Like velvet worms, jawless fish were more diverse hundreds of millions of years ago than they are today. Some would say that lampreys, along with their even more primitive, slimy parasite relatives—the hagfishes (*Myxine*) that even lack vertebrae—are degenerate survivors from ancient times, protected from extinction by their vile and parasitic habits. If a geologist splits the right rocks of late Silurian and early Devonian age, a wide variety of agnathan fish can be recovered in some localities, as in Scotland and the Appalachian Mountains. None of them were parasites; they probably grazed algae and small invertebrates. Many of these extinct types of

jawless animals had a bony covering over the head, with holes for
the eyes. *Cephalaspis lyellii* was described from fossils from Scotland
that would be instantly recognisable as belonging to a fish, for a distinctly fishy posterior equipped with a muscular body and a swishing
tail extended behind its bone-covered head-shield. This particular
species was named for Charles Lyell, the pioneering geologist who
established several fundamental principles of that science, and gave
Charles Darwin the gift of geological time. Time, and lashings of it,
was a necessary ingredient in formulating a mechanism for evolution. Both Darwin and Lyell would have recognised that the curious,
porous tissue known as *bone* was already present on the early Scottish jawless fish. Bone is made from calcium hydroxyapatite; it contains phosphate (of course including the element phosphorus) and
is a tissue type unique to chordates. Bone offers distinctive proof
that all chordates descended from a common ancestor. It makes
our internal scaffolding; precisely, it props us up.* The earliest bone
probably dates from the late Cambrian, and is 500 million years old,
so its invention inserts into a low branch on the tree of life. Without
bone, no lungfish could have made its first strides landwards; without bone, there could have been no teeth, and teeth have put the bite
into evolutionary history. Agnathan fossils suggest that bone started
as an external covering for the more vulnerable head regions of the
animal. Only later was it recruited internally to stiffen the backbone,
and provide the skeleton that we recognise as a sign of our kinship
with beast, fowl and fish, the skull beneath the skin. We might take
a journey back into a still earlier branch of the evolutionary tree
leading to the vertebrates. Further out to sea from the muddy flats
where I collected the brachiopod *Lingula* (see p. 127), the lancelet
(*Branchiostoma*) still lives in large numbers. In my student days we
called it amphioxus, and I have stuck with this habit of naming. It
is a flattened, transparent, but scarcely fish-shaped little organism
about the size and design of a willow leaf. It lives buried in sandy
sediment with its head end projecting, where it filters out tiny edible

---

* When the capacity to produce bone fails in the genetic disorder known as
brittle bone disease, the body collapses. Ivar the Boneless was a ninth-century Viking
king with this condition who had to be carried around on a shield, which apparently
did nothing to diminish his bloodthirstiness.

particles from passing currents with the help of some tentacles and a series of "gill slits," something like those of the lamprey. It has neither fins, nor skeleton, nor much of a brain. But it does have a bit of a tail and a notochord running down its back. It also has muscles along its flanks arranged in blocks, providing a close match with the agnathans. In 2008 its mitochondrial DNA was sequenced, which confirmed (for once) what my biology master told me when I was still a youth: amphioxus is related to the vertebrates; it is like a vertebrate with almost everything subtracted, a half-sketched blueprint. It resembles the lamprey larva even more closely than its adult, the former also has tentacles at the head end and filter feeds. Amphioxus also has several peculiarities of its own: its reproductive system is different from that of agnathans, and the notochord runs into the head region, for example.* Evolution has not allowed even this simple organism to stay unchanged: for natural selection applies to all creatures, great and small. Dramatic confirmation of the seminal position of these animals came with the discovery in China's Chengjiang Cambrian fauna (Chapter 2) of fossils that were relatives of *Branchiostoma*. *Myllokunmingia fengjiaoa* is an inelegant name for what is unquestionably an early manifestation of the amphioxus. It looks as if the little animal had been pressed delicately onto a shale surface with the care of a printer, and it even shows the muscle blocks lined up along its flanks. Thus with a swipe of flesh on stone we carry our vertebrate pedigree back 525 million years. We have pursued our own trajectory for as long as molluscs or brachiopods or arthropods. A silvery sliver of a thing took us away from trilobites and snails onto our own special path: one which led to land, to *Tyrannosaurus rex,* and *The New York Times.*

We know that the history of our most distant relatives proceeded from jawless to jawed. How that happened was recognised by studying transformations in the arches that supported the gills of jawless fish and their relatives: their basic anatomical elements were reassembled to become moveable parts, the struts and levers of jaws. One of the heroes of this triumph of the science of comparative anatomy

---

* Hence the scientific name for the group to which amphioxus belongs, the Cephalochordata; *cephalos* is Greek for "head."

was a Swede called Erik Stensio (1891–1984), whose magnificent and meticulous work in the 1920s and 1930s laid the foundation for modern understanding. I met him as a very old, wizened little man in the Swedish Natural History Museum in Stockholm. When I looked surprised to see such a famous figure still working, he squinted at me humorously. "Aha," he chortled, "you see before you a living fossil." I would have preferred the words "living legend." Two arches are invoved in making jaws: the mandibular arch, forming the jaws themselves, and the hyoid arch, which acts as a support for the mandibular arch. The earliest jawed fish, an extinct group of fishes known as acanthodian "sharks," have a hyoid arch that still largely resembles the old gill arch. So it seems that both of the arches involved in making jaws did not transform at a stroke, but by a number of smaller steps. Nowadays, we can study the expression of developmental genes during growth, which supplies a way of showing how complex organs are assembled, a window opening onto history. So zoologists are going back to the classical nineteenth-century study of embryology, to see how arch structures develop from cells in the neural crest of lampreys, the same potent area studied by Jean Joss and her colleagues in the lungfish. The way genes are expressed during very early development casts light on evolutionary events that happened 440 million years ago. They may also help us to understand medical conditions still affecting people today that have a genetic cause. Our bodies still read out ancient instructions. Isn't it extraordinary that the comprehension of the remote past might inform the understanding of the future? Every student of these problems must thank the higher being of their choice that the lamprey survived, for the fish is no mere curiosity; it is a fortuitous key both to distant history and therapeutic developments to come.

I hardly need to add that sharks are *Jaws* incarnate. Marine predators supplied with a never-ending conveyor belt of terrifyingly efficient teeth, sharks are also among the long-term survivors in the ocean. They scarcely require a special field trip to see them. I have walked among small sharks on coral reefs, and swum far above them off the Galapagos Islands. The latter did induce a queasy sensation, although my guide assured me that only other fishes were of interest to that particular species. Sharks survived, as head honchos of the food chain in the oceans, through the times when dinosaurs relinquished that position on land. They range from small, hungry fishes to giants made for Hollywood hyperbole. *Carcharodon megalodon* was the largest predator of them all, and survived until about 1.5 million years ago, which is comparatively late on the geological clock. At twenty metres long, it would have dwarfed even the great white. The largest living species, the whale shark, has secondarily become a peaceable browser on the plankton. Sharks are related to rays and skates that have crushing teeth and make quick work of hard-shelled molluscs. In their Palaeozoic past, sharks and their relatives were even more diverse; several kinds invaded fresh water, and others were covered externally with tooth-like "scales." Fossils of sharks are usually comprised of their jaws or teeth alone, because the rest of the skeleton is not made of bone. Instead, tough cartilage supplies the material for the mechanical engineering in the shark body, and such gristle does not have substance enough to be preserved in sedimentary rocks. But it is tough enough to support the motion of their threshing tails through all the vicissitudes of geological time, from their deep origin even before that of the lungfish until now. Humans are likely to exterminate some species of shark, because of our appetite for their fleshy fins. Sharks deprived of these necessary stabilisers are not even killed, but are thrown back to corkscrew helplessly to their death in the depths. I am ashamed of our species: Is this what evolution has brought us to, so smug on our bipedal pedestal? Do sharks deserve such an ignominious dispatch after 400 million years as kings of the sea?

We might with some relief move onwards from ancient seas and the lungfish to the terrestrial branch of the evolutionary tree. This is a story of progressive freedom from water. Amphibians need to return to water to breed; their eggs are laid there and larvae grow up there. Reptiles lay larger, yolky eggs surrounded by a thick membrane that protects against their drying out. If these "amniote" eggs can be placed in a suitable place to incubate, there is no need to return to fresh water at all. Babies can hatch out as miniature adults, ready to take on the world. Some advanced reptiles, especially snakes, have taken the whole process a stage further and give birth to live young. Both reptiles and amphibians carry within them a legacy of their aquatic life, as they cannot regulate and maintain their body temperature by internal metabolic processes; they are cold blooded (poikilothermic). Frogs and toads shut down and hide away in cold weather. Lizards in cool climates have to bask in the sun to get going in the morning. Once they are fully operational, they can move with great speed. One advantage of this basically slow machinery is that reptiles and amphibians require less food less often; indeed, crocodiles can starve for months. Hot-blooded creatures demand regular suppers.

I cannot point to a living species of amphibian surviving from the base of the tree of life. It is my major missing survivor. Today's amphibians with their familiar tadpole babies and their delicate skins are all descended from a common ancestor long *after* the tetrapods crawled out of the water. Most of them evolved following upon the mass extinction at the end of the Permian Period 250 million years ago. There has only been one 290-million-year-old Permian age fossil "ancestor" claimed for them so far. All these frogs, toads, newts, salamanders and their relatives are referred to together as Lissamphibia. The classification of more ancient amphibians known only from fossils is a complex and unresolved matter. I am not going to pursue this here, embroiled as it is in many technical arguments among specialists as to which of them are, in fact, more closely related to stem reptiles rather than to younger amphibians. It's a matter of "bone wars" among palaeontologists, I am afraid, and such campaigns can claim innocent victims. Instead, I am going to cheat. I shall briefly give what my American friends call "a visual" of the early tetrapods by looking at the living giant salamander (*Andrias*). It may not be a member of the early amphibians, but it makes a plausible attempt to look like one. For a start, it is huge: some individual giant salamanders are as long as a man. Small and delicate amphibians, like frogs in cloud forests, were not the first to evolve. Then the salamander looks as if it has been squashed flat by a heavy weight pressing in from above. It is a great, flattened grotesque of a thing with a broad, smiley mouth. It has a passable fin on its back end. In fact, to a moderately tutored eye it looks generally like some of the reconstructions of the lumbering early Devonian tetrapods from Greenland, *Ichthyostega* and *Acanthostega*, or one of the Carboniferous creatures we used to call labyrinthodonts before "bone wars." Its legs seem inadequate for the job; they stick out like some kind of failed robotic appendages. In short, it resembles a terrestrial animal put together as a prototype by a rather inefficient committee. No question, it is a treasure. In detail, it shows odd features that prove it is no generalised "ancestral" form after all. For example, it retains gill slits (as well as having lungs) as an adaptation for aquatic life. It is related to the hellbender (*Cryptobranchus*) of North America, a black and crinkly animal ready supplied with a name for a horror movie. Anglers expecting a plump trout recoil aghast when they land one of these wriggling grotesques instead. These animals continue to make a good living by eating crustaceans in rivers and streams, although the far eastern species are under intense habitat pressure in the wild. The Chinese giant salamander can live up to seventy-five years in captivity, so, like the Australian lungfish, it is a creature that likes to take its time.

No cheating is necessary to visit a reptilian survivor. The tuatara (*Sphenodon*) is superficially unchanged since Triassic times, 200 million years ago. It has endured in New Zealand alongside several of the organisms described in this book. It cannot thrive in competition with recently introduced predators, like rats and foxes, and for this reason populations have been established on islands around New Zealand to ensure its survival. To find it I visit Somes Island (*Matiu* in Maori), just offshore from Wellington; a small pimple of land, out in Day's Bay. Notwithstanding its small size, Somes has had a colourful history, one that has been told by David McGill in *Island of Secrets*. It was used as a quarantine camp for would-be settlers in the nineteenth century, and many people died there. A sad monument records their names, including those of children. They must have been already sick on arrival in New Zealand; this stark list is the only record of "Baby Arthur Aged 17 months." Judging by the roll call of deaths, the years 1873–1875 must have been a dreadful time. Isolation ensured the survival of other individuals. On a little islet off Somes called in Maori *Mokopuna*, a "Chinese leper" lived in isolation: he was fed by a basket that was winched to him daily, until his death in 1904 (in the end, he may only have had tuberculosis). At the outbreak of the First World War, Germans living in New Zealand were rounded up and dumped

58. (ABOVE) The brook lamprey *Lampetra*, a surviving jawless fish, showing the series of gill openings that reveal the primitive condition in vertebrate evolution. River Lambourn, Berkshire, UK.

59. (RIGHT) Brightly coloured fire salamander (*Salamandra salamandra*), an amphibian that is not uncommon in Europe, found here by the author in Italy.

60. (LEFT) The hellbender (*Cryptobranchus*) is a giant amphibian that offers a close visual parallel for some of the early amphibians that lived in the Carboniferous coal swamps, although it is not particularly closely related.

61. (ABOVE) The tuatara
(*Sphenodon*) of New Zealand is
the one surviving sphenodont
reptile. Its resemblance to a
"lizard" is superficial.

62. (LEFT, TOP) A giant
leatherback turtle (*Dermochelys*)
returns to land to lay its eggs in
the sand in northern Trinidad.
Turtles survived the mass
extinction at the end of the
Cretaceous.

63. (LEFT, BOTTOM) A giant
fossil turtle *Archelon ischyros*
from the late Cretaceous rocks
of South Dakota, preserved in
the collections of the Natural
History Museum in Vienna. It is
5.25 metres wide.

64. (ABOVE) The egg-laying mammal echidna out for a walk in Kangaroo Island, South Australia, in search of invertebrate fare.

65. (LEFT) The Australian duck-billed platypus (*Ornithorhynchus*) is the most extraordinary of our survivors. The "bill" is used to grind up the crustaceans and other invertebrate prey on which it feeds. Like the echidna, it lays eggs in burrows.

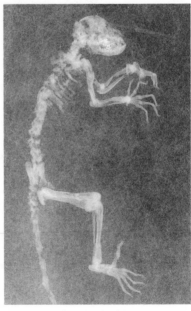

66. Ida (*Darwinius masillae*), a controversial fossil primate from the Messel pit near Darmstadt in Germany (Eocene, 47 million years old); it is 58 cm long. The fossil as it appears on the rock.

67. The miraculous details of the X-ray portrait.

68. Colourful reconstruction of a "dinobird" (Cretaceous) in the Vienna Museum. The boundary between dinosaurs and avians is becoming less distinct as more transitional fossils are discovered, especially in China.

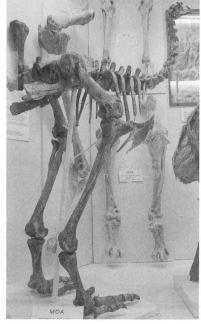

69. (RIGHT) The skeleton of an extinct moa preserved in the Bell Pettigrew Museum at St. Andrews University, Scotland. The extinction of these giant birds in New Zealand is almost certainly related to human hunting.

70. (ABOVE) The tinamou in its native habitat, Colombia, South America, is possibly the most primitive living bird.

71. (RIGHT) The little ferreret, or Mallorcan midwife toad, at home in a small patch of permanent water on the island of Mallorca. Its survival is typical of the problems faced by island animals.

72. (BELOW) The Chillingham cattle, Northumberland. They have several features that suggest they are "wild" cattle, but they are probably not direct descendants of the auroch.

73. (RIGHT) My colleague Paul Smith took this photograph of a group of ice age survivors, musk oxen, in Greenland. They survived while many of their Pleistocene contemporaries did not.

74. Mountain ibex photographed at alpine altitudes in northern Italy.

75. (LEFT) Ibex were deftly sketched by our ancestors on cave walls at Niaux, in France. They also drew animals that failed to survive the warming at the end of the ice age, or subsequent human maltreatment.

76. (BELOW) Noble ice age survivor: bison in Yellowstone National Park. This animal only just survived persecution at the hands of our own species during the nineteenth century.

77. (LEFT) Huangshan ('Yellow Mountain') eroded granite peaks in Anhui Province, China. The flanks of the mountains are home to several survivors.

78. (BELOW) Huangshan inspires a classical concept of the picturesque in Chinese culture; this is a cheap tourist item, which nonetheless remains true to the tradition.

79. (LEFT) A garden variety of *Magnolia* in bloom. Wild magnolia can be seen on the slopes of Yellow Mountain.

80. A bonsai specimen of *Ginkgo biloba*. It would be sad if the survivors described in this book continued their existence only in cultivation or captivity. Natural time havens merit conservation.

81. One of the great survivors—indeed, almost indestructible—a fossil cockroach compares closely with one of its living relatives (BELOW, 82).

in Somes. Major Dugald Matheson was appointed Camp Comman-
dant and recruited a disreputable bunch of guards that a Wellington
detective described in *John Bull's Register* as "a worse gang could not
have been leagued together." The camp was certainly cruelly run, but
when on 30 July 1918 four inmates escaped on a crude raft of three
kerosene cans, the same journal noted: "The Hun is a crafty and cun-
ning animal . . . he will stop at nothing to gain his desired ends." The
island then functioned once more as a quarantine area during the
flu pandemic of 1919. In the Second World War it took on a mis-
cellaneous bunch, including fanatical Nazis from German Samoa
together with communist "agitators," and any unfortunate Japanese.
Herman Schmidt was a communist who had helped to expose New
Zealand Nazis, but who was incarcerated willy-nilly along with the
President of the Auckland Nazi Club, Jonathan Blumhardt, and its
Treasurer "who had drawn attention to himself by driving around
with a Nazi flag draped on the front of his car." A more enterprising
inmate started a successful business making ornaments from *Paua*
(abalone) shells, a skill which has left descendants in every tourist
shop today. It seems appropriate that Somes received yet another
fugitive, although this fugitive is from the past: the tuatara.

After we arrive by ferry, our bags are searched for stowaway rats;
there certainly aren't any in mine. The warden, Rob Stone, guides us
along a path that skirts the perimeter of the island, winding its way up
and down through the scrub. A profusion of shrubs tumble together
over the sea cliffs beneath us in a dense scrum. I recognise *Coprosma*,
Pittosporum and *Hebe;* the latter are British garden favourites. Tuatara
must be hiding here somewhere, if only I could see it. The thickets
provide plenty of cover for native birds, and Somes Island has been
recruited to help conserve a rare species of red-capped parrot alongside its ancient reptile. More geologically recent reptiles are everywhere: little greenish skinks that sun themselves on the path side or
scuttle through the dry leaves, grabbing flies wherever they can. No
sign of cold-blooded torpor in these elegant, sharp-eyed sprinters.
They surely dodge even the hungriest mammals, except there aren't
any now, as before the tuatara could be introduced the island had
to be cleared of all warm-blooded pests. A population of the reptiles introduced in 1999 appears to have become well established,

after a failed earlier attempt. There are two species of *Sphenodon,* and the rarer of the two was the object of the introduction on Somes. "Tommy" the tuatara was brought from the South Brothers Island in the Cook Strait, where he had apparently attained maximum size. But after a few years in the more spacious accommodation on Somes, he grew both longer and heavier. Tuataras evidently live for more than a century. They go through life *lento* compared with the skink's *allegro con brio.* But they must still be fast enough to elude me now.

The path winds onwards around the island; there is no sign of our quarry. Rob Stone is not worried. Tuataras are creatures of regular habits, and we will soon get to one of their usual haunts. And there it is! A tuatara sitting on a log apparently oblivious to the human intruders nearby, looking as if it were resting after a stroll from the Triassic. It is not a small animal by any means, perhaps three-quarters of a metre long. It clearly displays the zig-zag crest along its back that is recognised in its Maori name. The dozing reptile looks generally similar to a greyish, rather slack-skinned lizard. This is a misleading resemblance because the tuatara is only distantly related to our more familiar reptiles. This particular example does not do very much for a long time. In fact, it did not do anything at all for a *very* long time; I am hoping it would at least lift a leg or something to indicate that it is alive. Instead it just sits there, enduring through geological time. At night tuataras are more vigorous, and go in pursuit of very large, succulent, flightless insects related to crickets known as *wetas.* I did discover an unpleasantly spiny, discarded *weta* leg as proof of their tastes. Male tuataras are territorial and chase rivals with a threatening open mouth, while raising the spines along the back. It is not too fanciful to imagine the same displays happening under ginkgo trees 200 million years earlier. Rivalry is not just a characteristic of hot-blooded suitors; it is as old as Pangaea. The teeth of the tuatara are primitive; they are just outgrowths of the jawbone. Nor are they replaced during their long life, so gradually become blunter. Aged tuataras, like some human nonagenarians, have to subsist on softer food, such as worms and slugs. Tuataras have a single row of "teeth" in the lower jaw, and a double row in the upper jaw; they occlude together quite tightly. Many other features of the skeleton point to the fact that these strange reptiles fit onto an evolutionary tree below lizards and snakes, with which they ultimately share a common ancestor. Their alliance with these reptiles, and more broadly with the extinct "sea lizards" like ichthyosaurs, dinosaurs, crocodiles and flying reptiles (pterosaurs) is suggested by a pair of large openings in the sides of the skull. These openings are called fenestrae, a term derived from the Latin for "windows"—which is exactly what they look like.* Their hatchlings have a "third eye" (pineal eye) in the centre of the forehead, which is probably another retained primitive feature. Tuataras have even been taken (incorrectly) as a model for a basal reptile: a creature apparently transported in a time machine to the present day. But 200 million years ago it had a variety of other relatives, belonging to several different families, some more snake-like, others probably living in ancient lakes. These are classified together in an Order Sphenodontia, of which the tuatara is the only living representative. It is probably no coincidence that this sole survivor is found in New Zealand with other relics of the ancient supercontinent Gondwana. Maybe isolation in those islands protected the tuatara from competition, as if they were a Somes writ large; the tuatara was allowed to waddle at its own, very slow pace into a younger world.

Nothing is ever quite that straightforward. We have seen before that there is no such thing as no change at all, and the tuatara has changed in subtler ways. In 2008, a study published on the DNA obtained from bones of tuataras known to be eight thousand years old allowed for an estimation of its rate of change at the molecular level over millennia. It turned out that the rates of change in the tuatara genome were *higher* than the rates characteristic of many other organisms. Lack of change in the shape this animal presented to the outside world was no indication of what might be happening at the most fundamental level of all. Beneath its stolid exterior, gene sequences were churning away as fast or faster than in animals higher in the tree of life. This was despite the fact that tuataras have

---

* Referring to this structure, these reptiles together are classified as "diapsids" ("two openings")—as opposed to "anapsids" ("without openings")—which have traditionally included turtles and tortoises together with some long-extinct marine animals like mesosaurs. Recent molecular work, however, has suggested that turtles were secondarily derived from a diapsid ancestor. If this is true, the anapsid skull of the turtle is another example of parallel evolution.

the lowest metabolic rate of any reptile, and are also the most "cold blooded" with a 16–21°C optimal body temperature. This is why they can hunt at night. One might almost say the Somes castaway was warming up from being a fish. As for the tuatara's prospects in the longer term, it may be secure for now on its islands-off-the-island, thanks to human intervention. It would be more than unjust to let it die; it would be an insult to the virtues of endurance. Biologists at Victoria University in New Zealand have shown that there is another possible threat to the old campaigner. Its sex ratios are controlled by temperature. This fact was established by altering the temperature of developing eggs to show that above 21°C only males are produced. During the climatic vicissitudes that must have happened in New Zealand over millions of years, the tuatara could change its preferred latitude, or even move up and down mountains, to secure its future in an appropriate temperature regime. If global warming becomes a reality, such options may not be possible on a tiny speck of rock like Somes Island. Can you imagine the last males of the tuatara, centenarians all, vying with one another to secure a meaningless territory, a barren estate?

The first fossils recognised by palaeontologists as reptiles are Carboniferous in age, about 318 million years old. Footprints made by these remote second cousins of the tuatara crossed mud flats where early relatives of the horseshoe crabs still scuttled beneath rotting vegetation; where creeks wandered through the hot and humid swamps of coal forests; while far above, giant dragonflies skirted tree-sized relatives of *Huperzia*, or rested briefly on horsetails or seed ferns. The Bay of Fundy in Nova Scotia, Canada, preserves both footstep and fossil in its sea cliffs. When fires raged through the ancient forests, stoked into ferocity by higher levels of atmospheric oxygen than we have today, hollow tree stumps provided the briefest haven from the heat for fleeing vertebrates, before becoming both their tomb and archive. It is a long journey through time and space from such early reptiles to their living relatives. If the fossils of animals such as *Hylonomus* look generally like lizards (or tuataras, come to that), which they do, that is only because being designed like a lizard is a general option for a reptile that probably hunted insects and other invertebrates, just like the little skinks on Somes Island do today.

Many of the skeletal features of more advanced living reptiles were lacking in the very first of their kind. These egg-laying animals* prospered, and radiated into a range of ecological niches on land and eventually in the air; some became "ruling reptiles," while others returned to the sea. Many did not survive the mass extinction at the end of the Cretaceous Period 65 million years ago. But a few groups passed through the lethal portcullis into the Tertiary and survive today. There is a recent obsession with swimming with animals that is slightly perplexing. Friends of mine who return from holidays abroad smile with satisfaction when they describe the high spot as "swimming with dolphins" or "swimming with sharks." Those who have travelled further afield swim with whale sharks, or blue-footed boobies. It could have something to do with being nearly naked. It may signify a return to an imagined Eden of direct communication with animals, before the Fall, to a time when we could roam unsullied and unclothed in a paradise of vertebrate equality. In Hawai'i I swam with turtles, so I have an inkling of this primal frisson. If I am candid, I do not believe that the green turtle worried one jot about this particular advanced mammal (with his incompetent diving moves) that hovered nearby as it nibbled algae for breakfast. I had also seen them lug themselves onto a beach of black volcanic sand to laboriously dig holes for their leathery eggs to be incubated under the sun. Distant ancestors of these same turtles probably performed similar rituals 100 million years ago. Tiny hatchlings are already miniature turtles— no namby-pamby nursery stage for these creatures. They must run the gauntlet of bird predators to the sea, just as once they probably dodged small (and possibly feathered) dinosaurs looking for an easy

---

* My more rigorous colleagues would expect me at this point to note that reptiles are not a natural group in modern zoological understanding. One group of reptiles gave rise to mammals, and another to birds, which *are* both natural groups descended from common ancestors, as touched on in the next chapter. This means that reptiles are a set of branches in the tree of life from which some derived branches have been "lopped off"; so they are not the complete ticket. They are united by retaining the general characteristics of laying amniote eggs and having neither fur nor feathers nor hot blood (but even that may be no longer true; see below). So if we were being persnickety, we might always refer to "reptiles" in quotes; but as Stephen J. Gould pointed out some time ago, nearly everybody has a good idea of what the term means, so let's stick with it!

meal. Once in the sea they have a certain elegance, using their flippers to join a relatively leisured class in a world of darting opportunists.

These living reptiles are also fugitives from the end Cretaceous extinction. A magnificent fossil specimen of the extinct turtle *Archelon* is displayed in the Vienna Natural History Museum, an animal that would have been a contemporary of the dinosaurs at their peak. In a gallery dating from the later years of the Austro-Hungarian Empire, with an ornate ceiling supported by extravagant carytids, and painted scenes lining the walls, a mere fossil would need to be something special to attract attention. But *Archelon* was the largest turtle ever to have existed, and grew up to four metres long: it is a show-stopper. The skeleton looks at first glance like some sort of archaeological relic, like the ribs of an upturned boat. The visitor then notices the spread-out flippers and the massive bony head, and its biological signature is obvious. There is no sign of any "windows" in the sides of the skull, as in the tuatara and its relatives. The ribs of the skeleton are fused to make the body armour, termed the carapace above and the plastron below. *Archelon* is related to the living leatherback turtles, where the shell is not entirely solid, but the bones support tough organic armour functioning rather like the hide shields used by medieval foot soldiers. Other turtles, like their relatives the terrestrial tortoises, are bony nearly all over and virtually impregnable as adults; at least until mankind appeared to apply his ingenuity to a particular gastronomic challenge. He wiped giant tortoises out from mainland South America, appreciating the virtues of a meal coming in a ready-made casserole dish. Since large tortoises and many turtles live for over a century there were no quick replacements ready in the natural larder, and such species were doomed.

The evolution of the body armour might seem to present particular problems. Is it possible to have half a turtle? The discovery in 2008 of *Odontochelys semitestacea* in Triassic strata from Guizhou, China, went a long way towards solving them. These wonderful fossils show that the plastron evolved first: the armoured underside came before the carapace. Maybe ocean-going tetrapods were particularly vulnerable to attack from beneath by sharks. The driver of evolutionary change was likely to be an arms race between the hunter and the hunted, and in nature, no potential meal ever goes unchallenged.

There is never such a thing as a free lunch, unless you are a human arriving for the first time on an oceanic island stocked with giant tortoises, and even here the restaurant is soon exhausted. The species name of the fossil "*semitestacea*" means "half shelled," while the genus name "*odonto-*" refers to the presence of teeth, which are lost in all living species, with their horny beaks. If the turtles and tortoises evolved from a more conventional reptilian ancestor, we would *expect* their early representatives to retain teeth, and here they are. It has become unfashionable in some scientific circles to claim the discovery of an ancestor, such intermediates having too often been mere hubris in the past. But in *Odontochelys*, we have one fossil animal that indeed seems to show all the right intermediate features at just the right time. Creationists will have to cross off the origin of turtles from their list of the allegedly inexplicable.

"Swimming with crocodiles" is not an appealing option, especially in Australia. I have heard tales of my colleagues sprinting for their lives from hungry saltwater crocs while on geological fieldwork in the Northern Territories. The reputation of the largest predator among living reptiles is splattered with bloody stories, even if recent research points to a measure of parental care unusual among their cold-blooded relatives as evidence of finer feelings. Mother crocs help their babies to the water after they hatch and protect them in a kind of "crèche." Crocodilians—alligators, crocodiles and gharials—are all survivors from the "age of reptiles." The living species descended from a common ancestor some time in the Cretaceous, so are approximately the same age as *Archelon*. However, several ancestral groups related to crocodilians go back further to the Triassic, so they join the tuatara and the turtles as part of a great expansion in reptilian variety at that time. The surviving animals need the heat of the warmer parts of the world to thrive, but have turned their reptilian qualities to advantage. They eat rarely, but can swallow almost anything they catch, so they can bide their time. They live long lives, much of it spent in a kind of alert torpor. Waiting. Crocodilian jaws have tremendous strength. Their eyes and nostrils are situated high up so that they can peruse the world safely hunkered down in creeks and backwaters. Their oddly crooked toothy grin is actually designed to grasp and not let go. Crocodiles are distinguished from alligators because, among other things, their lower teeth are still visible even when the mouth is closed. Lewis Carroll exploited this difference when he described the Nile crocodile:

> *How cheerfully he seems to grin,*
> *How neatly spreads his claws,*
> *And welcomes little fishes in*
> *With gently smiling jaws!*

Crocs and gators lived alongside dinosaurs even while those aggressive and most innovative giants pushed out many competing reptiles for dominance of the land. The most primitive living crocodilians are called gharials. The Ganges gharial (*Gavialis gangeticus*) is a greatly endangered, fish-eating species with a much more slender snout than the crocodile. Recent molecular work relates this animal to another very rare reptile, the false gharial *Tomistoma*, that lives in remote areas of Borneo and Sumatra. Until comparatively recently this reptile was more widespread in the Malay Peninsula and Indochina, and has close fossil relatives in south China. *Tomistoma* does not indulge in parental care, unlike its more advanced relatives. Its youngsters take pot luck, like any tiny turtle. Even while the dinosaurs strutted their stuff over dry land, the muddy riverbanks and swamps probably always belonged to the crocodilians. Where now they grab wildebeest in Africa or kangaroos in Australia, they once grabbed small dinosaurs. Those big jaws have served them well.

The undergrowth probably already belonged to the snakes, another reptile group with a fossil record in the Cretaceous, and one which slithered through the hard times at the end of the period that eliminated so much of the competition. An early and well-preserved snake fossil (*Pachyophis*) found in Cretaceous marine strata proves that some of them soon returned to the sea, because the thick ribs of this animal testify to a relatively rapid derivation from a terrestrial ancestor. Snakes underwent a great diversification alongside mammals after the disappearance of the dinosaurs in the early Tertiary. There is evidence that many furry animals (including primates) are genetically hard-wired to fear snakes. After all, these particular reptiles offer the choice of crushing the breath from your hot body or poisoning you, before swallowing you whole by unhitching their jaws. If Eve had followed her genetic bidding, she would have run a mile from the sibilant tempter in the Garden of Eden. But to be fair to the Suborder Serpentes, they are not all venomous. The evolutionary relationships of living snakes is still far from settled, but many authorities point to a group of worm-like blind snakes (Scolecophida) as the most primitive living snakes. They live a harmless existence under logs and in leaf litter feeding on small animals. Blind snakes are distributed almost worldwide in the tropics and subtropics, including South America and Australia. By now this will be a familiar tale, suggesting that they might have already been present at the time of the ancient "supercontinent" of Pangaea. Since lizards and snakes are both offshoots from the main branch that also includes the tuatara nearer its base, it seems logical that these animals originated while the continents were still married together.

I did not have to make a special excursion to discover snakes in the field for this book. After working in desert climates for many years, I have had enough close encounters to bore my companions over sherry in the Traveller's Club. In Oman, I once lifted up a slab of rock in the middle of the Huqf desert to reveal a greyish asp curled up: fortunately, it shot off across the sand rather than up my trouser leg. In the very centre of Australia, I climbed a series of rock ledges one by one in search of fossils, only to peep into the eyes of a large king brown snake all curled up as I heaved myself upwards onto its private place; I reversed back down in a state best described as silent funk. In Nevada, I climbed high into the Cherry Creek Range while my companion assured me that we were now too high for "rattlers." Almost immediately there came a strange sound of desiccated maracas, or shaken beans: a western rattlesnake was displaying the end of its tail while sunning itself into activity on an exposed log—just in case we got any fancy ideas. Strangely enough, at these moments I did not wonder at the ingenuity of the evolutionary process to dispense with legs and substitute rippling musculature. Nor did I reflect on the reasons why snakes might have joined Crocodilia to cheat mass extinction at the end of the Cretaceous. I was interested in being the survivor closest to hand, and getting the hell out of there.

8

Heat in the Blood

Sugar gums are the prettiest trees with their clustered crowns of leaves resembling so many stacked, pale green parasols. All along the Rocky River the landscape is airy and full of light, because eucalyptus has dangling foliage that does not challenge the sun: the shade made by the tree is always dappled. Groves of trees are open to the sky and often underlain by low shrubs to waist height. The spiny Christmas bush takes advantage of the partial shade provided by the gums. In March it has tiny, oval bright green leaves running along its thin branches between ranks of spikes, but it will look more like a bundle of dead sticks in the dry season. Nearby, a group of wan-looking, drooping trees have thin branches apparently with no leaves at all. These are native cherry trees (*Exocarpos cupressiformis*) and a parasite on the roots of the eucalyptus. Like *Gnetum* (see p. 179), it is a plant that has traded in leaves for photosynthesising directly from its green twigs. A naturalist on Kangaroo Island must try, and often fail, to grasp the puzzling variety of eucalypts, which have evolved so spectacularly as natives of Australia. The sugar gums, *Eucalyptus cladocalyx*, are just one of the easiest to recognise. Along the riverbank they are joined by pink gums (*Eucalyptus fasculosa*), with light grey and white trunks, and particularly narrow leaves. Identifying eucalypts makes any observer particularly sensitive to the textures and colours of bark. These can be smooth, furrowed, blotched, peeling or patchy according to species, or shining white to matt grey. There are too many subtle combinations. Away from the river, the scrub "heaths" on Kangaroo Island had been blasted by a great fire in 2007. Skinny black burnt poles are all that remains of the former bush. But shrubby mallees, another category of eucalypts, are specially adapted to just these circumstances by readily regenerating from their tuberous roots. Three years on, the open bush is a mass of fresh green leaves around every mallee stump with daisy bushes in between, and the occasional *Banksia* shrub sporting dense yellow flowers looking like so many bottle-brushes. Dense, spiky-leaved tussocks of grass trees (*Xanthorrhoea*) also survived the fire. Now their black flower spikes, often taller than a man, rise from the middle of each plant like giant black pokers. They used to be called "black boys," but for obvious reasons are currently known as "yaccas." This open bush is not exactly beautiful, but it has a resilient grandeur about it, a defiant persistence in the face of fire and fortune. It is a special habitat, where a remarkable survivor can be discovered.

I am on Kangaroo Island, a short distance off the coast of South Australia beyond the Fleurie Peninsula, south of Adelaide. No bigger than a small island in the Mediterranean Sea, it has been protected by its isolation from the effects of introduced predators. It lives up to its name with bouncing wallabies and kangaroos being a common sight. Unfortunately, many more of them are also lying dead by the side of the road: they have no road sense at night. The interior of the island is farmed, with small vineyards producing excellent, deep red wines made from the Shiraz grape; elsewhere there are sheep. The western end is taken up by the Flinders Chase National Park, where the Rocky River creeps between stately gum trees. A few areas escaped the big fire around here. The great explorer Matthew Flinders' name can be attached to many things in South Australia, parks and mountain ranges among them. In Kangaroo Island he lends his title to a new Visitor Centre staffed by charming enthusiasts, but with virtually no books. We follow one of the paths leading off from the Centre into the bush. One of these soon leads into regenerating mallee country, where the vigorously sprouting scrub is interspersed with the occasional surviving big tree. The sounds of bells and whistles keep us company. Birds with gently curved bills, honeyeaters, flit through the branches, calling as they go. There is always a species of *Eucalyptus* somewhere in flower to satisfy their thirst. It is hot, and there is a kind of dreaminess about the walk, a feeling I remember from wandering along hill paths on Greek islands. A feeling anything could happen.

What actually happens is that an echidna trundles purposefully across our route. It is almost too easy this time, as this is the special animal that we had come to find in the wild. It was such a casual meeting, as it might be with an old but sprightly gentleman out for a stroll. On first impression the animal looks like an animated ball of long, pale brownish spikes with pale tips. It is surprisingly large, much bigger than the European hedgehog it superficially resembles. It has a strange little dark nose projecting pertly out of its spiny ball, which looks as if somebody had stuck a cigar out in front of it. While not exactly terrified of human intruders, once it realises who we are, the echidna rapidly heads for a tussock of sharp-leaved grass and pokes its head deeply within it, leaving us to contemplate a singularly spiky rear end. It seems to puff itself up to give the spikes maximum value, and the result is something like a dangerous punk haircut. Nobody volunteers to disturb the creature, which seems perfectly content to wait us out. After a while there is nothing for it, but to go on our way. Doubtless, when all is still, the echidna will continue its quest for the invertebrates such as worms, ants, termites and beetles that make up its diet. There is plenty to eat in the bush if you have a sensitive nose to sniff it out. The common name of "spiny anteater" hardly does justice to its range of tastes. The echidna proves to be quite common in the western part of Kangaroo Island. We see five or six of them at different times, travelling alone. Often they potter along the side of the road on their determined way, but always perform the same trick of protective half concealment when an attempt is made to get the perfect photograph. It is rather frustrating. Hedgehogs tend to be more obliging.

The echidna is a mammal, but it is a special mammal. It suckles its young with milk just as we do. On a brief acquaintance it is hard to judge its intelligence, but it is warm blooded and clearly has the instinct for self-preservation. It is also one of a very few mammals that lays eggs. Its young are born at a very early stage of development, no more than tiny, wriggly things, a cipher of an organism. Although marsupials also give birth to tiny, incompletely formed babies, the echidna does not have a well-developed pouch to nurture its young, as do kangaroos, wombats and possums. Echidnas do not even have mammary glands unless they are lactating. Hedgehogs, by contrast, are regular placental mammals with wombs and mammary glands, and babies that even look a little like hedgehogs. Their resemblance to echidnas is a fine example of parallel evolution in response to a generally similar mode of life: in this case the steady pursuit of invertebrates. However, in detail, the "snouts" of echidnas and hedgehogs are completely different structures. Echidnas are classified along with the duck-billed platypus (*Ornithorhynchus*) in a group of egg-laying mammals known as monotremes. There are four species of echidnas in Australia and New Guinea, three of the long-beaked echidnas (*Zaglossus*) confined to the latter, and the *Tachyglossus*, or short-beaked echidna, in Kangaroo Island and widely elsewhere Down Under. The last species to be discovered in the New Guinea highlands was named as recently as 1961 *Zaglossus attenboroughi,* in honour of the famous naturalist. That makes a grand total of five species for all the monotremes; they are special mammals indeed.

I went off to see Peggy Rismiller to learn more about "spiny ant-eaters." Peggy has been studying *Tachyglossus* for many years, particularly at Pelican Lagoon on Kangaroo Island, where she runs a sanctuary for the creatures. I have noticed before how researchers develop a proprietary affection for the animals they know about. This may be because there is never an end to the fascinating details that can be learned about an animal's biography. To twist Socrates' famous phrase "the unexamined life is not worth living," one could justly say that every life is worthy of examination, and the deeper the examination the more worth is revealed. Almost everything about the echidna causes raised eyebrows. Peggy likes to shock her interlocutors in a gentle way, and she whips out of one of her folders a picture of the echidna penis with four openings. "There," she says triumphantly. "What do you think of *that*?" It looks like a small pink succulent; apparently, the animal urinates from the cloaca and the penis sits to one side. Almost every aspect of echidna anatomy is equally odd, and many of its features can be interpreted as primitive: for example, the bones in the shoulder have a distinctly reptilian cast. On the other hand, echidnas do not have a hinged jaw, and the lower jaw is reduced to a couple of struts, a unique feature. The palate carries a coarse cartilage grating that is used to mangle food against the tongue. The spines themselves are modified hair and are shed and renewed. Echidnas have poorish sight but good hearing, although they don't have ears in the usual mammal sense. Instead, they have big ear slits and can move their spines around it to amplify sound, which comes in useful for detecting scuttling insects. Their body temperature at between 31–33°C is the lowest operating temperature in Mammalia. It would perhaps be over-imaginative to suggest that echidnas were arrested in the process of "warming up" from the reptilian state. Unlike most mammals, they neither pant nor sweat so they are vulnerable to overheating. Occasionally, when it gets too hot, they disappear into caves, where they are capable of going into a state of suspended animation for several months. Spiny anteaters can live for up to fifty years, which is a very long time for an insectivore. To contrast a mammal from my own garden, shrews can go through their whole life cycle in a year. Echidna's nails grow continuously, which is useful for scratching a living, but these handy tools can eventually become old and horny. Their scat (or droppings) is a brown, dry cylinder with a fine mucous membrane around it, so that no body water is wasted.

The echidna is equally distinctive in its reproductive biology. In winter, male echidnas begin to pursue females, with one female taking the lead and a train of hopeful suitors following behind: usually three or four, but Peggy has seen up to eleven. An egg is laid about every three years, and the mother carries it around for about ten days before the baby emerges. The tiny hatchling weighs about 0.3 grams, which is almost nothing at all. At this stage its nostrils are not open, so it respires through its skin. The tiny creature hangs on to its mother's stomach with its minuscule front feet, and her soft belly flesh develops into a kind of "pseudopouch" to hold the baby in. After five or six days the baby's nostrils open. The milk patch has already come into action, stimulated by the presence of the tiny offspring: about 120 tiny pores exude milk. It is nutritious stuff, for growth is rapid. After thirty-five days, the little creature develops fine, fur-like peach fuzz. It has begun to present its mammalian credentials (we are all brothers under the fur). When the little animal is able to leave the mother, after about fifty days, she constructs a nursery burrow for her youngster that can be up to two metres long. Whenever she quits it, she backfills the entrance so that it is hard for a human, or any predator, to find. The milk must be rich stuff, for the mother returns to the burrow for only two hours every five days. The milk changes during the development of the baby echidna, starting off watery, then becoming progressively thicker and full of good things.* The small echidna is charmingly known at this stage as a "puggle." It becomes really round and fat like a little balloon after the maternal visit: it can take in up to 30 per cent of its body weight in one feed. So the young life continues for seven months, while the fur changes into spikes. Only then is the echidna ready to start an independent existence.

The short-beaked echidna is the most intensively studied species, but only in Kangaroo Island have individual animals been tagged and studied over the long term. It has been proved that young animals can move up to forty kilometres, so they are quite adventurous, at least for a monotreme. The long-beaked species of New Guinea are much less well studied, and at least two have "endangered" status. Nobody has ever seen their eggs or young. One wonders who will study their puggles, and who will ensure their survival in the wild. Such wonderful animals are worth protecting.

I chose to visit the echidna as my example of an egg-laying mammal, but I did also make an effort to see its more familiar relative, the duck-billed platypus—another burrower. Its favoured habitat could be seen further down the Rocky River, where a deep pool fringed with vegetation offers just the right protection to make the animal feel secure. The platypus prefers to excavate its breeding holes in soft riverbanks adjacent to stretches of water. Here it can pursue crustaceans and other invertebrates using its sensitive bill as a probe and a trap, and its webbed feet to propel it through the water. Sadly, I failed to catch sight of this remarkable creature on Kangaroo Island. Perhaps I did not wait long enough. I had visited the Warrawong Sanctuary near Adelaide on a similar mission some years before. A high fence surrounds the reserve, designed to keep out introduced predators,

---

* Peggy tells me that the richness of the milk is attracting research into its use as a "formula," for example in treating eating disorders.

most particularly *Felis domestica*. Its founder, John Walmsley, is a raffish and exuberant Australian who was in the habit of wearing a Davy Crockett hat made entirely of pussycats. The tail hung down his back, all ginger tom. His rant against our European introductions was splendid in its undiluted passion. According to Walmsley, our favourite nocturnal hunter has had a catastrophic effect on the populations of native marsupial mammals. Inside his protected park, they flourish once more. At dusk there were throngs of delightful paddy melons, bettongs and potoroos to prove his point. I should add that these animals were fed artificially; so one could argue that their populations were much larger than they would be in the wild. Whether Warrawong is artificial or not, the disastrous role of introduced predators in Australia is not in question. A night tour around the park through eucalypt groves led finally to a pond designed to appeal to the duck-billed platypus. Our guide motioned the small party to be completely silent. After a few minutes in the gloom listening to the sound of our own breathing, there was a distant "plip" somewhere across the water. "That'll be him," she whispered confidentially. Everybody glanced across the inky surface of the water. Just possibly, a ripple may have been the wake of a swimming platypus. It was hard to tell in the dark. It is certainly far easier to get a good view of the living animal in the Sydney Aquarium, down by the harbour. A display tank there has swapped day for night, so the visitor can watch the platypus busying itself happily in an artificial current. Its flattened tail helps it twist this way and that in the water, as it probes for edible morsels with its versatile bill. There is nothing half-hearted about its aquatic adaptation; it looks as lithe as a seal.

Platypus and echidna are so different that they must have diverged from their common ancestor a very long time ago. Indeed, the platypus is so utterly unlike any other mammal on earth that this alone argues for an independent history running to many tens of millions of years. Unlike the echidna, it has no ecological counterpart in the other hemisphere. Platypuses and echidnas do share a very curious feature. The male platypus has a spur on the hind foot that is venomous. He has to be handled with care, because the spur can deliver a very painful wound. The female has the spur also, but it does not develop into a venomous structure; the same applies to both sexes of echidna. No other mammal has such a peculiar appendage and, given the additional similarity in egg-laying habits, this does support the idea that the two living monotremes are related at a deep level, despite their many differences as adults. The fossil record does not help definitively in this case. There is a well-known fossil giant platypus in Australia, perhaps 20 million years old, and already a typical platypus. A few other tantalising fragments have excited much debate. In 2008 a re-examination of a 100-million-year-old Australian fossil called *Teinolophos* led to its assessment as an ancient relative of the platypus rather than the echidna. This pushed the split between these two kinds of animals still further back in time; but, as usual, not everyone agrees with this interpretation. A previous estimate based on molecular divergence had already yielded a maximum of 80 million years for the date when platypus ancestors separated from their echidna kin. So if *Teinolophos* is correctly interpreted, the monotremes as a whole have to be still more ancient. They assuredly earn a place in this book.

A great geological age for echidna and its relatives seems to me to be a reasonable conclusion. Very early mammal fossils are rather rare. However, it is now an established fact that they lived alongside the dinosaurs, as small, insect-eating specialists scuttling in the undergrowth, unnoticed by their towering reptilian contemporaries. Only after the giants had died out at the end of the Cretaceous 65 million years ago were they released from their bit parts and allowed to take on the most dramatic roles, as predator or herbivore. They did this with great dispatch: it was another of those great evolutionary "explosions." However, early mammals did not look so different from the shrews in my garden; these little animals with their long snouts do indeed represent an ancient lineage. Most old insectivore fossils are often only known from teeth, but fortunately teeth are informative. Specialisation of mammal teeth into different kinds for different functions— like incisors or molars—is a feature that takes the group beyond most reptiles. A few early mammals are more completely known; I should mention ten-centimetre-long *Megazostrodon* dating back to the Triassic, more than 200 million years ago. Unfortunately, as fur does not readily fossilise, there is no proof yet of the origins of this useful insulation, which would presumably have accompanied the development of warm-blooded metabolism. If being warm blooded required more food, well, there it was in abundance, as succulent insects fed on plants and decaying wood and, presumably, reptile dung. When flowers entered the forest, new insects found a new living, and insectivores a new prey. Everything fed back into an ever-growing richness. The mammals waited for their moment to arrive.

There are interesting matters to think about in this scenario. If, as is generally accepted, the mammals evolved from one of the "mammal-like reptiles" (therapsids), then it follows that mammal ancestors must have laid eggs, like virtually all reptiles. Nowadays, among the thousands of living mammals, only the monotremes lay eggs. Commonsense then suggests that they are survivors not just for 100 million years, but from such an early phase in mammal evolution that neither placental womb nor marsupial pouch had yet evolved—which almost certainly dates before 200 million years ago. This pushes our Australian animals further still down the tree of life. Time moved on for the other mammals, but echidna and platypus stayed with ancient ways. If this were true, it means we have an even bigger hole in the fossil record relating to their early history. There is another way to look for ancient history: by sequencing the genome. In 2008, Richard Wilson of Washington University and numerous colleagues showed that the platypus genome was indeed "a fascinating combination of reptilian and mammalian characteristics." The platypus genome is about two-thirds as large as ours, and they have no less than ten sex chromosomes (most mammals, including humans, have two), of which the X chromosome compares with that of birds. Since birds *also* descended from reptiles, this is not as surprising as it may seem. Although 80 per cent of platypus genes were shared with other mammals, there were many that were shared with reptiles. To give one example, the curious and venomous platypus claw always seemed to have more to do with reptiles than with mammals (who, with the exception of critics, entirely lack venom). The same genes appeared to be involved, although the authors claim that platypus "invented" venom independently. There was nothing discovered in the platypus genome to falsify the idea that here is a heroic survivor from the earliest days of today's top dogs—which, of course, are barking mammals.

The poor platypus! It has had to endure a barrage of derogatory remarks. Articles I have read describe it variously as a "freak of nature" or begin by saying that only a mother platypus could love it. There are references to the fact that when it was first reported in Britain in the eighteenth century, nobody believed that an animal with a bird-like bill, fur and webbed feet that also laid eggs could possibly exist. The type specimen that proved its reality is still kept safely in the Natural History Museum in London. I find platypuses rather beautiful, especially at home in water. The echidna did not have such an ambivalent reception, but is probably as remarkable a survivor. Among other Australian native animals, marsupials also have a fossil record extending well back into the Mesozoic. They are often considered "primitive" relative to placental mammals, but they have evolved into a remarkable range of creatures, and fossils tell us of an even greater variety in the comparatively recent geological past. Carnivorous kangaroos, marsupial "lions" and lumbering herbivores all evolved from marsupial ancestors on the Australian continent in its splendid isolation.* The variety of marsupials living today is still considerable, ranging from marsupial "mice" to badger-like wombats and not-truly-bearish koalas. The kangaroos and wallabies are among the best-adapted herbivores on earth, the match for gazelles over the long haul. One really should not describe such a wealth of diverse creatures as primitive, although it has to be added that if one gazes into the eyes of a wombat, there does not seem to be much of a spark inside.

What of the antecedents of the animal that is supposed to be the brightest of all, *Homo sapiens*? Most mammals, such as carnivores, whales and their relatives, larger herbivores, and bats had their main evolutionary radiations after the dinosaurs and their reptilian

---

* Although this is broadly true, the break-up of Gondwana was not a simple split, and the history of populating Australia with animals was more complicated than I have stated. Marsupials were much more widespread millions of years ago; opossums are still very successful marsupials that have spread northwards from South America through the Isthmus of Panama to colonise America as far north as Virginia. The ease with which wallabies establish themselves in the wild in Europe as escapees proves that they can compete perfectly well with placental mammals. However, marsupial encounter with European predators in their Australian homelands has resulted in many species extinctions.

relatives had died out at the end of the Cretaceous. Some high-falutin' terms have been used to described this process, such as "ecological release," but what it comes down to is that mammals filled almost every vacant part of the world in an exuberant burst of evolutionary creativity. The invention continued, off and on, over the next 60 million years, as one group of mammals after another rose to prominence in different continental parts of a now-separated Pangaea. The group to which we belong together with our evolutionary relatives, the primates, was initially rather an inconspicuous part of this radiation. There are a few fossils that suggest that the group rooted down into the Cretaceous, so primates are by no means among the most recent of mammals. My colleague Alan Cooper believes that based on molecular evidence they may have originated as long as 90 million years ago. The most primitive living primates are small, shy, nocturnal tree dwellers, and if this were the case in the past, one would not expect them to be common fossils. As we have seen several times before, placing animals on one branch or another at the base of an evolutionary split is often controversial. The primates are usually divided into two major groups, one leading to the lemurs and the other, simian branch leading on to monkeys, apes and ourselves. It hardly needs to be said that the latter excites the most public interest. Lemurs had their own isolated, evolutionary radiation on the island of Madagascar when it split away from Africa, where a number of marvellous species hang on precariously in the face of deforestation and human population growth. It would be tragic to lose them. Their Asian relatives the lorises are equally under threat, not least because their huge eyes, adapted for life in the deep night, provide an ingredient for traditional medicine. I am frequently led to doubt the *sapiens* part of our species name. The living animal closest to the base of the simian branch is now considered to be the tarsier, an appealing little animal comprising a number of species living in the forests of Borneo, the Philippines and Sumatra. Fossils show that they were once more common and widespread, and have changed little in 40 million years, so they are definitely on the list of survivors. To the casual observer, they look not unlike lorises and lemurs, and it is not surprising to learn that their classification has divided scientists in the past. All these animals are rather sweet looking, with huge eyes and

The tarsier, charming survivor from the base of the primate lineage, now confined to South East Asia. The fossil known as Ida looks generally similar, and was found in Germany.

sensitive long-fingered hands like those of a concert pianist. However, certain characteristics have pushed tarsiers onto the tracks leading along the simian line. To select one interesting example, tarsiers, in common with all the other species on the monkey clade, have lost the ability to manufacture their own vitamin C, something that lemurs and their relatives continue to do. This means that we humans have to eat fruit for the good of our health, as do our simian cousins; and, of course, tarsiers. This may seem trivial, but it is indeed a reflection of fundamental biochemistry. It's a pointer showing the path evolution has taken.

It is not to be wondered that classification of early primate fossils can be problematic. One 47-million-year-old specimen attracted much attention when it appeared in public for the first time: on display was a remarkably complete example of an early primate, *Darwinius masillae*. The discovery was made from a famous fossil site in Germany called Messel, where a wonderful variety of delicate fossil mammals (including bats) and even fossil insects preserving their colour had been known for decades. Although the fossil primate species had been recognised since 1983, the completeness and beauty of the specimen published in 2009 was something special (see image 66). It was acquired on behalf of the Palaeontological Museum in Oslo, by a charming, extrovert palaeontologist called Jørn Hurum. I went to look for it in 2010; even though it was not a living survivor like *Peripatus,* I felt it should be on the list. It was a different kind of survivor—an almost perfect fossil that survived to tell us of our early history. The Oslo Museum sits in the middle of a lovely botanical garden, and is a serious building of granite, built to last. I arrive to the usual hubbub of children that accompanies any dinosaurs on display. Below the public area lies the hidden world of the curators, a world I know well from my working life. It is a place where casts of ichthyosaurs lie next to former exhibitions, in an amiable jumble. There is dust about, but it is dust that is meant to be there, the dust of ages. And there is *Darwinius* on its slab, neatly laid out as if it wanted to be discovered just so. I may be no primatologist, but I appreciate a beautiful fossil when I see one. To my untutored eye it does look monkey-like, almost frozen in motion, startled by its sudden death and burial, recording a moment in time and a vanished world. Its graceful hands are splayed in mortal surprise. Even if it were a lemur, no living relative is found within ten thousand miles of Germany. It is too easy to take for granted the changes wrought on the world by geological time.

Jørn is not only personable but a persuasive entrepreneur. He coordinated the official publication of the animal with press releases, a television programme narrated by the doyen of public natural history, Sir David Attenborough, and the publication of a book, *The Link,* written by Colin Tudge. The fossil got a nice, short memorable name, Ida. It was a big hit. Jørn raised the necessary large sum of cash to buy Ida for science from a private owner. He lobbied the right politicians so that they would have felt embarrassed if they had not come up with the money. The brouhaha surrounding Ida was based on a claim that this particular fossil lay at the base of the simian branch of the evolutionary tree. This is another way of saying that it ultimately lay at the root of us humans. It was "the link." A cute fossil at the right time proved the truth of evolution, or at least that was the newspaper headline. It did not take long for the counter-claim to arrive in the shape of an article by Jonathan Perry and his American colleagues in *Nature,* published in October 2009. No, they said, Ida was actually a rather familiar animal somewhere on the "stem" leading to the lemurs, and no more than a sideshow in the circus of human evolution. It was related to a number of similar primate fossils and, to be frank, no big deal. The American study was based upon an analysis of Ida's skeleton (that includes the teeth) relative to the features of other relevant primates. A cladistic analysis of this kind is a comparatively routine way of investigating relationships, based on evaluating the shared, evolved characters of a group of species under study. A computer sorts out different arrangements, eventually "deciding" on a tree diagram, or cladogram, that portrays the relationships of the species concerned in a way that is consistent with the most characters—arranged in the best way. It all sounds very cut-and-dried. But we also know that organisms close to fundamental branches in the tree of life often show conflicting combinations of characteristics: some are "left over" from their ancestors, and might be very similar to those on an adjacent branch that may eventually lead on to very different creatures. Other features typical of later members of a particular group may not yet have appeared; there was a time before leopards had spots and seals had flippers. Other characters again are simply ambiguous in their interpretation. Any scientist who has worked with cladograms knows that they change with experience, and from worker to worker. When I mentioned the American cladogram to Jørn Hurum, he snorted with some vigour, and noted that a new analysis with the characters "properly" coded supported the original interpretation. "Just wait and see," he said. It is currently only the end of Round One in this particular sparring match, and I would not like to predict the outcome. What I *can* say is that wherever Ida fits on the tree of life, neither this ancient animal nor the other living members surviving from near the base of primate history have the large brain we like to associate with ourselves. This is undoubtedly a property of our close relatives among the great apes and chimpanzees. Cerebral magnification was one of those characters that would be acquired later. But our short face, forward facing, acutely sensitive eyes and useful fingers were already in place. Just give it time.

Internal temperature regulation in mammals was important; it allowed for everything to speed up. Burning food to keep the organism active even when the sun is set may have proved expensive for the metabolism, but it allowed for many new activities. Nocturnal habits are typical of many species of mammals, allowing them to live in deserts, or hide away from predators during the day to busy themselves at night. Fur coats and big energy intake allowed for colonisation of the high Arctic regions. Greater and more continuous activity led naturally enough to greater neurological complexity. That old "arms race" between the hunter and the hunted, predator and prey, was racked up another notch. Teeth became more specialised: to grab or chew, to masticate at leisure or to delicately nibble. To the faster and smarter came the prize. The more sensitive, swifter gazelle lived to breed and pick at leaves another day. The hunter that could collaborate with its fellows more effectively landed the larger prey. Noses and whiskers twitched with enhanced effectiveness; larger brains more quickly evolved new strategies for attack or defence. That fed back once again into favouring still larger brains. Many small mammals opted to "live fast, die young"; but elephants took the opposite course. Mammals like the whales returned to the seas from which their distant ancestors had emerged 300 million years before. Another group of mammals took to the air: the bats. Mammals became almost ubiquitous, occupying all the niches vacated by their reptilian predecessors. There is probably no entirely objective measure that would satisfy a critical scientist, but I am sure that the biological world in the Oligocene was generally smarter at the top of the food chain than was the world in the Permian. Heat in the blood led to brilliance. But the mammals were not alone. Other animals learned the trick of internal fire. They prospered alongside the fur-clad adventurers. Their eyes were as sharp; their brains were as alert. They were birds.

So we must go in search of an avian survivor. Along the eastern flanks of the Andes in Ecuador, lush cloud forest spreads up steep slopes weathered from volcanic rock, richly wrapping the landscape above the Tandayapa Valley. The Mindo cloud forest looks virginal, but even in this remote area loggers have passed through to reach the more accessible hardwoods. I am north of Quito in an area that probably has the most diverse bird fauna of anywhere in the world. Because the trees grow up steep inclines, I am able to see into the forest canopy from above. This is a better area to appreciate how the forest "works" than the vast area of Amazonia to the east, where the canopy is a hidden realm carried high above on stilts of towering trees. In the early morning in the Mindo, I can see clearly the conical peaks, stretching green-clad in ranks to the far distance. The forest itself looks like clouds at this hour: the canopies of the trees are puffy or pleated, piling one upon the other on the slopes like cumulus. The silver cloud-shape of the *Cecropia* tree is most obvious, making a crown of coarsely fingered leaves, but then there are a dozen shades of green forming mounds or pillows undulating into one another, and every now and then a bright cloud where a tree is in flower. Perhaps it is a daisy tree, dappled yellow, or maybe a burst of gorgeous pink blossom, species unknown. By the afternoon, wisps of real clouds creep up the valley like white smoke, and then they unite together; and as this happens the light begins to dim imperceptibly. Quite suddenly, I am enveloped in a cool, damp mist. The vista vanishes, and 99 per cent humidity makes all the special organisms feel at home. Ferns and liverworts feel as if they were back in the Carboniferous Period. The trees become silhouettes, dramatically draped with lianas against a dimming light. The forest acts as sponge, absorbing fine water droplets into an orgy of growth. Where the trees have been cleared by importunate loggers, bare gashes of rockslide reveal the volcanic tuffs and ashes weathered to clay and gingerbread-soft mash that underlies the regenerating forest. From hidden glades, whistles and trills betray the birds that happily dance and feed among the apparently endless trees.

This is a country of lianas and epiphytes. Every tree trunk is decked with climbing plants, and there are further plants upon those plants. Anything that can gain a roothold hangs on for dear life. Epiphytes cling on to the thinnest branches, feeding on sunlight and water. A profusion of Spanish "moss" makes everything greyish and cobwebby. Some of the epiphytes dangle sumptuous flowers. I immediately recognise the brilliant, frilly extravagance of orchids. Other epiphyte flowers belong to the bellflower family, or are related to *Fuchsia*, or are bromeliads with their characteristic rosettes of leaves; they might even belong to the family that includes the heathers and heaths. To a naturalist like me, the flowers are a source of continuous delight. I can't even count the species there are so many, and so many unfamiliar to me. I notice that many epiphyte flowers have a long tubular shape that suggests that they are pollinated by specialist creatures with matching long tongues. They might be moths, hummingbirds or even bats. The cloud forest boasts dozens of species of hummingbirds, which buzz about like giant iridescent bees. I know that the birds favour bright red flowers, but darker colours may be attractive to crepuscular bats. Some of the arum lilies smell of rotting flesh to attract flies. Lianas are as varied botanically, although when they dangle like strings or ropes from high branches, I would never imagine there were so many different kinds of plants with this habit. I see a bamboo climbing up trees, and the familiar pot-plant arum, *Philodendron,* a member of the pepper family (*Peperomia*), and even a climbing fern (*Diplopterygium bancrofti*). If there is a job to be done in the natural economy, it seems that there are many candidates that will do it *con brio.* Nature may abhor a vacuum, but in Ecuador any void becomes stuffed to bursting. The variety of trees creates a sense of bewilderment. A few old trees have not been felled, including the "Sangre de Grado" (meaning "blood of the dragon"). This tree is part of the genus *Croton.* The name refers to its reddish sap said to be effective against *Staphylococcus aureus.* Palms like *Wettinia* rise up through the forest; I am told that the particular species has to be identified by its fruits. *Cecropia* trees with leaves like giant, silvery spread hands are typical of secondary forest: they grow where old giants have been felled. *Melastoma* has a trunk that looks as if it has been badly wrapped in brown paper; the bark peels off and discourages epiphytes. The usual way to locate a tree species is by spotting a strange flower lying on the path, as it provides the only indication of what is happening up in the canopy. Flowers reveal the biological label of the trees in a way that leaves do not. So many cloud forest trees have similar simple leaves with pointed "drip tips," which is just an efficient shape for the purpose, and tells us nothing about the plant's relatives. By the crude tracks that wander between the trees, every clearing shows shrubs and herbs belonging to the coffee family with leaves of this kind. Different species have very different flowers: *Gonzalgunia* sports hairy-looking spikes; *Palicouria*, lilac, tube-like flowers in a branching arrangement that leave green berries; *Psychotria* has dull little green flowers and fruits. I am fumbling with a catalogue, overwhelmed by profusion. My taxonomic background bids me name and identify, but there is too much, and it is everywhere. In dim corners lurk some of the more persistent survivors mentioned in this book, liverworts and filmy ferns on damp banks, and *Selaginella* creeping along where nobody walks. It is all here somewhere, if only I can find it.

Birds are no exception to this rule. I am looking for one bird in particular, the tinamou. The cloud forest is a famous destination for bird watchers: those who have that kind of mind can tick off more species in a short time than anywhere else on earth. But birds are much harder to study than plants. They lurk within bushes and canopies, most species are secretive, and many species are also rare. Hummingbirds are attracted to artificial nectar feeders, so they can be drawn out from the forest. Close up, I can see their long tongues and readily imagine how a long flower might have evolved to become their dedicated food resource in return for "pollination services." Most of the other inhabitants of the thick leafy glades are more readily heard than seen. Particularly early in the day, the slopes ring with plangent piping, hoots, trills or penetrating whistles. It is hard to tell exactly where the calls are coming from. My binoculars reveal the flash of a primary feather, the flick of a tail. Or I glimpse the brilliant blue of a tanager for a few moments, or a tyrant flycatcher in a quick aerial display. Usually, the bird stays hidden. I learn to recognise a few, distinctive species from sound alone. Hidden in the undergrowth, Spillman's tapaculo makes a prolonged chirring noise like a football rattle. I catch a brief glimpse of one of these little brown birds for about half a second, so whoever Spillman was, he was certainly observant. Within the forest, my overwhelming impression is once again of endless variety. The old adage "birds of a feather flock together" does not apply to cloud forest inhabitants. Rather, small numbers of very many species are scattered through the thick cover, a few here, a few there, too few to flock. A young American called Joel is my skilled guide to help me join names to birdcalls; how I wish I had a better musical ear. What I actually *see* is a matter of luck, but every item is a thrill: a mottled potoo sitting motionless on a dead tree pretending to be a piece of wood is apparently a very lucky sighting. A glamorous toucan barbet with a big bill designed for eating fruit, a more predictable one. I cannot just summon a bird as I might a waiter; the best way to see a bird is actually to *be* a waiter, and then wait for a very long time. Would that I had the rest of my life!

High on my "shopping list" of survivors is a tawny-breasted tinamou. Could that mellow double hooting sound from deep in the undergrowth be my quarry? Joel tells me the call is just about right. But my attempts to observe the musician fail, the undergrowth is so impenetrable out there. The definitive guidebook to the *Birds of Ecuador,* by Ridgley and Greenfield, says of the tinamou: "more often heard than seen." I may have heard my object of desire, but failed to catch it in my binoculars. On my way around the jungle paths, a medium-sized, brown squat bird pops out of the undergrowth, and bounces ahead of me along the way. I hope against hope that it is my subject. But no, says Joel, this is the giant antpitta: rather a perky bird, compact, speckle-breasted, with longish legs, a stout bill, and evidently reluctant to leave the ground. When I return to base camp and tell my story, a fellow guest, a Dutchman, jumps up and down with frustration. He has been waiting for days to catch a glimpse of the giant antpitta, which is rare and local, and he has failed repeatedly. I dare not ask him whether he had seen the tawny-breasted tinamou. Both of us had learned much about the habitat of the birds we sought, but we had also learned that attempts to observe a particular species in the wild often lead to disappointment. Well, I *had* heard it.

Tinamous are almost certainly the most primitive living birds. There are several dozen species, all living in South America. They are ground-dwelling creatures, preferring discretion to display. Like several birds that choose to live this way, their plumage resembles that of the tweedier variety of English country gentleman: chestnut browns and speckles and occasional soft greys. It is easy to imagine them disappearing into the jungle and evading the binoculars of the eager birder. When threatened, they creep away and stay motionless among leaf litter or in holes at the base of old trees. They look a little like guinea fowl, with the same compact, almost dumpy mien, although they are not closely related on the avian evolutionary tree. Like guinea fowl, tinamou peck around for buds, seeds, insects and little snails on the ground. They are generalist feeders at the opposite end of dietary faddishness to the exquisitely fine-tuned humming-birds. Their eggs are of an unrivalled glossiness: blue, green, yellow or chocolate-purple, according to species. They are laid in a shallow, somewhat perfunctory nest on the ground, with the male perform-ing the incubation. Freshly hatched chicks are able to run around almost at once. Tinamous can fly, but they do so with reluctance, and even then not very well, unlike guinea fowl, which can take off very fast when under threat. Tinamous possess small hearts in relation to their bulk, so they cannot exert themselves for long. I was happy to have seen exactly where these survivors had continued living their unspectacular lives in the company of gaudier colleagues from higher in the tree of life. I shall never forget the profusion of life in this rain forest.

The special place of the tinamou in bird evolution was recognised long before modern techniques of molecular analysis confirmed it. Charles Darwin's champion, Thomas Henry Huxley, had suggested as early as the 1860s that the South American birds were related to a group of mostly large, flightless birds including the ostrich, emus, cassowaries and kiwis. These are feathered bipeds that have given up on the air, though some of them run like the devil to make up for it. They do not have the deep keel or carina on the breastbone that all flying birds possess—this is the flat, blade-like bone that divides the middle of the meaty chicken breast. These flightless birds are known, collectively, as ratites, and they don't need powerful breast muscles, nor the bone to which they attach, that are necessary to allow lift-off. Their wings are either reduced or used for purposes other than flight. Tinamous, by contrast, do fly and do have a keel. However, they share with the ratites a special and primitive construction of the bones of the palate. Tinamous and ratites are grouped together in a larger group, called the Palaeognatha, the name deriving from ancient Greek, meaning "old jaws." The significance of these head structures sparked many an argument among academic ornithologists in the twentieth century. Some, following Professor Alan Feduccia, have claimed that these features may have evolved more than once; others asserted that retained primitive features should be irrelevant to classification whatever the case. It was clearly time for an independent arbitration on the relationships of these birds, to be revealed by a tree based on molecular sequence similarities, particularly in the mitochondrial genome. Several analyses have been published since 2000, and they do bring the tinamous and the ratites together in a single group. This is further supported by studies on the microstructure of the eggshell. Nearly all the analyses, whether based on morphology or genes, show the tinamous emerging as the basal branch on the tree. These unspectacular birds root back deepest into history. However, there are several competing versions of the relationships of the flightless birds *within* the ratite group itself. Kiwis emerge high on the tree in some analyses, or as basal ratites on others. Having completely lost their wings and with modified feathers, not to mention a unique "nose," kiwis have clearly been evolving along their own track for a long time, wherever they ultimately prove to fit.

Thinking about the tinamou's place close to the root of the avian evolutionary tree allows us to draw up a picture of some part of bird evolution. First, it can be deduced that the loss of flight in kiwis, emus and their relatives is a *secondary* feature. Birds originally flew, and later took to wandering over the plains on two legs. The tinamou is a better representation of an early bird than either a kiwi or an emu. Once the power of flight had been lost, the hefty musculature connected with flight was no longer required, and the breastbone became reduced. The resulting large flightless birds are tough creatures. Emus are little more than feathered stomachs borne on mighty legs and ruled by a tiny brain. If an emu wants one of your sandwiches, he will get it, and then run away. He cannot help you with your sudoku. Second, tinamous have distinctive but not elaborate calls: music seems to be a defining feature of bird-ness. Subsequent evolution would play with this gift to produce the elaborate musical wonders of the perching songbirds. This leads us to wonder whether musical calls of a sort were also present further back into bird ancestry. Third, some extinct ratites belong within the same narrative. I briefly mentioned moas when I was describing the velvet worm in New Zealand. They were the ultimate large flightless birds, magnificent and stately; but this did not protect them from extermination, most likely at the hand of humankind. The elephant birds of Madagascar (*Aepyornis*) met a similar fate, even though they were recorded as alive and well as late as the mid-seventeenth century. Elephant birds looked like massive ostriches; they towered over humans. In my place of work, the Natural History Museum in London, one of the few exhibits that survives from the early displays features birds' eggs of the world, ranged in size from the smallest to the largest. The tiniest belongs to a hummingbird, like one of those I saw in the cloud forest; the largest, with a volume equivalent to 150 hen's eggs, belonged to the elephant bird. Small children peer with disbelief at this monster; at a metre circumference, it must have fed a family and all the relatives at one sitting. Think what commercial possibilities there might have been had the animal just survived for another century or so! Sadly, *Aepyornis* joined the auroch (Chapter 9) as one of the "near misses" in my story.*

One emblem of extinction, the dodo, was a flightless bird that survived on the island of Mauritius about as long as the elephant birds did on Madagascar. At first glance it could be assumed that the dodo was another ratite, a stunted version with a different kind of massive beak. No complete specimen exists in a museum collection. At the time when the population of the dodo was shrinking rapidly in the wild, the concept of a museum as a permanent repository was in its infancy. Nobody thought to collect one. Specimens on display in the Natural History Museum in London are actually bogus. Similar reconstructions in many museums have been based upon paintings like that by Jan Savery (1651), showing a very plump bird, like a turkey ready for the table. Recent research suggests that this painting might have been based on a captive bird that grew unnaturally

---

* I should note here that some recently published (2010) molecular research suggests that the ability to fly may have been lost on separate occasions in ratites. Several birds may have independently achieved ground living habits after the extinction of the dinosaurs, stimulated by the absence of competition and danger. It is therefore still possible that the shape of the bird tree could be redrawn (again), although the tinamou will doubtless remain near its base.

rotund. Slimmer versions are likely to prove to be more true to life. Fortunately for science, a mummified head of a dodo survived in the Oxford University Museum. This specimen was acquired by John Tradescant for his seventeenth-century "Ark" in London in the Borough of Lambeth, south of the Thames, which has a claim to be considered as the first public museum in Britain. Tradescant's collections later became incorporated into Elias Ashmole's founding collections for the Museum in Oxford University, and so the mummy survived to tell its story. Recent study of DNA preserved in this specimen confirmed that the dodo was no ratite, but was actually related to the pigeon family. It's a giant, flightless dove—and an evolutionary side branch. The dodo provides another example of a bird losing the necessity for flight because of a long period of isolation on an

The extinct dodo has become an emblem of Mauritius. Here it is approximately portrayed on the ashtray of the Hotel Maurice.

island habitat; and the Oxford specimen of the extinct bird provides an unimpeachable case history demonstrating the importance of museums.

However diverting the dodo, it is also a diversion. The tinamou is evidently closer to the main trunk of bird evolution. The fossil bird *Archaeopteryx* is late Jurassic in age (146 million years old), so a thread of descent must connect the tinamou with the final flight of a long-extinct bird across a warm, limy lagoon in what is now Germany, the same lagoon in which horseshoe crabs were trapped for posterity (see p. 17). A handful of fossils of "the first bird" show feathers and a poorly developed carina, as well as retaining teeth from their reptilian ancestors. Recent CAT scans of the inside of the skull have confirmed that the fossil had a brain similar to that of a bird; and if it thought like a bird, and flew like a bird, then, by golly, it *was* a bird, although it was unlikely to have been a very efficient flyer. Curiously, the feathers that would once have clinched the assignment are no longer so critical. In recent decades a variety of feathered dinosaurs have been discovered, especially in China. The inelegant term "dinobirds" has been coined to describe them. Their feathers were initially not used for flight, but rather for insulation and, quite possibly, decoration. I love to think of striped and speckled dinosaurs, precocious plumage on speedy little runners decked out like fighting cocks. New images from Bristol University have proved that the feathered dinosaurs were indeed colourful. All the new discoveries have served to bolster the notion that birds descended from lightly built theropod dinosaurs with relatively long arms. When I first wrote about this theory two decades ago, it was still controversial, but now it is almost conventional; such is scientific progress. As so often in evolution, a feature that appeared for one purpose was recruited for quite another use, and a feather is not something that is likely to have appeared more than once in evolution; it is just too complicated to be invented by nature on a whim. Some lightly built dinosaurs with feathers on all their limbs have been shown by experimental modelling to have been effective gliders, affording a kind of halfway house to true flight. Then, with the apparent perversity which is also common in evolution, flight was secondarily lost in many birds, so now the ostrich strides once more in the fashion of its antecedents, and the horny feet of the emu leave prints that would not disgrace an *Allosaurus*. Birds survived the mass extinction at the end of the Cretaceous 65 million years ago, the event that put paid to their reptilian relatives. So dinosaurs did not die out, they just flew away. Or even walked.

The journey from *Archaeopteryx* to modern birds is still a long one. Living birds comprise a group of feathered, dinosaur descendants typified, among other things, by high body temperatures and a four-chambered heart. It could be argued that they are now the most successful terrestrial vertebrates. After all, the perching birds (passerines) alone have twice as many species as that most diverse of mammal groups, the rodents. Warm blood unites creatures of fur and feather, which is why they find a place together in this chapter. The fossil record of birds related to those still living extends back into the Cretaceous, so we can be sure that there were ancestors of extant birds flitting through the trees while large dinosaurs lumbered through the landscape. Birds, taken as a whole, are survivors. It is considered that the palaeognath birds split from the rest of the birds about 100 million years ago, so when we hear a tinamou call, we are truly eavesdropping on ancient history. The distribution of the ratites bears the signature of the old Gondwana continent, as they are found in South America (rheas), Africa (ostrich), Madagascar (elephant birds), New Zealand (moa) and Australasia (emu, cassowary). Like several other survivors in this book, they root back to a time before the fragmentation of that former world, and were carried to their present localities on the back of moving continental fragments. That still leaves tens of millions of years between *Archaeopteryx* and the origin of our living avian fauna. During that time in the Cretaceous Period, many kinds of birds prospered that left no descendants. Large toothed birds like *Hesperornis* successfully colonised oceanic habitats, became specialised in diving and catching fish, and lost the ability to fly. They resembled gannets in a general way, and made a living as gannets do, grabbing fish under water. Because of their oceanic habits, it is not surprising that fossils of *Hesperornis* are among the commoner examples of ancient birds. Just as one imagines penguins will be readily entombed in the rocks at some future date, and exposed to view when alien palaeontologists finally reach earth from the Planet Tharg, sediments welcome marine animals into their clammy embrace and keep them forever. Enantiornithines are a different matter: they, too, were mostly toothed birds, although they could flex their wings, but many species are known from just a few bones. They varied in size from little birds the size of sparrows to large predators almost as big as eagles. They ranged over seas and forests, estuaries and ponds. It is a pleasure to use this particular technical name to describe them because it was coined by a colleague of mine at London's Natural History Museum, Cyril Walker. He recognised the distinctive features of the articulation between the scapula and coracoid bones that defines the group. Cyril belonged to a class of naturalist who have become as extinct as *Hesperornis.* He was largely self-taught and therefore untrammelled by convention. He worked with fragmentary remains that many other scientists had overlooked, and he drew his own conclusions. The sheer variety among enantiornithines proves that birds had already evolved a great array of different lifestyles while the dinosaurs were still head honchos in the terrestrial biosphere. The birds were ahead of the mammals, as the latter only came into their own *after* the dinosaurs had relinquished their grip on the higher tiers of the pecking order. Indeed, perhaps that last term is inappropriate, as it could be claimed that you cannot really have a pecking order without a beak to enforce it.

Could it be that the feathered dinosaurs sang like birds? The thought of the Cretaceous landscape ringing to their calls is appealing; one could imagine an array of voices to match their precocious plumage. Sadly, it seems unlikely. The tinamou's call, though musical, is a simple one, although since separate species do have different calls, it is certainly useful. The intricate melodic signalling of songbirds is likely to be a later evolutionary feature intimately connected with the elaboration of the syrinx, an organ at the base of the trachea where complex music is produced. Since warbling also requires a receptor in the inner ear to receive and interpret it, the complexity of the latter structure indicates whether or not it was receptive to elaborate songs. The CAT scan model of the inner ear of *Archaeopteryx* does resemble the bird structure rather than the reptilian equivalent, but it is no match for the more complex structures of the passerines (perching bird), the real songsters. It is possible that there were simple calls made by the ancestors of birds, but there was probably no morning chorus. This does not altogether rule out Cretaceous music. The most primitive living passerines are New Zealand "wrens." Of these, the rifleman bird is still relatively common on both islands, for all that it is rather a feeble flyer. Several of its relatives became extinct after humans arrived, and one species, the bush wren, breathed its last as recently as 1972. The rifleman is a tiny, discreet green or brownish bird, which is probably why settlers from England called it a "wren"; actually, it is not closely related to the European bird at all. Instead, like the velvet worm and the tuatara, it is one of New Zealand's survivors. I did catch brief glimpses of this charming little animal on a walk through native forest as it flitted after tiny insects disturbed by my footsteps scuffing the litter. Its brief churring song could scarcely be described as highly musical. Some authorities believe that the position of the New Zealand "wrens" at the bottom of the passerine evolutionary tree suggests a Gondwana history comparable with that of the ratites. Under this theory, they would have been carried off to their island redoubt at the time of the break-up of part of the supercontinent perhaps 80 million years ago. That would imply that there were already passerines living during the Cretaceous alongside the dinosaurs. Other authorities point to the apparent absence of fossils of passerines in rocks more than 50 million years old as evidence for a younger origin of the group, an origin long after the extinction of the dinosaurs. Then again, it could be argued that fossils of small birds are always rare and hard to find as their bones are so fragile. Maybe scientists have just not yet struck the "pay dirt." Only if the early origin were correct would little perching birds have flitted among the treetops while dinosaurs browsed or fought below. Only then would birds have greeted the dawn with the first music in history, while the largest terrestrial animals ever to have lived awoke to stride across the plains. I wish future palaeontologists good fortune with their geological hammers to harden this vision into history.

9

Islands, Ice

Now we are approaching the top of the tree of life, a period about a hundred thousand years ago, during which our own species emerged from one of a group of hominin ancestors. It was as eventful a time as any other mentioned in this book, for the same period saw the world gripped in a great ice age, obliging many animals to evolve to cope with the new challenges posed by freezing climates. Elsewhere, protected from the direct effects of ice, evolution continued its creative work on islands scattered around the world that had been isolated for millennia. Wonderful species appeared in these unsullied Gardens of Eden. But humankind has now violated all that was once remote, so there are other survival stories to be told of animals coping with unprecedented invasions by those who sought new lands, and who changed the face of the earth.

The ancient city of Las Palmas on the island of Mallorca spreads out into an interminable strip of apartments and hotels adjacent to the Mediterranean Sea. Each block of five or six floors is equipped with a balcony to catch the breeze and the precious sea view. This creates the paradox that the view from the sea is an endless wall of balconies and terraces with a few umbrella pines between them. Every so often a bay intervenes, crammed with yachts, and backed by bars. It is a kind of holiday suburbia-on-sea, decked with flowers. Some parts are posh; others are notoriously vulgar. Airplanes bring more people in almost every minute. It is something of a relief to escape into the countryside, where an older Mallorca soon asserts itself. Sad stumps of little windmills dot the landscape. They were used to bring up water, not to grind grain; now engines do the job more efficiently, at least for the time being. The landscape is dominated by limestone. The older buildings are made of it, and blend in well with the natural environment. Much of the centre of the island is flat and cultivated, but agriculture is no longer as important to the economy as yachts and burgers; half a century ago there were 125,000 farmers, now down to fewer than 10,000. The island must have looked very different to early tourists. A small, locally printed handbook, *Trim's Majorca Guide* (1954), written by the eponymous cocktail-swilling lounge lizard, paints a picture of a louche and slightly eccentric crowd fleeing the conventional resorts of the bourgeoisie. Trim gives us an amusing, but I suspect bogus, glimpse of a rural economy before industrialised tourism. Almonds are "the farmer's friend," says Trim, and "once a year you go and beat his branches with a long pole to knock the almonds off. You pick them up and take them home. That is almond farming." He continues: "Majorca also has lots of olives. Once a year you go out to the olive trees, and beat the branches with a long pole to knock the olives off. Majorca also produces algarrobas, trees that bear long beans, which horses and mules eat. Once a year you go out with a long pole and beat the branches to knock the beans off." The almonds, olives and algarrobas are still there, scattered through fields behind roughly piled-up stone walls. I suspect they require more attention than Mr. Trim suggests. Mules have vanished completely now that everyone drives everywhere. Today, the richer ground on the plain is used to grow new potatoes for the European market, but small vineyards still create local wines. The island has changed more in fifty years than it had for centuries. Farmers who do a bit of this and a bit of that are probably regarded as living fossils.

The northern end of the island is dominated by the Sierra de Tramontana, making wild and mountainous country and, with the exception of a few obvious tourist traps, is still largely untouched. Relentlessly winding roads lead up through fragrant native pine forests to passes from which abandoned terraces can be seen on the hillsides beyond. The farmers have gone elsewhere. The ancient olive groves survive with many of the trees hollow, riven or twisted and tortuously bent, but still gamely supporting a canopy of tiny grey-green leaves. They are here for the long haul. Even if they have been butchered during road widening, their grizzled stumps send forth new shoots; their roots go down forever. Carob trees—Trim's algorrobas—have much darker canopies with bigger, divided leaves, so they make welcome shade as well as long beans, which hang down all knobbly above the loiterer's head. Where footpaths take off into the wilds under the pine trees, shrubby bushes of lime-loving heath *Erica carnea* testify to the underlying geology, and many plants carry spines to discourage grazers, while on the dry slopes holly oaks look almost black in the bright sunlight. One particular rock formation makes a huge yellowish cliff that dominates views of the south side of the range, and in medieval times supplied the obvious locations for fortified castles and monasteries. On the high road beyond Lluch Monastery a grey limestone has been weathered into a series of fretted and fluted pinnacles providing proof of torrential winter rains. Water runs quickly off the mountains by way of "torrentes" that are completely dry by May. Eventually, all rainfall disappears underground. And when water finally arrives, at depth, on the plains, it is pumped up again to put ice into Mr. Trim's cocktails.

However, in a very few places in the Sierra de Tramontana water does linger. This can only happen in streambeds floored by limestone, where the surrounding rock is unusually massive and lacking the cracks and joints along which the precious fluid usually seeps away into underground drainage. The floor of the stream needs to be deepened into a pothole to provide a natural pond, preferably partly shaded by overhanging rocks to slow the rate of evaporation. Such deepening usually only happens where a boulder has been trapped inside the hollow and scours the bed round and round as it is whirled by the winter rains. These pools are little islands of permanent wet within the other, greater island far out in the Mediterranean Sea. They also provide the habitat for the Mallorcan midwife toad (*Alytes muletensis*), a living fossil.

For once, there need be no qualification in using this term. For this particular toad species was actually scientifically named as a fossil *before* it was discovered as a living animal. It was originally described by J. Sanchíz and R. Adrover in 1977 from bones discovered in Muleta cave, near Soller on Mallorca, in a site where the fossil bones of a strange and extinct relative of the sheep called *Myotragus* were particularly numerous. However, when comparisons were made with some previously collected museum specimens of living toads, it became clear that the "fossil" had survived the extinction of its mammalian contemporary five thousand years ago and lingered on in secret places. It was rediscovered in the wild in 1979. It was time for me to visit the toad before it met the same fate as *Myotragus*.

Samuel Pinya is a Mallorcan whose father is a renowned expert on cooking the wild mushrooms of the island. Samuel is in the middle of research for a thesis on the toad, which is fully protected on the "red list" of species threatened with extinction. Like all of the people mentioned in this book, he works out of enthusiasm for his subject rather than for financial gain, and strenuously monitors the sites where the toads still survive, after his day's work as a conservation officer. He is very fit, because it is always necessary to scale up steep limestone slopes to discover the toad's remote redoubts; he is certainly in far better shape than me. A traditional farm at the foot of the Tramontana provides the gateway to discovering one of these sites, although even this apparent survivor is not quite what it seems, for the landlord is a Swiss banker with ecological sympathies. Never mind, it is good to see a few pigs happily grubbing around under the carob trees, and horses nibbling at hay in another stone-walled field, and gnarled olive trees everywhere. Samuel shares a few incomprehensible words with the farmer in the old Mallorcan language. The streambed we follow into the slopes is, as expected in June, entirely dry. A few flowering myrtle bushes with white flowers like tiny feather dusters line its flanks, their deep roots doubtless tapping into some hidden source of moisture. Pale limestone boulders flooring the "torrent" are large and rounded, proving that when the waters are in their brief spate, a furious flow will lift and tumble improbably large pieces of stone, rounding off their edges in the general jostling. Soon we are obliged to leap from boulder to boulder, and then as the gorge becomes narrower, the climbing begins. Our guide leaps expertly from one side of the narrow cleft to the other, grabbing on to rocky projections and gaining height rapidly. His followers blunder about with considerably less elegance, but somehow cling on, panting and slithering. Limestone country is relatively friendly to the climber, since it is full of cracks and fissures that afford a foothold, which is fortunate. Samuel gestures as we climb, pointing out neat cushions of interesting species of endemic plants clinging to the slopes. Dwarf palms apparently sprout straight out of the rock. The ascent seems to go on and on; large crows come to have a look just in case there are any corpses to be picked at. But then Samuel signals to us to stop and listen.

We hear a distinctive "plink" sound, reminiscent of the noise made by one piece of metal striking another. Then there are several of these musical but percussive notes, all slightly out of sequence, so together they sound like some outré pieces of contemporary music. "Ferreret," says Samuel: the local name for the midwife toad. The name refers to a blacksmith, and it is not difficult to imagine the sound of tap-tapping at an anvil in the ferreret's repetitive calls. All of a sudden we are upon the pool whence the sounds originated. The pool is only a few metres across, not very deep, and tucked deep into its miniature gorge, it comes as a surprise to see standing water making a green pond high up in these parched hills. The chirruping continues in our presence, and not a croak to be heard, anywhere. Peering down into the pool, an astonishing sight meets our eyes: the largest tadpoles imaginable. Some appear to be longer than my thumb, with great black bulbous heads propelled by the usual undulating but relatively skinny tails. They are apparently busy nibbling the algae and bacteria that clothe the side of the pool, and one or two of the biggest ones seem to be taking "sips" of air from the surface. The only other inhabitants of the pool are very large black water beetles, and they come to the surface regularly to refresh the air bubbles that allow them to dive for up to three minutes at a time. These beetles are predators, and it was soon clear what happens to any dead tadpoles lying on the bottom of the pool—they are fought over by hungry beetles. Carcasses are tussled over and bitten to bits. The little pool is a miniature ecosystem with its own producers, grazers and predators, the smallest world that could be imagined.

Samuel is poking around in the cracks under the rocky overhang looking for dark crevices hiding the adult ferrerets—and they prove to be tiny, just a few centimetres long. They seem out of proportion to their very large tadpoles, at least to those accustomed to the life cycle of *Bufo bufo,* the British toad. They are pretty too, sitting so diminutively in Samuel's palm all mottled yellow and brown, with fingers as delicate as sparrow's feet. They have a way of looking surprised and resigned at the same time. The markings of each toad are as distinctive as a human face, and part of Samuel's research entails keeping track of individual toads by photographing them after weighing them, a mere two to three grams, and referring back to his toad identikit files. That way it is possible to find out how long toads live, and perhaps work out the main threats to their continued existence. Males and females are very similar, although the tympanum—the vibration-sensitive patch behind the eye—is larger in the former. As adults, they hide out in their cool crevices during the heat of the day, coming out at dusk to search for insects and woodlice.

The "midwife" part of the toad's title comes from the fact that the males (scarcely the "wife"!) carry the fertilised eggs around until they hatch as a way of offering additional protection from potential predators. Midwife toad species on the mainland of Europe carry up to fifty eggs, but the Mallorcan species bears only a dozen or fewer, very big eggs, and keeps them on board for two or three weeks. The males hide safely in crevices during the day, and return to the pool at dusk with their precious load. This special parental care allows relatively few eggs to be produced with a higher chance of success. Eventually, the tadpoles leave the security of dad and move out into their little watery world. Samuel has proved that adult ferrerets do not live for more than four years, but the second year is the critical time for reproduction since most individuals do live that long. However, some of the tadpoles live for two years, which must partly account for their huge size. Slim pickings in the tiny pools mean they have to take their time, so they are not like our mainland species that seem to progress from prolific spawn to tiny toads all in a rush. There was one pool in which a seagull died when apparently the tadpoles gorged upon it and grew to an enormous size, and fast. The older tadpoles metamorphose into adults at the end of July, and after a little growth begin clinking away to attract their mates. The pool we were examining might support two hundred ferrerets. Samuel is studying eight other pools, all in the Sierra de Tramontana, to which the species is now confined. There were originally eleven pools with ferrerets, but with new introductions these now number about thirty-five. Tadpoles outnumber adults about ten to one. The main mortality occurs after metamorphosis but before the adults have had a chance to breed. The total population is about three or four thousand, which is about the same as a large English village. This is not a great number to keep a species alive.

And the ferreret is not without enemies. The most implacable is the viperine snake, which was introduced to Mallorca by the Romans. This particular snake likes to lurk in water, and if it discovers a pool in the hills full of ferrerets, it takes up residence there for several years until its destructive work is complete. Samuel says that if a snake gets into a pool, the tadpoles stop wriggling and drop to the bottom; but that does not save them. The adult toads do not have the nasty taste in the skin of their relatives on the Iberian Peninsula. Fortunately, individual pools are far apart and hard to find, even for a snake. This, too, might prove to be a problem. Recent genetic work has shown that ferreret populations along a single torrent course are related, but that there is little genetic exchange with populations elsewhere. Some pools are, quite literally, gene pools. Such relative inbreeding may make them more vulnerable to infections or an unhelpful mutation.

There has been unwitting damage done by humans with the best of intentions. Shortly after the ferreret was discovered in the wild, a captive breeding programme was set up in Jersey Zoo, an organisation well-known for its conservation of rare species. This proved successful, so much so that the toads were reintroduced into the wild. However, this also introduced a nasty chytrid fungus into Mallorca that attacks the thin skins of frogs and toads, and has been blamed as the cause of massive declines in vulnerable species elsewhere in the world. The Jersey Zoo was entirely innocent in this accident, as the chytrid in question was not discovered until 1998, long after the programme had come to an end. Fortunately, because of the disjunct distributions of the ferrerets in the wild, the chytrid was not able to spread widely among the natural populations, and one of the two reintroduced populations affected has survived to "plink" into the future. This history offers an important lesson about how careful we must be when we interfere with nature.

Fossils indicate that the ferreret was formerly more widely distributed on the other Balearic Islands; its current range is much contracted. There are three other species of midwife toad on the Iberian Peninsula. Separation of the Balearic Islands from the mainland happened about 5 million years ago when the straits of Gibraltar were breached, and the Mediterranean Sea attained something of its present shape. Sea levels fell again at the height of the Pleistocene ice age much more recently, when passage to the islands would have been possible once again. During their long spells of isolation, the Balearic Islands were worlds unto themselves. Other endemic animal species evolved there: shrews, dormice, *Myotragus*. Of these, only the ferreret and some special Balearic lizards survive.

I have described the little ferreret in some detail, not just because it is a living fossil in the most pedantic sense, but as an emblem of all those species that have evolved in comparative isolation on islands. The ferreret was derived from one of its mainland relatives, no doubt after becoming isolated on the nascent Balearic Islands when the sea level rose. Several of its features point to a rather sheltered existence in its new home, with few predators: the loss of the subcutaneous glands that make its relatives such an unpleasantly garlicky morsel, the very few yolky eggs, the long-lived giant tadpoles. The midwife toad evolved to match its environment, as organisms always do, a strategy that worked well until the viperine snake arrived. Doubtless, some of the island's birds of prey took ferrerets when they could grab one, but this snake was a specialist. As humans cultivated the land and extracted water, the available habitat for the ferreret gradually shrunk, until only the Sierra de Tramontana remained to give it lodging. It could so easily have been known from fossils alone. But the ferreret was a survivor—just.

The story of the ferreret could be repeated in its essentials in a hundred islands around the world. The more remote the island and the longer its history of isolation, the more endemic species will have evolved. There is also a direct relationship between the size of the island and the number of species that can "fit in"; broadly speaking, the bigger the island, the more the species. For remote volcanic islands like Galapagos and Hawai'i, the founding species that arrived at the barren rocks on the winds or on the back of floating logs determined what happened later. Evolution got to work on whatever material there was to hand. In Hawai'i, a common fruit fly produced dozens of spectacular endemic species. From Darwin onwards, scientists have been marvelling at the different life habits adopted by endemic finches on the Galapagos. They are small brown birds, superficially no more remarkable than the common house sparrow, and all evolved from one founding ancestor. Yet they have evolved on one island after another to cope with their specific challenges: dry climates, maybe, or fluctuations in a common food source. Peter and Rosemary Grant have spent a lifetime tracking these changes from their base at Princeton University, while remaining as British as teacakes. I heard them describe how one finch had recently learned to peck the breasts of sitting seabirds to get a vital bloody dose of protein. Islands are indeed special places.

But they are also vulnerable. Species that have evolved in isolation have not encountered everyday threats from cats or rats or rampant weeds. Some islands in the Galapagos are already compromised, covered in quinine bushes and alien brambles, their natural experiment in evolution violated by strangers. When Hawai'i was reached by questing Polynesians perhaps two thousand years ago, its virginal diversity was breached forever. The golden plumage of its endemic honeycreepers soon made magnificent robes for great chiefs. Rats came later, and all the Hawaiian endemic flightless birds that laid their eggs on the ground were doomed. Some misguided meddler then introduced mongooses to control the rats, and they also turned on local creatures. Alien trees displaced slow-growing native stands. Much of what evolution had created in isolation, natural selection destroyed in communion with the rest of the world. What I have described on Mallorca happened on an enormous scale in Hawai'i.

There are some scientists who respond to this extinction with a shrug: "too bad," "*tant pis.*" It would have happened sooner or later, they say; this is just what you expect to happen in islands. Why grieve for the Fijian longhorn beetle, the second largest beetle in the world, for who will notice when it finally becomes extinct? And it is true: the disappearance of a species from a Pacific island does not make much impact on the rest of the world. It does not threaten ecosystems in Asia, nor does it tip the scales in some delicate ecological balance in the African continent. The fossil record tells us that many island species have always been unlikely to survive in the long term: think of *Myotragus* and its Mallorcan contemporaries. So let's throw a few more into oblivion. Who cares?

The answer to such a crass challenge is to be found in the narrative of the ferreret. The animal is no mere statistic: it has its own, fascinating and unique story to tell. I did not know its story before I went to Mallorca, and the chances are high that you did not know it either until you read this book. I would assert that *every* organism that has evolved its place in nature has such a tale to give us, if only we knew it. E. O. Wilson's notion of "biophilia" posits an innate bonding of the human species with the natural world that might (or might not, according to others) be a deep reality in our makeup. I cannot disagree with his statement that "Humanity is exalted not because we are so far above other living creatures, but because knowing them well elevates the very concept of life." However, I am not in sympathy with the idea that what matters about a species is how we humans react to it, which seems allied to a view that nature is only validated by observation from this particular hominid. Organisms are entitled to their own unique narratives—I would like to use the term *bio*graphy in a literal sense if the word had not already been hijacked for humans. We don't reckon the worth of a species by the "damage" its extinction would do to other ecosystems. We cannot rank the products of more than 3 billion years of evolution in utilitarian lists. The richness of the biological world is the most wonderful feature of the biosphere, and every story is worth the telling no matter how humble, or indeed insular, is the organism concerned.

In his recent book *The Greatest Show on Earth,* Richard Dawkins has made a telling point about islands. He remarks that one can look at the whole surface of the earth as a patchwork of islands. For example, if one mountain range is separated from another by a deep and wide valley, this feature may be as impassable as an ocean for the organisms that live there. The separated mountains are, in a sense, islands. I have already commented on mountainous New Zealand as a huge isolated island when describing the survival of the velvet worm and the tuatara. In an analogous way, a mountain chain may serve to divide two large land areas—as the Himalayas isolate the

Indian subcontinent from the rest of Asia; or the Pyrenees, the Iberian Peninsula from Europe. The areas thus defined could be thought of as (admittedly very large) islands. It is certainly true that species subsequently evolve that become typical of these sub-continental "islands." For example, the Iberian Peninsula has a whole series of endemic birds, reptiles and plants, not to mention three different midwife toads. It is a good point, but it does not really acknowledge the special nature of *oceanic* island animals and plants. When they encounter organisms that have evolved on large mainland areas, such true island species always seem to come off worst. To turn it the other way round, I can discover no example of a species that is known to have first evolved on a small oceanic island subsequently becoming a major coloniser of the nearest mainland by out-competing their inhabitants. Yet the list of extinctions of island faunas and floras following human contact—from the dodo onwards—is apparently endless, and continues to this day. Almost every issue of my magazine from the Royal Society for the Protection of Birds contains an appeal sponsoring efforts to ensure the survival of some brightly coloured island avian endemic. Flightless rails (though not colourful) and other ground-dwelling birds are obviously particularly vulnerable to being caught by domestic cats, but many songbirds and small parrots seem unable to adapt to aggressive invaders to their island home, from the common sparrow upwards. Even island insects can be bullied into oblivion, or starved by changes in the flora. And what do we do to protect New Zealand endemics? We send them to smaller islands off the islands. I do not think it does justice to the wonders of evolution to simply regard the demise of these island species as "collateral damage" to human spread, even if the root cause is rather obvious. The story of the ferreret concerns one of the lucky ones, one whose place in the scheme of things has been recorded and valued. How many more *life histories* have been thrown away, grieved by nobody?

Before we leave the ferreret in its mountain refuge, it has another surprise for us. It may originate on a very high twig on the tree of life, because its appearance does not predate its geologically recent isolation on the Balearic Islands. It is also a member of a family called Discoglossidae (Alytidae in some sources), which is an extremely

254 HORSESHOE CRABS AND VELVET WORMS

ancient group of animals. It is not even correctly referred to as a toad, but lies close to the junction where frogs and toads diverged from a common ancestor, and most authorities seem to think it crawls into the froggy side of the divide. This group of amphibians sharing a disc-shaped tongue (disc, obvious; plus tongue: *glossus*) has a fossil record going back 150 million years to the Jurassic. Tadpoles belonging to the ancestors of the ferreret matured in pools waded through by dinosaurs, and may have provided food for pterodactyls. So although the ferreret is a high twig, it is connected to a low branch in the evolutionary tree. Imagine if at some point in the future the chytrid that is causing so much grief to the amphibians around the world were permanently to remove its discoglossid (ancient "frog") relatives from mainland Europe and northern Africa. Then the unassuming little ferreret would be the last representative of its ancient lineage, protected by its isolation, rather like the tuatara on Somes Island. It would be a living fossil by any definition.

It remains true that there are many more examples where a small island has proved to be a permanent graveyard. None is more poignant than the case of *Homo floresiensis*. There was initially a measure of scepticism when a fossil discovery made in October 2004 was reported in the world's press: the remains of a diminutive human had been found on the Indonesian island of Flores. The little hominin was only about a metre tall; it did not take long for it to be christened "The Hobbit" in the media, thereby immediately identifying it with some of the cosier inhabitants of Middle-earth. The scientific press, on the other hand, went into its customary debating mode. Some anthropologists claimed that the remains were those of a pathological modern human, while others defended the notion of a hitherto undiscovered human relative. A further surprise came with a date as recent as 12,000 years before the present, for the strata with which the remains were associated. Since our own species had already reached Australia before 40,000 years ago, and possibly a lot earlier, there is no doubt that modern humans and this diminutive biped lived at the same time. As more and more fossil discoveries were made on Flores, it became impossible to maintain that the fossils were pathological modern humans. The Hobbit showed a number of distinctive features: relatively large, rather flat feet and curiously primitive wrist bones among several other peculiarities. It began to seem plausible that the little "man" had evolved on a line entirely separate from that leading to modern humans. Since it was already clear that our own ancestor *Homo erectus* had left Africa about a million years ago* and moved eastwards to Java, one possible explanation for the origin of *H. floresiensis* was as an offshoot from that early migration. A population of early humans became isolated on Flores by a sea level rise and in time developed their own distinctive features. Like the ferreret, they became smaller, and perhaps more defenceless. Since stone tools dating back as early as 800,000 years have been found on Flores, *Homo floresiensis* could have pursued its independent history for tens, or even hundreds of thousands of years, and certainly until modern humans colonised the islands for the first time. Indeed, local myths still include stories of "little men" in the hills. There is something peculiarly tragic in imagining tiny people nervously contemplating the species that would displace them, perhaps hidden wide-eyed in the secrecy of the bush, maybe ineptly brandishing their crude stone tools, or more likely fearfully withdrawing to the safest and most inaccessible place they knew. Unlike the ferreret, we shall never be able to read the full *bio*graphy of these small people, but what a story it would have been, and what additional chapters it might have added to our own history.

I first became a professional palaeontologist when I carried out fieldwork on another very different island. Spitsbergen, the largest island of the Svalbard Archipelago, lies well within the Arctic Circle, and has permanent ice sheets covering the elevated regions of its interior that flow down and reach the sea in great glaciers. Months of my life were spent camped between an ice sheet called Valhallfonna and the frigid sea in pursuit of trilobites from the ancient rocks that cropped out there. The bleak raised beaches seemed to support virtually no vegetation, but close to the ground were small patches of lichens, moss

* The timing of a migration "Out of Africa" event, or events, of ancestral *Homo* species is still controversial, with some claiming a first movement of *H. erectus* out of the continent as early as 1.6 million years ago.

and occasional patches of yellow Arctic poppy and purple saxifrage, tucked into hollows where meagre soil accumulated. Not for the last time, I wondered at the capacity for life to survive almost anywhere. Arctic terns made their nests on the empty shore, enjoying the summer plenty of the sea, which briefly erupts into glorious life every year when continuous light makes plankton bloom and crustaceans thrive. Spitsbergen is a favoured habitat of the polar bear—the ice bear, or *isbjorn* in Norwegian. The huge, shaggy creature feeds mostly on seals that come ashore on the ice floes that surround the island, but it will grab protein in avian form, or even human, if it gets the chance. It is famously a specialist for life in high latitudes, toughest of a tough family. Having glamour on its side, the polar bear is a species that is a cause for concern in the context of climate change. The Arctic ice cap is getting thinner, making the platform from which the bear pursues its prey more treacherous; the bear can move fast on solid "ground," and frozen water is perfectly adequate for the purpose. Although it is a strong swimmer, the bear cannot match prey like seals whose mastery of the sea is virtuosic. If the ice support gives way, the bear flounders. Hungry bears, or bears that have failed to rear cubs, are becoming more common. They cannot roam the Arctic wastes as once they did. Television crews are dutifully there to record their plight for the ten o'clock news. Bears are photogenic enough to attract attention.

The high Arctic can be considered as another kind of island, united by low winter temperatures with perpetual darkness. It's like a frigid Australia walled off at the top of the world by warmer climes to the south. Animals and plants that learn to live there are confined by their special living conditions almost as effectively as if they had been contained by an ocean barring the way to lower latitudes. Yet this "island" differs from any other encountered in this book by its capacity to change shape quickly. This it has done repeatedly over the last 2 million years. The Arctic ice cap has waxed and waned several times throughout the Pleistocene Period, reaching its maximum extent 20,000 years ago at the Last Glacial Maximum (LGM). At that time massive ice sheets spread down across America as far as the Ohio River, and almost all of Great Britain and northern Europe was similarly deeply covered. The roof of the world was perpetually frozen.* Nothing could survive atop this pitiless mass of glaciers. Particularly around the Asian edges of the icy wastes, vast areas of cold marshy tundra and grassland provided peculiar conditions in which specialist species happily prospered. So a time of ice was also a time of opportunity for some, just as it drove more temperate organisms southwards towards their comfort zone. But the quantity of water locked up in the ice sheets meant that global sea levels fell to a dramatic low-stand. As one result, the LGM also marked a time when the Indonesian archipelago formed a continuous land bridge into Asia, allowing humans to migrate towards Australia; perhaps the little men on Flores contemplated their voracious big-brained relatives for the first time. Cold-blooded reptiles were, naturally enough, banned from high latitudes, but it was a great time for animals that could make wool: mammoths, rhinoceros, cave bears. As anyone who has spent time in a tent in Spitsbergen will know, surviving cold climates is mostly a matter of having enough food to fire the internal metabolic engines, and enough insulation to keep out the worst of the frost. Large size also helps to supplement the work of the woolly coat, as it reduces surface area relative to volume; short-faced bears and mammoths accordingly became the giants of their kind. For the same reason, the latter also became the logo of a French hypermarket chain. There was even a super-sized beaver. Bones derived from huge numbers of woolly mammoths in Siberia testify to the productivity of the ice age tundra. The Pleistocene "megafauna" of North America is even more diverse, with sloths, tapirs, dire wolves, sabre-tooth cats, mammoths, mastodons and many more. Away from the ice, animals evidently adapted and evolved to cope with a rich and special world of open grasslands. When calibrated against the tree of life, nearly all of these species were late innovations. Like humankind, many Pleistocene mammals lay at the end of the last sprigs on the top of the crown of the evolutionary tree. Almost the last deal of the genetic cards, these mammals responded to a time of climatic fluctuations when the spoils might belong to the most adaptable, like the Arctic fox, or the most adapted, like that latecomer, the polar bear. Now

* There was also a massive extension of ice sheets in Antarctica and elsewhere in the southern hemisphere, which contributed to the globally low sea level at the LGM.

we have to ask whether the Arctic specialists are all doomed, as the snowy "island" shrinks under the influence of climate change.

I have seen one survivor of the Pleistocene ice ages on the Arctic "island," albeit from a distance: the musk ox (*Ovibos moschatus*). This shaggy messenger from the height of the North American glaciation hangs on in Arctic Canada and Greenland. It has been transferred successfully back to some of its former haunts on the Taimyr Peninsula at the top of Russia and on Svalbard—which was where I saw a hairy mass in the distance looking like a large animated tussock perched on a soggy swamp. The musk ox is not actually an ox at all, but belongs to the same family as the sheep. It looks massive from the front, where it is indeed strongly built, but much of the bulk is made up of semi-*bouffant* hair. Long, water repellent hairs dangle rather scruffily almost to the ground on the outside, and are underlain by a furry, very dense coat that can withstand sub-zero temperatures with aplomb. The musk is the sexual attractant of the male, and doubtless females of the species find it very pleasant (unlike anyone else). As animals go, musk oxen do have a rather depressed appearance. This is partly their long face leading to a whitish muzzle, but droopy horns contribute to the effect, being of the same shape as the pendent moustaches of old-fashioned stage Frenchmen who inevitably "'ave ze secret plan." When threatened by bears, musk oxen line up to present a united wall. They are good at dealing with adversity. In the summer they eat grasses and rushes. During the long, harsh Arctic night, they are able to scratch through the hard covering of snow with their robust cloven hooves and dig out grey lichens. In Greenland they hang about around Thule Airport bumming almost anything they can get their tough teeth into, paper bags included. They may look a trifle gloomy, but they are survivors. They may look primeval, but they are latecomers to the story of this book. However, it is difficult to imagine them outlasting the melting of the Greenland ice sheet. Their shaggy coats would look *de trop* in a warmer world.

Many of the large animals of the Pleistocene have already passed away forever from the earth. Some can still be seen in the graphic drawings made on deeply hidden cave walls by our ancestors; a handful of virtuoso lines in charcoal or sienna capture a life, even a species. The entrance to such a cave might seem rather mundane now, just another large hole in the side of a cliff along a steep-sided valley through limestone country, but the cavern provides access to a time capsule. The art is nearly always hidden deep in the cave system, testifying to its magical significance for the original draftsmen. Many of the caves are now closed to public view: the drawings are just too fragile to endure streams of visitors. At Lascaux in the Dordogne, a black mould has attacked many of the famous and wonderful images; a mould that may have been brought on the breath of well-meaning tourists. Several famous caves follow the line of the Pyrenees across France and Spain. I was fortunate to visit the Altamira caves in northern Spain some years before they, too, had restricted public access. I shall never forget a ceiling covered with huge bison sketched with perfect grace and economy of line, caught as exactly in movement as in repose. Fine bulls of the auroch (*Bos primigenius*) in their prime were portrayed elsewhere in the cave; this species is considered to be the ancestor of the domestic cow—no less than a giant, and rather shaggier version of humankind's most useful animal. What could have been smudges of black and red in lesser hands became portraits full of character in the hands of these Magdalenian artists. La grotte de Niaux in the French Pyrenees is one of the few caves where time travel back twelve thousand years is still permitted, but then only under strict supervision. The art is so delicate that too much exhalation might destroy it. The images on the cave walls are vulnerable to carbon dioxide and moisture breathed out by visitors, who are let into the dark innards of the mountain only a few at a time, filing past ranks of stalagmites to reach a hidden chamber. Everybody in our party tried not to breathe too hard in the caves. Even so, a few children brought along by their parents for a different kind of day out bawled lustily in the dark, scared by the weak infrared torches employed to pick out the drawings. Here is a horse, caught tensed to sprint; there a bison, static and dignified; above, an alpine ibex, its long horns lovingly rendered. These drawings are survivors of a different kind from any others described in this book, for a mind has been at work here, and a skilled hand. They are fossils in a sense, but more delicate than bone, shell or amber: they are little more than a smear of paint flung in the face of time.

As for survival to the present day, most of the animals portrayed on cave walls scraped through, but only by the skin of their hooves. The European bison, or wisent, disappeared from one European country after another during late medieval times, gradually retreating to a last redoubt in the wild in the old royal forests of Poland. Its horns were in demand as drinking vessels, irresistibly calling to mind the more traditional productions of the operas of Richard Wagner. The animal also provided good meat and leather. A few years after the end of the First World War, only fifty wisent were left, all of them living in zoological gardens. More enlightened times have seen this splendid species carefully bred back to a healthy population from the small band of survivors and, quite recently, reintroduced into the wild. The mountain ibex (*Capra ibex*), known to the French as *le bouquetin,* has a not dissimilar story to tell: by the early nineteenth century it had been hunted almost to extinction. *Le bouquetin* was saved by being the supreme alpinist, able to climb where few enemies could follow—even ones on two legs equipped with rifles—and by being able to survive on sparse vegetation of all kinds. The wild horse (*Equus przewalskii*) retreated to the steppes of central Asia, having a narrow escape from extinction. In time, the horse was adopted and modified for speed by our own species. This changed the nature of warfare and the course of recorded history, not to mention equine morphology. The horse became the preferred military mount for a thousand years. Perhaps the "wild type" seemed dispensable at that time, but it survived, just, to tell us its story.

The auroch was not so fortunate. During the early history of mankind, this huge wild bovid was widespread all across Europe and Asia. Doubtless, it provided meat for wolves and bears, and it was a favourite object of pursuit by *Homo sapiens.* A specimen of an auroch skull on display in 2010 at The Royal Society, London, carries a stone hand axe almost too neatly embedded in its forehead. It looks more like an offering than a fatal blow. Cave drawings of auroch bulls are touched with a sense of awe; the big beasts must have provided a worthy adversary, with their flared horns and muscular haunches. Across a divide of thousands of years, a comparable respect lingers on today in the same countryside among the bullfighting fraternity in Spain. On the Iberian Peninsula I have heard the sound of lips smacking in appreciation of a particularly gutsy bull, and it is tempting to visualise generations of human males steeped in bullish admiration passing back in time all the way to the shaman creeping into the dark cave to celebrate his mighty prey. The best bulls for fighting seem still to possess the elegantly curved horns of the auroch. The domestic cow was derived from tamed and penned aurochs about ten thousand years ago in the Middle East or India, but the wild animal endured. Like the wisent (or indeed the wolf), its range diminished as wild forests were felled and cultivated, and swampy meadowland was drained. The auroch had probably already gone from Britain long before the Roman occupation. Through medieval times, aurochs disappeared progressively from central European countries, probably lingering on in the more remote parts of Germany—the kinds of places that later yielded stories of wolves and warlocks to the Brothers Grimm. The royal forests of Poland again provided a haven for the very last of these animals. Since they were the property of the King, they enjoyed a certain measure of protection. But by 1564 an official inventory recorded only thirty-eight animals remaining, and the very last animal died in 1627. Had he a mind to, William Shakespeare could have travelled to see a living auroch. Had the Kings of Poland been more assiduous in protecting their herd, we could be admiring this splendid animal in the flesh rather than deftly sketched on the walls of caves. The auroch is one of the near misses in these stories of survival. The dodo lasted a few years longer.

This has not stopped attempts to "revive" this distant ancestor of Herefords and Friesians. In the 1920s and '30s in Germany, two brothers—Heinz and Lutz Heck—made separate attempts to "back breed" modern cattle by selective crossing to reproduce the characteristic features of the auroch. Today one can still find descendants of these animals in game parks and "rare breeds" establishments. They do look somewhat like their model, although smaller. Given that Germany at the time was a centre for now utterly discredited notions of eugenics and racial "improvements," there is inevitably something slightly creepy about these attempts to turn back history, however dearly one would like to see an auroch. There are other bovine claimants for primitive status. The Chillingham herd has allegedly lived in its relatively remote woodland park in Northumberland, northeast England, for more than seven hundred years, at least since the park was enclosed in 1212. The story has it that these beasts are descendants of "wild cattle" dating from a time when this area close to the Scottish borders was highly forested from coast to coast. They do retain interesting traits suggesting that they have not been completely domesticated. They would rather scratch for meagre fare in winter than take cattle cake, and the cows go off alone to calve, and subsequently have to be accepted back into a herd ruled by a dominant bull. Did I observe something proudly independent about the way their white breath flared up in the frosty morning air? I expect I was merely wishfully imagining some innate grandeur, for several facts militate against the animals of the Chillingham herd being direct descendants of the auroch. For a start, they are too small and they are white, although some have red inside the ears. Then there is just too much of a gap between the apparent disappearance of the auroch from Britain, and its alleged emergence into a medieval world. Through many recorded generations there has never been so much as a hint of a "Chillingham brown"; these cows do seem to be another breed of the domestic cow, even though long isolation may have allowed them to revert to some wild habits. There are other medieval records of "red eared cows," so maybe this particular breed was once more widespread. Perhaps those unruly fighting bulls in Spain are closest to the common root after all. I might reasonably conclude from these attempts to revive the auroch that it is easier to make history than to reverse it.

Some Pleistocene survivors are far less conspicuous than the musk ox, but are more important to the climate scientist. It was not just large, cold-adapted mammals that moved north and south with the waxing and waning of the Arctic ice sheets, but all manner of lesser beings from voles to beetles. Plants moved closer to, or away from, the pole according to their climatic preferences as the ambient temperatures passed through cold and warm phases, leaving behind the record of their pollen as a kind of fossil thermometer in the sediments of lakes that abounded away from the ice front. Since I have referred to the appearance of ice as a stimulus to the evolution of specially adapted cold-lovers, it is worth noticing that this may not be a universal phenomenon. The doyen of palaeo-entomology, Russell Coope of Birmingham University, has rather emphasised a *lack* of obvious evolution in the beetles he studies. Beetles leave behind their tough wing cases, or elytra, when the softer parts of their anatomy have decayed away, so they make relatively abundant fossils. He has found not one, but a whole cast of survivors from the ice ages. Beetles common in Pleistocene deposits in England live on in appropriate, and mostly chilly, places today. For example, a fossil beetle abundant in glacial period deposits in Worcestershire is found today on the Kola Peninsula in Arctic Russia. Coope has identified other British fossils identical to dung beetles still happily living in Mongolia. Beetles would, it seems, rather move than evolve.

This fact makes beetle fossils rather good as thermometers of the past. An average of the climatic tolerance ranges of living species can be combined to suggest an ambient temperature and environment in which a collection of fossil species originally lived. The precision of the method has been called into question by some scientists, but generally seems to match very well results obtained by other means, such as stable isotopes. Beetle elytra provide one way to tell the story of the land bridge that connected Alaska with Siberia across what is now the Bering Strait. Appropriately enough, this drowned land is known as Beringia. It provided a vital corridor of steppe landscape that stayed open even while ice sheets massed together further to the east. Cold-tolerant beetle fossils provide a line of evidence showing that the land was often above sea level, although it was periodically submerged during warmer episodes in the Pleistocene when sea levels rose again. At emergent intervals, large mammals moved through the corridor from Asia into North America (and then still further on into South America). This migration undoubtedly included our own species. At some time during the phase when Beringia acted as a bridge between the two continents, about 22,000–11,000 years ago, *Homo sapiens* followed his prey into a new promised land. Whether this invasion happened more than once is still controversial. Humankind may well have lingered in Beringia for several thousand years before pushing on southwards when conditions permitted. Until recently, a major invasion at about 13,000 years ago was believed by many scientists to be the unique populating event, a threshold that heralded the widespread appearance in archaeological sites of well-worked stone tools produced by the so-called Clovis culture. However, there were also several claims for much earlier sites associated with human occupation dotted almost along the length of the Americas. None seems to go uncontested.

In the last decade there does seem to be a gradual build-up of evidence that mankind may have penetrated into the Americas at a date earlier than 13,000 BP. In 2008, fossils of supposedly human excrement, politely known as coprolites, from caves in Oregon yielded a date indicating that humans had penetrated into North America about 1,500 years earlier than the commonly accepted time. A few years ago, Eske Willerslev was a post-doc in the Zoology Department at Oxford University* when the possibilities of fossil excreta were being understood. It was claimed that it should be feasible to obtain sufficient DNA from such material to identify the maker. As one might expect in a laboratory full of students, there was much crude humour to be extracted from the new technology (along the lines of "These results are shit!"). But the methods seemed to work. A sample allegedly originating from a Tasmanian "wolf," or thylacine, another species that became extinct in the twentieth century, might have proved that the creature was alive and well and hiding in the bush. Sadly, it was identified as emanating from the Tasmanian devil instead. By the time Willerslev had become one of the youngest professors ever to grace the University of Copenhagen, examination of such unconventional excretory sources was a perfectly respectable scientific technique. His team recovered DNA from permafrost cores in Siberia that probably originated from the daily evacuations of mammoths and other herbivores. So it would be wrong to pooh-pooh (irresistible, I am afraid) new results pertaining to Beringia. Whatever the time of the first colonisation of the Americas, the disappearance of some of the unique larger mammal species that had flourished there followed shortly upon the arrival of the Clovis culture. The culture takes its name from a site in New Mexico where the distinctive stone tools used by these people were first excavated

---

* At the end of the 1990s and the early part of the first decade of the new century, the Zoology Department had an Ancient Biomolecules Unit, run by Professor Alan Cooper, when I had a Visiting Professor position in that department. Much pioneering work on fossil DNA was carried out there. Sadly, the unit was disbanded, although all its members went on to prosper elsewhere.

in 1932. Their thin flint blades have a very particular elegance, and were doubtless unusually efficient in assuring the successful pursuit of the mammoths that were such an important part of the Clovis livelihood. Several persuasive scientists, like Jared Diamond and Tim Flannery, point to the advances in hunting technology of the Clovis people coinciding in a general way with the extinction of many large Pleistocene mammal species. These large animals might otherwise have contributed to our story of old timers; a partial list would have to include: long-horned bison, giant capybara, camel, horse, American cave lion, sabre-tooth cats, woolly mammoth, American mastodons, dire wolf, short-faced bear and several ground sloths. It is quite a catalogue.

Having already learned the stories of the wisent, the ibex and the auroch, it would be unwise to doubt that the human species could have such a profound influence on other inhabitants of the planet. Other explanations for the extinction of such a varied fauna in North America have been advanced, the most compelling of which is their coincidence with a sharp deterioration of the climate known as the Younger Dryas (c. 12,800–11,500 BP).* A consequent profound change in ecology would favour only those animals, like the musk ox, already skilled in surviving privation and extreme cold. In 2008, a claim was made that a meteorite impact may have been responsible for this climatic crisis, thus transferring to extraterrestrial causes a very earthly tragedy. A review of this evidence by a panel of distinguished scientists in 2009 found it wanting. This has always been the way with ancient history: it is more complicated than it seems at first, and final proof of a plausible explanation proves elusive. Nowhere is this truer than in considering the role of humankind in extinction. Whether we scientists like it or not, our explanations are muddled in with preconceptions about human virtue or vice, wisdom or idiocy. One particular view of "native" peoples is that they are in harmony with the natural environment. Such a view belongs to an intellectual lineage derived from Jean-Jacques Rousseau's concept of the "noble

* This is named after a sudden increase in abundance of the pollen of the pretty white-flowered, cold-tolerant plant *Dryas octopetala* known in the UK as "mountain avens."

savage." In the present context, an appropriate label is provided by a 1990 movie called *Dancing with Wolves* portraying the life of an indigenous American tribe; or we should perhaps refer to them as a group of descendants of the Clovis people. In this version of human natural history, hunting is accompanied by respect for the hunted animal, sanctified by appropriate dancing and ritual. There is wisdom in the tribe that appreciates the interactions between species making up the natural ecology. The human species in a "natural" state is, at root, benign and in harmony with nature. This view seems much at odds with views (even facts) about man-made extinctions elsewhere. Predecessors of the Maori certainly exterminated the giant moas in New Zealand shortly before 1500. That was not long before the same thing happened to the dodo, but in the latter case we know that was no "noble savage" at work. In Easter Island an internal collapse of human society instigated the devastation of natural ecology of a whole island, and eventually the elimination of the whole population. A case has been made for fatal human interference in the extinction of larger mammals during the early human occupation of Australia, and human "overkill" is a potent theory for major extinction after the Pleistocene in the Americas, both north and south of Panama. The list goes on around the world.

It is a legitimate question to ask whether tribal peoples are, or ever were, wise stewards of their environment, or rather relentless butchers, or perhaps some uneasy mixture of the two. This is obviously germane to the wave of extinctions happening today. Is this another example of the human species carrying out "business as usual," only distinguishable from similar slaughter in the past by its technological backup and increased capability?

Tribal peoples are, for wholly understandable reasons, often anxious to portray themselves in the *Dancing with Wolves* image. In the title of the 1982 cult movie *Koyaanisqatsi,* the Hopi word for "life out of balance" was employed to describe the contrast between modern industrial–technological mayhem and some deeper spiritual harmony, which was understood by a people more in touch with natural processes. The aboriginal tribal groups of Australia, who have suffered so much since the "colonisation" by Europeans, often refer to a kind of idealised society whereby ancient knowledge is used to ensure survival in a hostile landscape. Again, the interlopers do not understand the deeper harmonies. The contrast has been portrayed eloquently in *The Songlines,* Bruce Chatwin's semi-fictional account of how aboriginal peoples see and understand the great open spaces of the outback. It has been repeated in a score of movies from *Walkabout* (1971) onwards. At this point it is important to make a distinction between the history of mankind's inhumanity to man and mankind's attitude to nature. Nobody looking at the disgraceful history of the treatment of native "Indian" tribes in North America in the nineteenth century or aboriginal tribes in Australia at about the same time could doubt the former. However, the truth of the brutal imposition of Western "civilisation" does not by itself constitute proof of a wholly benign history between man and nature among indigenous populations. One can quite understand why tribal peoples themselves might favour the *Dancing with Wolves* version. It adds another moral dimension to a list of undoubted injustices. It is something that can be turned to political ends, with regard to land or hunting rights, for example. It plays well with those who would sentimentalise "the state of nature," and perhaps unconsciously follow the Rousseau-esque archetype of the wise primitive. But, as the song puts it: it ain't necessarily so. It may be rather more depressing to think of humankind following ancient foolishness rather than ancient wisdom. If the Clovis hunters could pursue species all the way to extinction, why should their descendants be immune from such short sightedness?

Perhaps the most likely extinction scenario is similar to that portrayed for the crisis and demise of ancient civilisations by Jared Diamond in his 2005 book *Collapse.* In the good times during the diaspora of our species, human populations grew, and there was plenty for all. But when periods of climate deterioration occurred, the capacity of the environment to support large numbers of prey species fell. The land became overpopulated. Hungry people became desperate, and with desperation, social order broke down. Anything that could yield meat was vulnerable. Among large animals, the most likely survivors of the resulting annihilation were those that could climb upwards beyond trouble, such as the ibex, or those that were tough enough to survive in even the most marginal sites, such as the musk ox. Sooner or later, the population sizes of both hunter and hunted fell sufficiently for the land to support the surviving species—but some animals did not come through the hard times. In North America, for example, giant sloths may have been particularly vulnerable, being both meaty and slow. There is indeed evidence that they survived longer in the Caribbean islands until humans eventually arrived there. The final causes of such environmental crises might be connected with fluctuations of the great ice sheets, which have been linked in turn in some cases with atmospheric changes induced by the eruption of a mighty volcano. The explosion of Mount Toba in Sumatra 74,000 years ago is often cited as the cause of one particular climatic crisis producing a population "bottleneck" through which the human species just managed to pass. Whatever the combination of circumstances, and whatever humanity's role in the process, the fact remains that the fauna of large land animals is now impoverished compared with what it would have been only a few tens of thousands of years ago. Many lush twigs from the top of the evolutionary tree of life have been pruned away.

I should finish with one survivor that only narrowly escaped the fate of the giant sloth and the auroch, and where the role once played by our own species is known in some detail. The American bison (*Bison bison*) is a close relative of the wisent. It is not closely related to the "buffalos" of the old world, despite commonly being known by the same name, which in turn has become attached to a city in New York State. Its ancestor joined those mammals that moved from Asia through Beringia, but unlike many of them, it survived into historical times as the great grazer of the plains. The plains bison thrived in its millions, and was the principal source of food, clothing and shelter for tribes of Native Americans such as Sioux and Cheyenne. These peoples doubtless developed as close an ecological relationship with their preferred animals, as did the Sami people with reindeer on the other side of the Atlantic Ocean. Nonetheless, hunting was central to their survival, and they developed various ingenious methods of catching and killing bison, such as inducing them to stampede over cliffs. Respectful ritual towards the big beasts was a central part of tribal lives, and this, coupled with photographs of their richly individual and dignified chiefs, has probably contributed towards today's

Dancing with Wolves version of a prelapsarian "Red Indian" society. Traditional ways of earning a livelihood did not threaten the survival of the "buffalo" prey species; as with many predator–prey relationships, a dynamic equilibrium was reached. After many centuries of balanced sparring between bison and human, the arrival of the horse and the gun with the white Europeans swung the odds dramatically in favour of the hunters.

The killing of bison accelerated. In the 1830s, the demand for bison skins by the American Fur Company led to a trade with the tribes in exchange for beads and weapons. The weapons in turn speeded up the kill. One might justly say that market forces prompted the exploitation of the plains herds. A quarter of a million bison were killed in one year by the Comanche alone. The slaughter increased even further as the plains were colonised by white men from the east after the Civil War. Culling the bison herds also deprived the native population of their traditional succour, and for a while this was actively encouraged from Washington as part of the solution to the "Indian problem." The supply of big beasts must have seemed endless: nobody seemed to cry "stop!" The new railways across to the West could further speed up the transport to market of the valuable skins. Some animals were shot for their tongues alone, for bison tongue was regarded as a delicacy. This may sound barbaric, but similar practices continue today: I have mentioned how sharks are being deprived of their fins and thrown back in the sea to die in the cause of gastronomy. Bison meat was cheap and nutritious. Bison bones could be ground for fertiliser. One of the most horrifying photographs I know was taken in the 1870s and shows a mountain comprised of thousands of bison skulls piled higher than a large house, with a couple of grinning gentlemen in hats for scale. If there was any notion of not "killing the goose that lays the golden egg," it is absent from this ghastly image. Rather, it speaks of getting the very last one. As with bluefin tuna right now, when an animal finally gets rare, the price goes up, and those same market forces push ever onward towards extinction. The very last one will be the most valuable. Nobody comes out of the bison story well, unless it is James Philip of South Dakota, who saved a small population of "buffalo" from the gun in the early 1880s, and kept them. Towards the end of the nineteenth century, fewer than a thousand

Skulls piled high at the vast Michigan Carbon Works, known as "Boneville." The American bison was hunted to near-extinction when cash was paid for bones, used to produce charcoal and fertilisers.

animals remained, where once they numbered tens of millions. The founders of the American Bison Society in 1905 sought to reverse this apparently terminal situation using, among others, animals from the Bronx Zoo—as far away from the plains as one could get—to breed back a stock for nature reserves. Founder members of the Society included Theodore Roosevelt, who went on to play such an important part in American conservation. President Roosevelt was, among many other things, a hunter. He made the connection between ensuring the continuing survival of species and the capacity to shoot them for sport. Paradoxically, animals slain by the gun were eventually saved for the gun. Those who assert the inevitable benefits of market forces might like to think about the role played by such allegedly inexorable factors in opening up the West and destroying the bison.

The only truly wild bison population survived in Yellowstone, Wyoming. High up and hidden in the interior of its protected mountain basin, even the most intrepid hide hunters failed to oust the tough bovid from its last redoubt. Legal protection of the few tens of remaining animals was enacted just before it was too late. I looked down upon a bison herd in the Lamar Valley while snow still decorated the flanks of the surrounding hills. For a moment, the vast numbers that once wandered the Great Plains could be conjured up in the mind's eye. They were in no hurry, a few dozen of these big grazers, ambling along the wide valley bottom. Their massive heads tossed from time to time, but mostly butted the ground searching for a tasty tussock, or a green shoot some other member of the herd had missed. Their horns seemed insufficiently serious for such a powerful animal, looking more like a pair of pert handlebars than a method of defence. Like the musk ox, the bison is all massive muscle at the front. This comes into its own during the implacable winter in Yellowstone, when the bison can drive its massive head down through the thick snow cover to munch on meagre lichens and buried herbs. Numbers of animals fluctuate naturally within the Park, as they would in nature, but some culling is necessary to maintain the health of the herd. Since a few of the bison have become infected with brucellosis introduced in cattle from Europe, animals cannot wander freely out of the Park into the adjacent ranching state of Montana. Complex rules concerning culling and immunisation have been drawn up to protect agricultural interests whilst preserving as much of the wilderness intact as possible: a practical compromise. Once more it seems that the past, once lost, can never truly be recovered. The half-tonne beasts have been taking quiet revenge on their human oppressors in the Park by mooching in line along the roadways at an infuriatingly slow pace as if deliberately to hold up the traffic. Those who visit the Park are far too ecologically aware ever to toot their horns.

In the winter, bison take refuge in the thermal areas where bright, hot, slimy slicks defy the cold. When the roads into Yellowstone are first open, some beasts still linger close to the deceptively inviting bright blue pools. At Black Sand Basin early in the frosty morning, the breath from a bison cow and her calf rises in passing imitation of the steam swirling up from the vents beyond. The bison ignore all imprecations not to walk upon the bacterial mats, for they have been tramping here for a thousand years. Wispy grasses and lichens seem to provide little sustenance for such big animals, but they find something worth picking at on the damp ground. Around these hot springs, the two ends of the tree of life are drawn together. Archaea, bacteria and algae spilling out of the pools in their coloured oozes speak of a time when life and the planet were young, when the land was bare; when wrinkled and dimpled surfaces sported the first communities that converted energy into replication. These organisms have not remained unchanged—nothing alive on earth has remained *completely* unaltered—but they still reproduce according to ancient rules, still grab what they need from ancient sources. Sulphur can be food, and scalding heat can drive the motors of life. Unseen within the stomachs of the bison, other single-celled organisms drive the digestion of the plant food that gives the animal permission to survive on poor rations; without these benign helpers, the grasslands would be devoid of grazing herds. A last relic of the ice age, a late shoot on the tree of life, would have been unable to make a living. The dignified bison and its unseen ruminative helpers are all survivors in their different ways, although the fate of the former could have easily been damned by simple twist of history; just a few more gunshots or just a little less natural resilience, and there would have been no bison in my story.

I like to think of the double helix of the DNA that proves the shared ancestry of the simplest of prokaryotic organisms with the bison and her baby as a kind of plaited twine weaving through the tree of life and binding it together. All the organisms I know around the hot pools in Yellowstone National Park are threaded together as part of a common narrative: the coal-black raven eyeing the duckboards for the possibilities of a dropped snack; the scruffy conifers that survive poisonous gases and chilling winds; the humble mosses that fill in the gaps between tentatively flowering yellow daisies; a thousand kinds of tiny creatures that make the soil in the surrounding forests alive. But bound together by the twine of descent as they are, every one of these living beings still has its own biography, and every *bio*graphy could be as interesting as that of any organism I have selected for this book. No ant is too small to fail to deserve our attention, no microbe too hard to understand, no fungus too obscure, nor any flower too evanescent.

10

Survivors Against the Odds

Local newspapers always have ready copy when a member of their community reaches their hundredth birthday. A cosy article is accompanied by a photograph of the smiling centenarian surrounded by hordes of children, grandchildren and great grandchildren. The feature invariably finishes with a question about the secret of longevity. Assuming the old person is up to framing an answer at all, the wise advice is always different. One senior will attribute great age to teetotal habits and eating once a day; another will inform the reader confidentially that "a little of what you fancy does you good"; another might mention the ability not to take life too seriously; yet another praises the virtues of hard work and crossword puzzles. Few attribute long years solely to good luck or favoured genes. Old campaigners who have survived particularly bloody battles are usually more forthright about the role of brute chance. A bullet straying to the left rather than to the right might have "had my name on it," or there was that bomb that failed to explode. Warfare makes the random aspects of survival more explicit. In general, though, it is difficult to talk about survivorship except in terms of human virtues. Self-made tycoons tend to regard their wealth as just repayment for hard work and brilliance rather than being in the right place at the right time; so longevity is always a matter for congratulation, even if not self-congratulation. "Isn't she marvellous?" an admiring relative will exclaim. It is hard to avoid similar language in describing organisms that have survived from deep time and, since we are all relatives in the genealogy of the tree of life, it would be surprising if some of the same admiring tones had not crept into my account of the echidna or the tuatara.

I believe that it is inadequate to sum up all the enduring prokaryotes, animals and plants that reveal the history of life as "just lucky." It would equally be a mistake to describe them as entirely adaptively superior. The phrase "the survival of the fittest" never truly belonged to Charles Darwin, but was coined by the Darwinian economist Herbert Spencer in 1864. Darwin took it up in his fifth edition of *The Origin of Species* in 1869. It has been misused more than once in a false conflation with natural selection. The same phrase can rather easily be turned around to say that the fittest must be the longest-lasting survivors. After all, they have been "fit" enough both to survive and then to survive a lot longer than most other organisms. Once again, a subtle value judgement inevitably creeps in. In this scenario, the animals or plants in this book would be rather like those marathon runners that come staggering in on the last day of the Olympic Games, so admirable in their powers of endurance. I prefer to use another metaphor from athletics: we should think instead of hurdlers on a challenging circuit, clearing one fence after another, while fellow competitors stumble and fall in the face of obstacles, even though they may have briefly led the field. The fittest in this specific sense are the ones that reach the post, and they are not necessarily the temporarily speediest. Survival is a matter of carrying on a line of inheritance through the rebuffs thrown up by history, where luck may well play a part, but will be of no avail unless the enduring organism has appropriate qualities. The animal or plant may not be superior in any absolute sense, if indeed there is such a thing, but they had features that served to ensure their survival at a time when it mattered: something that helped them to clear a hurdle. Such an impediment or obstacle to survival could be one of the ancient mass extinctions or a sea level change; the result of moving continents or an ice age. Every organism has its trajectory through history—its *biography*— a narrative all of its own. Some are brief, like that of the cave bear (*Ursus spelaeus*), which thrived in the cold of the Pleistocene ice ages, but could not prosper thereafter; or think of an animal that briefly flourished on an oceanic island only to founder along with its home. By contrast, some microbes may well have narratives lasting billions of years. Where the trajectories of different organisms coincide, or independently tell a similar tale, then maybe we can discern a pattern, something that moves beyond luck.

I have been dismayed from time to time by some portrayals of evolution as a kind of simultaneous advancement in which the traces of the past are continually erased. This notion is implicit when politicians utter such phrases as: "we must change; we must evolve new solutions before it's too late; we can't be dinosaurs!" On the contrary, we have seen that imprint of the past in organisms surviving from branches low on the tree of life is almost everywhere. Evolution has left clues of its past history in a ragbag of survivors, some of whose biographies I have explored in detail. As Richard Dawkins has said, we could trace conclusive evidence for evolution even if there were no such things as fossils. It is there in the genome, and it is there written in the comparative anatomy and development of living organisms. This is true, but it also misses the point. Fossils provide evidence of when, how and why history happened. How much more satisfying it is to have evidence that feathers preceded flight and that there were such creatures as "dinobirds." It would be a poorer vision of the world without trilobites or dinosaurs, though both are extinct. I can think of no other way of discovering that mankind's forebears walked upright *before* they acquired a capacious brain than by discovering early bipedal fossils. Without fossils, we would not know that the first movement of our tetrapod ancestors onto land happened 400 million years ago. We could, I suppose, have built up a picture of ancient Rome from the written accounts of Rome's poets and historians alone, but how much richer it is to have the archaeological record as well. Let us carry on digging! I have referred to fossils frequently, not just because I am a palaeontologist by training, but also because fossils add depth and meaning to the story of every organism. However, I have been rather cautious about using the tag "living fossil" too readily. Just as it is untrue that evolution erases the past, it is also the case that no organism remains *completely* unchanged through long periods of geological time. The horseshoe crab may carry an ancient carapace on its back, but it has still moved with the ages. Even the tuatara has evolved at the genomic level for all that its obscurely smiling visage seems to speak directly of the Triassic. I can use the term "living fossil" with a completely straight face to describe a few organisms that were found first in the fossil state, and later discovered alive: like the ferreret, Australian lungfish, swamp cypress, coelacanth and Wollemi pine. But it would not be correct to use the same description for organisms such as crocodiles, for all the endurance of their kind, nor yet for bacteria whose tenancy of the earth is of such extreme antiquity. What these examples provide is evidence for deep branches in the tree of life that have continued to put forth new shoots and leaves despite their antiquity. On this dynamic earth, nothing stays still: neither climate, nor life, nor the oceans and continents, although they all move to different rhythms.

The hurdles cleared by survivors have been referred to many times already but not anatomised. I should list the most important mass extinction events in a little more detail. These also serve as a brief résumé of the story so far. It will be obvious that the thread of life connecting the longer term survivors will have crossed more of these events than will organisms with a more recent origin. As with so much in history, the older events tend to be the more obscure, and I am sure that there are important new discoveries still to come in the vast compass of Precambrian time. The first crisis that should be revisited is the "great oxidation event" at 2.4 billion years ago. When cyanobacteria introduced a significant level of oxygen into the atmosphere, the effect on the hordes of prokaryotes that thrived only in its absence must have been devastating. The extremophile bacteria that had once owned the earth endured in hot springs and foul corners, but the atmospheric rules had changed forever, and would eventually allow for respiration and the dominance of modern animals and plants. In the biggest picture, change in numbers of one organism relative to another is as important in our planet's history as extinction by the complete obliteration of lineages. Then there were periods of extreme climate change. The "Snowball Earth," or Marinoan glaciation of 650 million years ago, was a time when—so its proponents believe—the earth was encased in a solid layer of ice. Enthusiasts for this kind of event recognise several other "Snowballs" in the

Precambrian, and if they are right, periodic freezing conditions were hazards that life had to negotiate. Even under the most punishing scenario, however, there must still have been islands of water around active thermal vents like volcanoes and hot seeps, which should have provided a conduit for survivors. Recall that most organisms were small at that time, so this is not like conserving the giant panda. Many scientists believe that the extreme Snowball is not correct, and that an equatorial region remained free of ice (or was "slushy" at most). Very few dispute that the influence of ice extended down into low latitudes during this interval, and must have had a profound influence. After the freeze, the flowering of the latest Precambrian, Ediacaran fauna followed, as I described at Mistaken Point in Newfoundland, where a strange undersea confection of soft-bodied but indubitably size-able organisms spread over the world. Life grew bigger. This is one biological drama that seems to have vanished without trace. Jellyfish were there, and probably sponges, but most of the Ediacaran animals remain more or less controversial, and do not include obvious "ancestors" of those that followed. Alas, they do not survive. It is hard to know whether it would be more accurate to describe the Ediacaran as the end of an era, or the start of a new one.

No such ambiguity attaches to the Cambrian Period that followed. It bears something of the stamp of the modern biological world. We are all children of the Cambrian, whether we know it or not. We have seen how vertebrates and invertebrates alike root back to this most creative time in our planet's history. A time when hard shells appeared in several groups of organisms for the first time allowing for protection from predators, and for the lodging of strong muscles. But "soft bodied" animals are as varied as shelled ones when suitable rocks to preserve them are discovered, as in Chengjiang in China, so it was not just a matter of shells making merry. Many of the animals in this book entered the stadium at this point: molluscs, arthropods, brachiopods, penis worms, peanut worms, velvet worms, wormy worms (see p. 130), hemichordates, chordates. Yet the Ediacarans had vanished, so there must have been a mass extinction of these enigmatic fractal organisms and their odd companions at the close of Precambrian times. Most palaeontologists agree that there was a time lag in the fossil record between the disappearance of the

Ediacara curiosities and the full "explosion" of Cambrian forms. In many places with otherwise excellent rock records, there was a withdrawal of the sea at just this time, so evidence of the critical period is missing. Martin Brasier has given an entertaining account of the discovery of such a "gap" in rock sections lining the Lena River in Siberia in his recent book, *Darwin's Lost World*. Where rock sections *are* completely preserved, there is evidence of the wide spread of anoxic seawater at, or at least close, to the critical time. The lethal water wells up onto the continental shelves overwhelming animals that need to "breathe" oxygen at moderate levels; under these conditions, the old, old bacteria briefly reassert their dominance. Shales turn black. Other life dies. Maybe it was this kind of event that removed Ediacarans like *Charniodiscus* and its relatives from the marine realm. Certainly, oxygen levels were higher in the ensuing Cambrian (542 million years +), although there were further oxygen crises later in the same period[*] that stressed animals living on the continental shelves, and encouraged those that could tolerate less benign conditions. A few trilobites became masters of the starved seas. Through all this, remember that survivors from lower in the tree of life were continuing to continue, even when they didn't announce their presence as fossils.

Onwards and upwards to the late Ordovician, when another glaciation was centred on the South Pole of the time, in northern Africa. 445 million years ago ice sheets grew once again, sea levels fell concomitantly, and the tropical zone contracted in sympathy. I have stood on top of scratches made by an ancient ice sheet in the Anti-Atlas Mountains of Morocco while all around a dry desert stretched away, and some bird sadly mewed in the sky for the least edible scrap. I was reminded how profoundly the continents move, slowly shifting climate zones, slowly transforming the face of the world. At the end of the Ordovician, geographic and oceanographic conditions were just right for the growth of ice sheets, and the world was plunged into a climatic extreme. Many animals that had flourished since the Cambrian went into decline. When the ice sheets melted, another low oxygen crisis was released on the world's oceans; many more

---

[*] One of these in the late Cambrian has been claimed as a mass extinction, but in my view its importance has been exaggerated.

invertebrates died out, including several trilobite families. The survivors clung on. After the crisis, life in the sea expanded once again. Ancient oceans were consumed as mountain chains grew and were eroded away. Plants and animals made the crucial moves towards colonising the land, and more of our potential survivors had entered the circuit of life. Yet 60 million years after the Ordovician glaciation, another crisis in the late part of the Devonian Period put paid to coral reefs and many more trilobites, brachiopods and jawless fish. Recent research supports the idea that this was an extinction event with several pulses rather than the single event (termed the Kelwasser Event) that was first recognised. Once again, the spread of oxygen-poor waters through the oceans played an important part in the stresses placed on marine life: another hurdle to clear.

All this was as nothing compared to the massive extinction that happened at the end of the Permian Period 250 million years ago. Life both on land and in the sea was devastated. The Carboniferous and early Permian Periods had seen amphibians, reptiles, horseshoe crabs, scorpions and many "lower" plants appear in the fossil record, often nurtured in the sweaty bowers of the coal forests. Even a long-lived glaciation centred on Gondwana only had the effect of pumping up biodiversity by inducing regional floras adapted to different climatic zones. How many of these potential long-distance runners would survive a crisis when, according to some of its leading students, 90 per cent of species became extinct? The continents were amalgamating into one to create the "supercontinent" of Pangaea; any trouble on land would have to be shared by all inhabitants in a kind of catastrophic lottery. Rock sections in China record completely the details of the tragedy as it played out in the sea. A vast volcanic eruption of the Siberian "traps" is thought by many to have been the proximal cause of the crisis. Possibly the greatest eruption of basalts in 500 million years would have spewed out millions of tons of sulphur and carbon dioxide into the atmosphere over a geologically short period of time. Even today, the black lava "traps" occupy an area as large as Western Europe, and they were originally much more extensive. Inside the unity of Pangaea, these volcanic piles were also much more centrally placed than they would be today. Poisoning of the atmosphere and rapid climate change would be inevitable consequences of massive gaseous emissions. One thinks of Friedrich Engels' description of a polluted Manchester in 1844: "how little air, and such air!" This was a world crippled by gases. Meanwhile, deserts had spread over the continental areas of Pangaea, and there is little sustenance to be found in sand dunes, even for bacteria. Terrestrial life was pushed to the edges: there were fewer shelters in which to hunker down through the bad times. In the sea, anoxic waters with their low oxygen levels briefly reigned supreme, spilling even onto the continental shelves that surrounded the huge Pangaea supercontinent. Few marine animals could cope with it. Many rock sections show a "dead" interval comprised of depressing shales with hardly a single fossil spanning the critical time. When animals that could leave fossils behind returned once more, they were stunted, sad little things lacking in variety; a few, short-lived, resilient molluscs surviving from the older world, perhaps, accompanied by a skimpy group of heralds for the new one. It must have been a ghastly episode for life, a time fit only for opportunists, and for microbes that flourished on others' discomfiture. It passed as these things do, but many casualties from the ancient world did not endure from the Palaeozoic into the Mesozoic Era (that is, from Permian to Triassic Periods), my trilobites among them. The tuatara roots back to the Triassic, as do living amphibians. There must have been links to their Palaeozoic past and the hot coal forests, but the fauna living today dates to a time that the biological world emerged battered but durable from the great dying, to make a new beginning. It was an event that reset all the clocks. Some of the tottering survivors from the end Permian event were culled subsequently at the end of the Triassic, some 200 million years ago. The causes of this particular mass extinction are still not settled, but the casualties included brachiopods and sea lilies that dated back to the Silurian. Had they just lasted through one more test, maybe, just maybe, they would be with us still. I grieve for them. I would dearly have liked to visit them. The same extinction event removed several potential competitors to the dinosaurs, and so ushered in the hegemony of every schoolchild's favourite "monsters" for more than 130 million years. The movie epoch had begun; Steven Spielberg could now begin to take notes.

It was not to last. Nothing lasts forever—unless it is bacteria that can feed on froth, poison and air. The next great mass extinction at the end of the Cretaceous Period removed the dinosaurs from their cock-of-the-roost position, and terminated the Mesozoic Era. When I wrote an account of this event fourteen years ago,* there were still many who doubted that the effects of the impact of a giant meteorite on the Yucatán Peninsula in Mexico were an important part of the reason for their demise. There are fewer doubters now. A majority of scientists accept chemical signatures from rock sections scattered around the world that betray a massive impact 65 million years ago, coinciding with the greatest mass extinction in the sea and on land since the end of the Permian. Some palaeontologists would like to add into the mix the effects issuing from another vast mass of basalts that erupted in what is now northwest India—the Deccan traps. Whatever the concoction finally proves to be, dust and gross darkness covered the earth, while acid rain on a vast scale produced an environmental catastrophe that saw the leaves wither on the vine, and the seas poisoned. Food chains collapsed. Herbivorous dinosaurs starved, and the carnivores that preyed on them were doomed. In the oceans, even single-celled plankton suffered a drastic reduction. Ammonites were just one of the more prominent invertebrate victims. These coiled molluscs had already weathered the Permian débâcle; they had made Jurassic rocks a pleasure for the palaeontologist. They finally met a hurdle that was too testing for them. The eventual recovery from the trauma allowed mammals and birds to thrive in a way they never had before, for flowers to brighten the landscape, for bees and butterflies to buzz and flit into their respective futures. All mass extinctions ultimately generate winners as well as losers. They help life move on. And so, to a climatic deterioration a mere 1.6 million years ago which brought on the last of the ice ages, in the Pleistocene. As ice sheets grew, they covered great areas in higher latitudes. If animals and plants were able to migrate away from the ice front, then that is exactly what they did; as long as a climatic gradient remained intact, extinction was certainly not inevitable. The

---

* *Life: An Unauthorised Biography* was published in America as *Life: A Natural History of the First Four Billion Years of Life on Earth*. Like some organisms, book titles change subtly when they cross the Atlantic Ocean.

Pleistocene includes a rich story of faunas moving to the north or to the south, in harmony with oscillations in climate and the waxing and waning of the ice sheets. A variety of specially adapted mammals, often woolly or huge, that evolved in response to frigid conditions were much more vulnerable in the long run, when climatic conditions eventually improved. However, even before the last glacial maximum some 20,000 years ago, one bipedal relative of the chimpanzee had already left his African home on a long trek around the world. Water locked into ice sheets produced low sea levels that helped to make bridges for this interloper. His tribes moved from eastern Siberia into the Americas. His numbers grew; he learned to dress for the cold and to plan for the future. He may have hunted edible mammals and birds to the point of obliteration even in his earliest days; he was the first species that deserves to be called the Terminator. Humankind is responsible even now for what biologists are accustomed to call the "sixth mass extinction,"* and there is no sign of it stopping. If the bipedal creature had evolved a conscience somewhere in its special adaptive package, then it has to be said that it is not effectively doing its job.

With such a catalogue of disasters to interrupt the continuity of life, it might seem a wonder that any organism at all survived from deep time. Even now, I have not been exhaustive. Recent research is finding evidence for sudden episodes of global warming associated with the melting of methane hydrates (methane "ice") normally sequestered safely within ocean sediments. Methane is the most effective—if short-lived—gas to produce climate change. Currently, the "ice" is stable below about 300 metres depth, but a warming climate may cross a "tipping point" causing the layer to melt at an alarming rate, accelerating global warming. Similar dramas have been added to the scenarios proposed for the five mass extinctions, and may help to account for some otherwise inexplicable smaller events. Then we have to remember that the surface of the world is ever on the move, with some tectonic plates moving apart, others converging. Island continents like Australia remain all alone; after the break-up of Pangaea,

---

* This is counted only from the Cambrian, so it does not take into account the important Precambrian events that preceded the end Ordovician mass extinction.

the Indian subcontinent was married to Asia after millions of years of autonomy, when its collision with the larger continent to the north threw up the Himalayas. The Antarctic was not always at the end of the earth, nor was it always deeply buried in ice. Only specialised giant penguins can survive for long on land that frigid nowadays, but there was a time when mammals lived unremarkable lives in the same place. The growth of ice sheets doomed them to extinction, but it was not their fault—just bad luck. Once upon a time it might have been possible to imagine a land that drifted away with a cargo of ancient animals, an Ark of survivors, as in *The Land That Time Forgot* by Edgar Rice Burroughs (1918). I wish it were out there, somewhere, anywhere, but it is not. Instead, we know that isolation often produces endemic bursts of evolutionary creativity. Recall the lemurs on Madagascar, a wonderful bunch of primates stranded on their own island, or think of the kangaroos and wallabies that make Australia so much its own place. There is no such thing as evolution standing still, no past world exactly preserved, no secret hideaway cocking a snook at the passage of time. Even Shark Bay is a simulacrum of antiquity, a fluke of circumstances mimicking a time before memory.

We must return to the survivors, and consider how they could have negotiated the series of hurdles presented by the mass extinction events. The first thing to say is obvious. Survival is about endurance of habitat. Every organism has a place where it fits in, earns a living, reproduces—a niche in nature. Some niches are very specialised. I think of those hummingbirds in Ecuador exquisitely engineered to taste the nectar of particular flowers, and those flowers in turn evolved to please only such dedicated pollinators. Or how Charles Darwin wrote on the perfect co-evolution between orchids and their insect pollinators in 1862, an elaborate *pas de deux* danced between one organism and another over millennia. It would be easy to imagine such a delicate balance upset: if one or the other partner suffered a population crash, for example, its collaborator would be doomed. If climate change affected vegetation, the results could be the same. These are examples of extremely narrow niches, and it is easy to understand how the organisms concerned might fail if fate tossed something unexpected their way. To survive a major earth trauma requires some continuity in earning a living from nature through the difficult times. Some plants may be able to survive through the worst traumas as seeds, ready to germinate when the bad times have passed; seeds buried with the pharaohs have famously retained their viability. But other organisms do not have such an option, and rely on an appropriate habitat surviving with them, buoying them through. Microbes have the least problems, because their small size means that the tiniest crevice is a world to them. There will always be an option; when things get bad, they simply hunker down and slow down their rate of reproduction. After all, they can live inside solid rock. Others prosper while misfortune rules the world: a poisoned ocean is a wealth of food for sulphur-loving bacteria. Briefly, they can re-establish a dominance that they had not enjoyed since the Archaean. When free oxygen poisoned the world for the earliest prokaryotes, there were always hot, sulphurous springs or fetid, slimy places where they could carry on their ancient trades. Mid-ocean ridges and hot and cold seeps have been present in the seas ever since the motor of plate tectonics began turning over 4 billion years ago. The arrival of a meteorite might be no more than a series of new opportunities for many of these anaerobic micro-organisms; after all, one bug's poison is another bug's meat. So we do not have to wonder at the survival of the early prokaryotes. Nor at the long history of cyanobacteria, which can appear in a glass of water left on a window ledge or the merest scratch in a rock ledge. For these simple photosynthesising organisms, where there's light there's hope. Habitat is not a problem.

Moving up the tree of life to larger organisms, it is more challenging to think about some of the invertebrates. Why, for example, does *Lingula* hold such a long-distance record among the brachiopods? In Hong Kong's New Territories, I held a spade full of muddy sand with the old animal lying among the sediment, and it felt like a sample that could have been easily matched in the Ordovician. But the biology of this particular animal has special features, which suggests that the habitat explanation might work rather well. First, it is not alone. Recall that *Lingula* is found alongside the peanut worms, a type of organism that we know, thanks to the Chengjiang fossils, roots back to the early Cambrian. This is a beast without a shell, but otherwise no more or less complex than a brachiopod, and another long-term survivor. Just offshore lives the primitive chordate amphioxus that once again has a history as distinguished. Now recall the ancient clam *Solemya*, which I encountered on "Straddie" in eastern Australia. Elsewhere on the same island there are the biggest specimens of *Lingula* living in the sediment that I have ever seen. They are as large as mussels. So here we have another pair of long-term survivors. The clam in question has gone into partnership with sulphur bacteria and can tolerate, indeed *needs* to live in a marginal site where few other animals can thrive. Elsewhere, priapulid, or penis worms can be found in a similar habitat; this is a group that was more diverse in the Cambrian than it is today. The argument about habitat can be threaded into the confluent *bio*graphies of animals with different evolutionary origins. Maybe this particular habitat survived as a package through the mass extinctions at the end of the Ordovician, within the Devonian, the big one at the end of the Permian, the smaller one at the end of the Triassic, and finally the trauma at the finish of the Cretaceous. All survivors came through. For a habitat to contain one survivor today might just be luck, but for it to include a whole cast list begins to suggest that convergent biographies are revealing a tale in common. I am tempted to return to the military metaphor: maybe this habitat was like a tunnel that simply went under the front line. The luck came in if you happened to belong to that special battalion with access to the tunnel. I must repeat again that this does not imply that evolution stopped; the species are not the same today as those in the distant past. The genes still moved with the times even if the genealogy were an ancient one.

The habitat in question lies low in the intertidal zone and extends into shallow subtidal habitats, on muddy-sandy shorelines. A simple thought experiment suggests that this could indeed be a favoured habitat for survival. It is reasonable to believe that when anoxic seas became widespread, as they did repeatedly during phases of mass extinction, this particular area was probably still oxygenated at the surface. After all, the wind still ruffled the waves on shore. All the organisms mentioned burrow into the sediment at low tide and feed on tiny particles when the sea came in. They can make do on little. If the sediments became depleted in oxygen at a shallow depth, as still happens, the burrowers were able to cope with it by keeping contact with the surface. In the case of *Solemya,* necessity became the mother of invention, and the sulphur bacteria naturally living below the surface took part in the mollusc's life history. It might be expected that natural selection would nudge out some of the survivors as more "advanced" organisms appeared, but it does not seem to work like that. Instead, as I saw in Hong Kong, recent additions lived in the same flats alongside longtime survivors: shrimps, snails and small fishes alongside *Lingula* and peanut worms. The habitat was a kind of jostling collection of different animals of all sorts of antiquity. There might be a special reason for this. Populations in many habitats are critically limited by the quantity of food available. However, in places such as mud flats food may not be the limiting factor: for filter feeding animals, a rich food store is carried into the area with every high tide, or is brought from nearby land during storms. The crucial thing is to find living space. The problem is not the food in the trough, but making a place at the stall. So if it can establish itself in its burrow, *Lingula* is as able to compete for food on equal terms with a later arrival (geologically speaking) like a shrimp. This habitat does seem like a good place to be for an organism with conservative tendencies. If its own place survives, then so will the beast. It is in the right place to weather mass extinctions that affect many other environments more severely. In one way it is "survival of the fittest," but of the "fittest" habitat with the right design specifications to offer long-term security. Stick-in-the-muds last longest.

Most animals are loyal to their habitat, but when climate changes, the habitat itself may shift its location. An organism might well be free to accompany its favoured niche to a new place. Other organisms move into, and adapt to, new habitats through the passage of millions of years. If those habitats are particularly specialised, these species will become survivors because they occupy special niches that are hardly contested. Scorpions evolved from originally aquatic habitats to become one of the few creatures that can prosper in hot and arid environments. They will doubtless see us humans into ultimate perdition. The little crustaceans sold as dried eggs in pet shops as "sea monkeys"—the brine shrimp *Artemia*—can exist indefinitely in desert salt pans until watered by rain, when they "come alive" and grow rapidly to maturity.* They are almost as antique as the scorpions. We have seen how the curious, long-lived plant *Welwitschia* is confined to an area of Namibia where very few other plants can thrive. As long as this habitat endures, so will *Welwitschia*. Like scorpions, velvet worms also began as marine organisms, but at some stage a terrestrial branch of the group became their sole survivors. The same logic applies to those unpleasant fishes that endure from the base of the vertebrate tree, the lamprey and the hagfish. They became parasites on other fishes, a niche from which they were never displaced. Early members of the jawless fishes were certainly *not* parasites: they did some of the usual jobs carried out by aqueous animals sucking up sediment to extract small animals. They were replaced in the ecosystem by a host of bony fish species, but their specialist cousins have continued to make a living around the world, and are still successful enough to be a nuisance to commercial fisheries. For them, the only interesting thing about a mass extinction was whether enough of their hosts survived to provide them with a good meal; evidently, they did. All these organisms still depend on the continuity of the special ways in which they earn a living, but even after hundreds of millions of years the chain could still be broken. Humans may yet accomplish what time and catastrophes could not if they continue to destroy natural habitats at the current rate.

For the many species that move around the world in tandem with their favoured habitat, there are two contrasting cases that it will be helpful to recall briefly. The first scenario is played out on large landmasses when climate change forces movement upon whole regions of forest or grassland, reshaping them into new configurations. In the last ice age, temperate forests in the northern hemisphere moved northwards during warm periods and southwards at times when the ice sheets advanced towards their glacial maxima. In appropriate areas, forests might equally move up or down a mountain range to maintain their preferred climatic regime. Only if a change was

---

* The tadpole shrimp (*Triops*) has comparable adaptations, and is another crustacean survivor, with a fossil record extending to the Carboniferous. Both *Artemia* and *Triops* are freely available in kits, to be grown in a jar on the window ledge, so these old timers can be visited in the comfort of your own home, and do not require a field trip to find them!

particularly sharp or drastic was there no time for adaptation. We have seen that during the Pleistocene, ice also helped to create a special frigid habitat in the tundra: sedge-rich grassland that supported specialist mammoths and predators. When the world warmed up, these animals had nowhere to go. It was easier for mammals adapted to a warmer climate to survive the cold than it was for cold-adapted mammals to survive the warmth. Even here, luck was not even-handed, since *small* cold-adapted animals could still find a niche in mountains where chill hung on. Unpicking the influences of luck, genes and history becomes complicated.

A second scenario is played out when the predominant influence is plate tectonics. Continents shift, carrying their biological cargoes with them; or maybe a volcanic island appears in the middle of an ocean, creating a *tabula rasa* on which evolution may draw up new species. Chance, of a kind, is the dominant influence in these cases, since the organisms themselves can hardly "choose" where they finish up. We have examined several cases where the present day distribution testifies to an early history on the "supercontinent" of Pangaea. Recall the velvet worm, the lungfish, the monkey puzzle trees, ratite birds. There are plenty more, each with its own story to tell. I would have liked to visit ricinuleid arachnids, which are neither spiders nor mites, and now live modestly in Africa and South America. These creatures have a history extending back to the forests of the Carboniferous. I should have discovered the primitive cactus *Rhipsalis,* found in Africa, Madagascar and Sri Lanka; cacti are now specialists for life in arid conditions in the Americas, but did they originate in damp forests in Gondwana? There is still so much to learn.

There are places in the world where survivors fetch up, just as some beaches are known for the precious flotsam that washes up on them. One could readily imagine species moving in response to climate change, clinging to their natural habitat, and being found today in special havens sequestered from further change by persistence of old environments. These sites will be best revealed if several survivors are found living together: another coincidence of *bio*graphies. One such place lies in the middle of China. Huangshan, the Yellow Mountain, in Anhui Province, is a collection of peaks that is now a tourist resort, and hugely popular among the Chinese. A forest of new high-rise hotels is growing at the base of the mountains to accommodate the hordes of visitors. The landscape above presents an archetype for all those classical Chinese paintings featuring precipitous mountains rising clear from a thick mist below, with the summits decorated with twisted pines, and hardly a figure in sight. Granite makes up all the mountainous area, but it is granite weathered in a way I have never seen before. In the British Isles and Europe I am accustomed to seeing severe, rounded masses making wild moors or, where scoured by ice, sheer cliffs and mighty crags. Granite is an unforgiving igneous rock, and yields grudgingly to the elements. In Huangshan a once mighty rock mass has been reduced to teetering columns, fluted and precarious, reminiscent of an extravagant Bavarian castle—but they go on and on retreating into distant mistiness. At this latitude trees can grow to the very tops of the peaks, leaning out at improbable angles, with their branches stacked like irregular plates; they gain a purchase even in the most inaccessible cracks. In former times Chinese poets would journey into the wilderness and be lost for days in rapt concentration. Now the area is served by cablecars that whisk visitors to the top, and even the most precarious ridges have been paved. Crowds of happy people troop to have their photographs taken against the backdrop of one of the famous vertiginously hanging trees. Tour guides bellow at their charges through microphones, hurrying them along. On the way up in the lift, I notice features such as perched granite boulders, left behind by uncountable ages of weathering, a last relic of some vanished rocky mass. I was familiar with similar rocks from the granite tors in Cornwall, western England, which is a place that was never reached by glaciers during the Ice Age, where time alone sculpted the landforms.

At 1,800 metres the trees have a comfortable look, not least because they belong to the same genera as one might find in England: pines (*Pinus huangshanensis*) of course, but also an oak (*Quercus stewardii*), beech tree (*Fagus engleriana*), hazel (*Corylus*) and a small shrub (*Enkianthus chinensis*), which reminded me very much of the strawberry tree (*Arbutus*) native to western Ireland. They may have been different species, perhaps, but this is a flora familiar enough to Europeans. However, in the warmth low down on the slopes, I am surprised to see some other trees that I recognise from home territory, though only in gardens. It is fortunate that I am visiting in April, because these small trees are in flower; otherwise, they might have lurked anonymously within the densely vegetated slopes. *Magnolia* announces itself boldly with huge pale pink flowers, just as it does in my garden. In the wild they look from afar as if a small flock of pale, untidy birds had become entangled in the branches. Much less conspicuous, but instantly recognisable, is the delightful shrubby tree *Hamamelis,* with yellow flowers decorating its bare branches, looking something like bright whiskers. Neither of these trees could be found higher in the mountains. Both of these plants insert into a very low position on the evolutionary tree of flowering plants, judged from the primitive structure of their flowers, and confirmed by molecular evidence. They are survivors from the Cretaceous that must have passed through the mass extinction that brought that period to a close. Along the road near the chairlift, a few huge planted ginkgo trees remind me that these Mesozoic survivors have their natural haven not very far away. Even the conifer *Cunninghamia,* which grows nearby on the slopes, is regarded as the living tree closest to the ancestor of the cypress family—more primitive even than the living fossil *Metasequoia,* according to recent molecular work. After all, *Metasequoia* itself was found not *very* far away. It begins to look as if a whole bunch of trees survived together from the era of the dinosaurs quietly to populate the lower slopes of mountains in Anhui Province in China.* Since habitat is the important control, it seems reasonable to conclude that there was continuity of the right habitat with the deep past; not necessarily in the same region, naturally, because the flora concerned was once much more widespread. But in this particular place, a collection of species was able to endure to the present, and their *bio*graphies all came together.

On the way up Huangshan mountain, a small museum explains something of the geological history of the area. The granites were

---

* The *Magnolia* family also has a distribution reflecting a Gondwanan history. It occurs with *Hamamelis* again in eastern America, where I visited a similar surviving flora to that described from Huangshan, but sadly not in flowering finery (the fall is reckoned the best time to make a call in those parts). A rather similar survival story might be applied there, since the flora is free to move southwards or northwards along the Appalachians as climate fluctuates, or up and down the mountains.

emplaced from hot magma during a long period of the Mesozoic Era. So the mountains were born deep in the earth at about the same time as the long-lived plants that now live on top, their eroded remnants now gracing the surface of the planet. To understand the vast span of time through which the plants have passed, imagine first the uplift of those granites, and then the slow erosion of the huge mass into the fretted relics that we see today. A century of weathering might produce the least roughening of the surface, a millennium a few faint wrinkles. How many millions to sculpt such a tough landscape, to carve valleys, and to define whole ranges? Climate changed repeatedly, mammals evolved, even the world's geography transformed as the slow abrasion of time worked ever onwards. Our old timers snuck around the northern hemisphere in harmony with their appropriate soil, temperature or rainfall. Eventually, they found their way to Huangshan. Then came a last insult: the Pleistocene ice age. In the museum, a monument to Professor Li Siguang (1889–1971) tells how he recognised glacial features in the Huangshan region. Since Professor Li was also a very senior Party member, it was probably unwise to disagree. The remarkable landscape with its perched rocks and grooved cliffs does not resemble the glaciated granites of Europe and North America in any of its details. Recent research is indeed questioning whether ice sheets penetrated this far south at all. Doubtless, there would have been a climatic cooling even if it did not proceed all the way to ice caps. The surviving trees would have been compelled to move down slope to a climatic regime that would have permitted them to continue to grow and reproduce, before moving back to their present altitude. In the last chapter we met ice age beetles that moved the other way, tracking the cold they preferred, eventually to finish up in the Arctic. Only humans can change their clothes if they need to stay abreast of climate.

I term the areas where survivors accumulate following the toing and froing of the tides of history *time havens*. They are a collection of "refugia" for individual species. Through millions of years these species have shuffled and accommodated alongside evolutionarily younger neighbours. Time havens are most easily identified in terrestrial sites where different ecologies can be parcelled up into more or less defined areas. In the marine realm, the planktonic larvae of many species enable dispersal to new locations, and this might be expected to blur maps, and to smear out any neat packaging. The shore I visited in the New Territories of Hong Kong is probably such a time haven. So, too, is the Queensland Plateau, where the pearly nautilus still bobs along over a sea floor carrying stalked crinoids and ancient sponges. The Daintree Forest Reserve has a host of primitive plants all living together. Nearby, there are places where the old clam *Neotrigonia* thrives; as does *Campanile,* a large sea snail and known fossil from Eocene rocks of the Paris Basin, where it is as long as your arm. Recall that the clam *Solemya* has a history going back much, much further, to the Ordovician 475 million years ago, but can still be dug up alive and well in Stradbroke Island. It has somehow survived all five mass extinctions. Such a region is not an Ark that somehow saved everything, but North Queensland *is* a gathering place for life with a long pedigree.

The history of species preserved in time havens can only be mapped out using fossils. The fates of old timers track similar paths through geological time. Ancient relatives of the molluscs *Neotrigonia, Campanile* and *Nautilus* include many European species that lived more or less throughout a vanished, warm sea known to geologists as Tethys. In the Mesozoic and Tertiary, its course approximately tracked the line of the Alps, and ran on through the Middle East and Himalayas. Tethys disappeared when the southerly continents eventually pushed northwards into Europe and Asia after the break-up of the supercontinent Pangaea. This damned seas to oblivion and forced up the present day mountain ranges that wind all across the northern hemisphere from Nice to Katmandu. This is a complex story, about which a great deal is now known. There were actually several successor seas that were consumed in turn, and the details need not concern us. However, it is most interesting that many Tethyan relics finished up in the Far East and Australia. The sponge *Vaceletia* would have felt at home in a warm Tethyan sea. Sea lilies, and even single-celled organisms called alveolinid foraminifera (*Alveolinella*) could join the list. My colleague Brian Rosen has shown that the diverse reef corals that today are found in the biologically richest part of the Indo-West Pacific region north of Australia were present far to the west in the Middle East during the Miocene Period 20 million years ago. There appears to have been a wholesale shift of marine faunas eastwards. Huangshan might be part of the same story on land, since the trees that flourish in the foothills were once widespread across Europe. New Caledonia lies across a tract of the Pacific Ocean opposite the time haven Queensland, at the very edge of this eastern trek. That's where the most primitive flowering plant *Amborella* lives, and the most species of *Araucaria*; it was formerly the centre of pearly nautilus populations and the only place to find *Allonautilus*. It begins to look as if ghosts of the Mesozoic haunt these quarters of the world in particular.

The Director of London's Royal Botanic Gardens at Kew, Stephen Hopper, and his colleagues have recently drawn attention to botanical biodiversity "hot spots." Such habitats are stuffed full with a huge variety of beautiful species of flowering plants, many of them (but not all of them) of comparatively recent origin. Famous examples are the Cape Province in South Africa, and the area of southwest Australia around Perth. Botanists jump up and down with excitement at the chance to visit such cornucopias of biological richness. New species are still discovered in these habitats with almost metronomic regularity. They become, as they should, conservation priorities, even though local property developers are notoriously cavalier about putting their own interests ahead of a special endemic species. As one Australian businessman put it to me: "Who cares about a bunch of bloody weeds?" Given this attitude, we need some criteria to decide which areas deserve to be given urgent legal attention. I think we should add time havens to the list even though they are different from, if not exactly the opposite of, evolutionary hot spots. The protection given to the Galapagos Islands recognises, as did Charles Darwin, that these geologically young islands are a proving ground for evolution in action in the recent past. The species there are mostly twigs at the top of the tree of life. At the other end of the temporal spectrum, time havens include species that tell deep tales of old events. They surely deserve protection, not least as a metaphor for the brevity of human history in the face of true persistence.

Now at last we can return to the question with which this chapter began: whether there are particular properties of the organism itself that may have helped with endurance. Qualities that can be added

to turns of fortune in geography or habitat that have helped an animal or plant clear extinction hurdles over the long history of earth.
Recalling one of those centenarians interviewed by the local newspaper, the answer will assuredly *not* be "eat only vegetables" or else the
reputation of the crocodile is on the line; and the echidna and tuatara
are protein eaters at a different scale. But some of the higher animals
and plants I have written about do share characteristics that may well
have been helpful in surviving bad times. Many of these species seem
to have long life spans. Crocodiles can lie and lie in wait for a good
meal; tuataras take their time; turtles and giant salamanders would
all make the local paper; *Limulus* takes more than a decade to mature,
which is a long span for an arthropod. Echidnas live five times longer than hedgehogs. Lungfish live life in the slow lane. Even molluscs
like *Solemya* and *Nautilus* grow slowly and live a long time. Among
plants, the bizarre *Welwitschia* is a real Methuselah, while *Wollemia*
and *Ginkgo* are in no hurry to complete their life cycles. Longevity
alone can serve an organism well in times of crisis; consider how the
Wollemi pine must have bided its time in its hidden redoubt during
the climax of drought in Australia. Some trees carry a cargo of primitive insects that might also be cushioned through a crisis, hidden in
their long-lived hosts. The kauri pine (*Agathis*) is food plant to one
of the most primitive small moths (*Agathiphaga*) that was probably
given free passage through extinctions at the end of the Cretaceous
along with its host plant. However, there is nothing magic about longevity alone. Large dinosaurs probably also lived a long time and that
did not save them.

Some bigger animals in for the long haul share another feature.
They tend to have relatively large eggs or few offspring. This is a way
of investing more in the survival of individual progeny. The opposite
strategy is to produce enormous numbers of eggs, seeds or spores and
cast them into the sea or into the wind on the chance that one will get
lucky and find just the right conditions to develop into a reproducing
adult. This technique has served ferns and liverworts well for hundreds of millions of years. They are still among the first plants to colonise after volcanic eruptions or floods. They get around, whisked in
upon the air. But lungfish and coelacanths and *Limulus* all lay large,
yolky eggs for their kind. This gives the babies a little grace before

they have to earn their way in the world. The alligator even looks after hatchlings for a while before they are loosed upon the waters. Large ratite birds lay few eggs. Even the echidna is sparing with its small puggles. All these animals take a relatively long time to grow sexually mature, but may well remain fertile for a long time thereafter. If one year fails, the next might succeed. This life strategy has obvious implications for their conservation: it cannot be hurried. Tempting though it is to build up the positive features of this lifestyle for exceptional endurance, it makes for an unsatisfactory theory. There are examples of animals, humans and elephants among them, that have a comparable reproductive system but whose longevity through time is yet to be proven. A close relative of the elephant, the woolly mammoth, was more in thrall to climate change than anything else (human interference possibly had a part to play). To use a hypothetical example, it is rather like discovering that centenarians tend to have blue eyes. It may have some genetic connection, but there are many more people with blue eyes who are not centenarians. Plants with long-lived seeds will be better placed to survive an environmental crisis, but only if there are the right conditions for germination when the worst is over. Curiously enough, the first fossils that are found in the rocks following mass extinction events are often huge numbers of shells belonging to a very few species. In the very short term, it pays to reproduce rapidly. However, these animals soon fade out when a normal marine ecology is re-established. Opportunism is not a durable strategy. Surviving in the marathon of life depends on a whole package, the body and life history an important part of it, but we must never forget about being in the right place at the right time.

The inescapable truth is that luck for old timers will eventually run out. It always does. In the very long term, we are all finally dead, even *Limulus*. Even the velvet worm.

EPILOGUE

It is always good to get back from a long field trip abroad, but I know at once that something is wrong. There is an unpleasant sweetish smell in the air. At first I think that a bird must have died in the house and gone through its cycle of decay, leaving behind a whiff, a memory of decomposition. There is something indefinably nauseating about the pong, something intrusive. I throw open the window and go into the kitchen. As light floods the room, a dense scuttling sound, a scraping, urgent, oddly dry noise alerts me to the truth. Ugh, cockroaches! Even now they are fleeing into the dark crevices between my cupboards and under my sink, and off into a mouse hole in the skirting board (I really must fix it one of these days). These insects shun light. I had no idea that they were infesting my premises. They must have proliferated in my absence. I accept insect visitors like tiny, flightless silverfish (Thysanura) with long bristly tails, as these are harmless little creatures; and anyway they serve as a living memento of a time before insects had wings. They are another old timer. But there is something singularly succulent and juicy about a cockroach that induces shudders, even in a professional arthropod man. Their antennae seem too long, their jointed legs unnecessarily spiky. They are not a good colour. When I attempt to hit it with a shoe, one fat specimen cornered on the floor hisses at me, but that does not save it. Then I open the cupboard door and discover a horrible jar of something heaving with cockroach nymphs of several sizes, wingless and twitching miniature versions of their parents: no maggot or caterpillar stage for these guzzlers. Peering into the cracks behind the sink, I see dozens of pairs of antennae, smelling the air. When I wake up the following morning, I realise a mouse has eaten the dead cockroach. That's ecology for you.

Cockroaches are not related closely to beetles, although they might look a little like them. My distaste for them is rooted in something irrational. After all, they do what they do very well, and they are remarkable old timers. Their relatives go back to the Carboniferous Period, so they have had 300 million years to perfect the art of living on almost anything organic. They have also come through two mighty extinction events, the one at the end of the Permian being the king of them all. "Alpha for persistence," as one of my schoolmasters was fond of saying. Cockroaches crawled on the damp forest floors while trees that would finally make coal grew all around. To most observers, the early roaches look much like the ones still living, but to those with a passion for Blattaria—and every animal group has its devotees—they had a way to go before they could move into my jam pot. They appeared 200 million years before the first moths and butterflies. Only the dragonflies match them in antiquity among winged insects.* Maybe out of respect I should not have squished that large specimen with my shoe. I should have helped it on its way.

Cockroaches are distributed worldwide away from cold latitudes, so there is nothing relict about them; they must have continued to scuttle over Pangaea when life was almost snuffed out 250 million years ago, and subsequently they were carried away to all parts of the world when the great continent broke up. They are numerous in Australia, where I saw really big ones clambering over eucalyptus trunks in the setting sun. I have watched with awe as they scaled bedroom walls in Malaysia. In their six-legged way, they do conform to some of the other survivors. They are long lived and can exist for several years in a tank. Many other insects have grubs that last for a long time buried in wood or living in ponds, but the adult usually only lives briefly after metamorphosis, when it is time to breed. Then, like the crocodile, cockroaches can go for a long time without food; this means something like a month, which is a long time for a small creature. Some species lay relatively few eggs, and one species even gives

* This is true if the Carboniferous dragonflies (Protodonata) are classified with the living ones. Some entomologists do not regard the earliest forms as very closely related to living species, which have a common ancestor in the Permian.

birth to live young, but others are as prolific as any bug on the block. They can endure high doses of radiation. Nobody seems to like them much, but they must earn a grudging respect for their durability and lack of fussiness. So the cockroaches will guzzle on until the last scrap of food is consumed when, it might be supposed, they will turn on one another. For some reason I am reminded of another animal that is too numerous, that seems to guzzle everything immoderately and may finish up turning on his fellows. D. H. Lawrence nailed him thus (although inspired by rabbits rather than roaches):

> *There are too many people on earth*
> *insipid, unsalted, rabbity, endlessly hopping.*
> *They nibble the face of the earth to a desert.*

It has been a privilege to get to know every animal and plant in this book. They were chosen from the millions of species on our planet because they had something to say about evolution. Old timers help to explain distant origins in strange worlds. I have no doubt at all that had I made a different selection of organisms to include in this book their *bio*graphies would have been every bit as interesting. Humans are only one species in the great inventory of the biosphere, but anthropocentrism rules just about everywhere. I have even met fanatics who maintain that wildlife was created as a source of food and entertainment for just one bipedal hominin, who deserves to have total dominion. I do not like to mix science with moral strictures, but it would not be possible to have made the journeys I have made without caring desperately that something as ancient as the lungfish should see another decade. After hundreds of millions of years, that is nothing less than justice. I grieve that the *Nautilus* that has survived the dinosaurs is declining because of a trade in tourist trinkets. Our "endlessly hopping" species is squeezing everything. The extinction event that is happening right now is the first one in history that is the responsibility of a single species. There's no meteorite this time, no exceptional volcanic eruptions, no "Snowball Earth," just us, prospering at the expense of other species. We have not nibbled the face of the earth to a desert yet, but if our human numbers go on growing it looks like a plausible end. Some time soon, it has got to stop. We can do something about it. After all, we are not cockroaches.

I am not worried about the survival of bacteria. They will be there to rot down the last bodies of the last humans, and then the wheel of life will have turned full circle.

GLOSSARY

While I have kept technical terms to a minimum, and attempted to define them where they first appear, it may be helpful to have definitions of some terms that are mentioned more than once, or require a little further explanation.

Ammonite. An extinct type of mollusc that did not survive the end Cretaceous extinction event. Familiar as elegant "rams horn" fossils.

Amoebocyte. A living cell looking somewhat like the independently living, single-celled organism known as an amoeba. These often perform a defensive role, engulfing "germs" in the blood of many invertebrate animals.

Anoxic. Describing a habitat lacking in oxygen.

Arthropods. The great group (Phylum Arthropoda) of invertebrate animals with jointed legs. These include spiders, scorpions, insects, crabs, shrimps and trilobites.

Bacteria. Minute, single-celled organisms reproducing by fission. With the archaea and eukaryotes, they comprise one of the three greatest divisions (domains) of living organisms.

Biofilm. A thin mat made by living microscopic organisms, such as bacteria and simple algae.

Brachiopod. An invertebrate with two valves, but unrelated to the molluscs, which lives by filtering minute edible particles from seawater. Phylum Brachiopoda.

Chelicerates. A group of arthropods including spiders, scorpions and the horseshoe crabs. A typical pair of head appendages (chelicerae) gives the group its name.

Chitons. An ancient group of molluscs having a series of plates covering the upper body.

Chloroplasts. The special bodies lying within the cells of typical photosynthesising organisms where carbon is "fixed" by the action of the green chemical chlorophyll.

Chordates. Animals with a cord running along the "back" housing the nervous system. This includes all the vertebrates with backbones, and groups of more primitive animals that lack bony structures but still survive, such as amphioxus. Phylum Chordata.

Chromosome. Chromosomes carry the genetic information encoded in DNA in our cells. Genes are laid out along the famous double helix of DNA, the blueprint for everything in biology.

Chytrid. A type of single-celled organism allied to fungi and living in moist habitats.

Cladogram. Diagram portraying the relationships between a set of organisms, often in the form of a tree summarising their relative similarities. Cladistic analysis is the general method for scientific classification of the natural world, and was formerly based upon features of morphology—like feathers, or fingers, or the hairs on a fly's legs. Similarities at the molecular level have provided a whole new wealth of data to a new generation of scientists, and classifications have been changed as a result.

Crustacea. The great group of arthropods including crabs, shrimps, woodlice and their relatives.

Cyanobacteria. Also commonly known as "blue-green" bacteria because of their characteristic colour—the result of chlorophyll. The first photosynthesising organisms to use light and carbon dioxide to grow (and emit oxygen).

Dating techniques. Thanks to the fixed rate at which radioactive isotopes of certain elements decay, it is now routine to use this natural chronometer to date rocks. Different radioactive elements have different rates of decay, so there is usually an appropriate candidate to use as a timepiece to date any particular sample. Uranium-lead, Potassium-Argon, Thorium and Carbon "clocks" have transformed our understanding of deep time. The age obtained by these methods is often referred to as the radiometric age.

Echinoderms. Sea urchins, starfish and sea lilies are perhaps the most familiar of these "spiny skinned" marine animals encased within hard plates of calcium carbonate (Phylum Echinodermata).

Endosymbiosis. Literally "living together inside." Referring to the discovery that single-celled organisms embrace and incorporate other small organisms within them. The "prisoners" subsequently prosper inside their hosts to provide new cell functions, working in fresh partnerships to produce more advanced species.

Eukaryotes. Those organisms having an organised cell nucleus (bounded by a membrane) containing genetic information: this includes all "higher" animals and plants.

Extremophile. Micro-organisms that favour living conditions which other animals might find extremely unpleasant—such as very hot water or strong acid.

Gondwana. The ancient supercontinent formed by the welding together of present day South America, Africa, peninsular India, Australia and Antarctica (and more fragments besides). Gondwana split up into the continents we recognise today during the Mesozoic.

Great Oxygenation Event. This occurred about 2.4 billion years ago, when oxygen was present in the atmosphere to a significant extent. It saw the end of the dominance of organisms that favoured anoxia.

Hominin. This term is used as shorthand for "mankind and his close relatives" (it has replaced "hominid" in recent books).

Hyperthermophile. Micro-organisms that thrive in extremely hot (indeed, almost boiling) conditions.

Last Glacial Maximum (LGM). The period of maximum extension of the great ice sheets between 26,500 years ago and approximately 19,000 years ago, during the last (Pleistocene) ice age. Not all ice sheets waxed and waned at exactly the same time.

Late Heavy Bombardment. The period 4.1–3.8 billion years ago when meteorites heavily blasted the earth and evolution of life in the conventional sense was improbable.

Lobopod. General term for the velvet worm and its relatives, with stumpy legs tipped with small claws and a ringed (annulated) body.

Mass extinction events. These are crises in the history of life. The simultaneous disappearance of many species proves that unusual circumstances prevailed. The most important mass extinction events in our narrative are:

1. "Snowball Earth" late in the Precambrian (Marinoan glaciation 610–590 million years ago) when the earth largely froze
2. End Ordovician (444 million years ago), another major glaciation
3. The near-end Devonian event (378 million years ago)
4. The end Permian "great dying" (251 million years ago)
5. End Triassic (201 million years ago)
6. The great Cretaceous-Tertiary (K-T) event, 65 million years ago

We also note the end of the Pleistocene ice age following the LGM 19,000 years ago.

Mitochondrion. An organelle within the eukaryote cell, often described as the cell's "powerhouse." It is the site of cellular respiration, where nutrients are turned into energy. Mitochondrial RNA is one of the molecules that has been sequenced most frequently by molecular biologists in making assessments of the evolutionary relationships between organisms.

Molluscs. Great group of invertebrate animals mostly carrying stout shells, including snails, clams, squid and the pearly nautilus. Phylum Mollusca.

Notochord. The uniting character of the Phylum Chordata, which includes ourselves. A flexible rod-like body (carrying the main nerve trunk) is found in the embryos of higher vertebrates, and is more obviously displayed in the primitive lampreys and lancelets. A chef in Latvia serves lamprey notochords as a gastronomic treat.

Onychophora. A phylum of living lobopods including the velvet worm, *Peripatus*.

Organelles. These are tiny bodies within the eukaryote cell which carry out particular functions in growth and reproduction. They are usually bounded by membranes. The endosymbiont theory contends that the organelles originally had an independent existence, and were "captured" by the larger cell to their mutual benefit.

Pangaea. Massive supercontinent existing about 250 million years ago, when all the present day continents were conjoined. Living plants and animals include some forms that still reflect this ancient geography. Pangaea split up into the continents we recognise today, when oceans slowly opened up between them. They are still drifting apart today.

Phages (or bacteriophage). Viruses that infect the cells of bacteria. They are probably the most numerous organisms on earth.

Plate tectonics. The geologist's explanation of how continents move and how mountain chains are formed by continental collision. The world is divided into a number of tectonic plates which slowly shift and change, providing a fundamental control on where animals and plants might live, and on the climates in which they find themselves.

Prokaryotic organisms. Minute single-celled organisms that reproduce simply by "splitting" and lack an organised cell nucleus. They are divided into archaea (formerly known as archaebacteria) and bacteria, which is one of the deepest divisions in the tree of life. Their simple external appearance (spheres or sausages are common) belies great complexity in their internal chemistry.

Radiometric age. See *Dating techniques.*

Ratites. Group of large flightless birds, generally reckoned to be primitive, including emus, ostriches, rheas and the extinct moa.

Saprobes. Organisms such as fungi that obtain their energy for growth from breaking down organic compounds made by plants—including wood and leaves. They are a vital part of natural recycling.

Snowball Earth. This happened late in the Precambrian when ice caps grew to a huge extent, even as far as the equator. This event must have affected the course of life profoundly.

Strata. Rock layers, especially referring to sedimentary rocks that accumulated beneath seas, rivers or lakes.

Supervolcano. Giant volcanoes throw out huge quantities of ash, such as Yellowstone 150,000 years ago and Mount Toba in Sumatra 74,000 years ago. Such eruptions affect world climate and cause extinctions.

Symbiosis. "Living together": a relationship between two organisms to their mutual benefit. An example is that between fungi on roots and the trees that host them: fungi scavenge nutrients for the tree and are rewarded with sugars for their own growth.

Tetrapods. Four-legged animals: those vertebrates that took to life on land, including the amphibians, reptiles and mammals and their common ancestors. *Homo sapiens* is a tetrapod, despite walking only on two of them.

Thermophiles. "Heat loving," especially as applied to primitive organisms.

Trilobites. An extinct group of arthropods that lived from the Cambrian to the Permian Periods, and rather more glamorous than dinosaurs. The author has spent much of his life studying these animals, so this assessment is, of course, entirely objective.

Vendobionts. This term is used to describe the curious, large organisms that flourished in the Ediacaran Period at the end of the Precambrian.

ILLUSTRATION CREDITS

The author and publishers gratefully acknowledge the following sources for permission to reproduce illustrations:

INTEGRATED ILLUSTRATIONS

Page

9   Trilobite *Calymene blumenbachii*. Photo © Derek J. Siveter.

16   Growth series of horseshoe crab, *Tachypleus tridentatus*. Photo © Kevin Laurie.

21   Scorpion sculpture by Gustav Vigeland, Frogner Park, Oslo. *Photo © Jackie Fortey.*

44   *Aysheaia pedunculata. Photo courtesy of Derek Briggs, figured in* The Fossils of the Burgess Shale, *Briggs, D. E. G., Erwin, D. H., and Collier, F. J., Smithsonian Institution Press, 1994.*

59   Map of Pangaea. © *HarperCollinsPublishers*

81   Endosymbiosis diagram. © *HarperCollinsPublishers*

85   Red alga *Bangia atropurpurea*. Photo © Juliet Brodie.

89   Testate amoebae *Bonniea dacruchares. Photo courtesy Susannah Porter, figured in* Vase-shaped Microfossils from the Neoproterozoic Chuar Group, Grand Canyon: A Classification Guided by Modern Testate Amoebae, *Porter, S., et al.,* Journal of Paleontology, *2003.* Used with permission from the Paleontological Society.

98   The Old Faithful Inn, Yellowstone National Park. *Photo © Jackie Fortey.*

105   Diagram of the bacterial cell. *Based on a diagram in N. A. Campbell and J. B. Reece, 2002,* Biology, *Sixth Edition, Pearson Education Inc. publishing as Benjamin Cummings.*

113 Cyanobacterium *Spirulina. Photo © David J. Patterson.*

117 Three Domains. *© HarperCollinsPublishers*

120 Yellowstone warning sign. *Photo © Jackie Fortey.*

133 Chiton *Ischnoradsia australis,* Bell Pettigrew Museum, St. Andrews, Scotland. *Photo © Jackie Fortey.*

138 *Scaphotrigonia. Figured in* Text Book of Geology, Vol. III. *Archibald Geikie.*

141 Ammonite *Dactylioceras. Photo © Jackie Fortey.*

165 *Ginkgo biloba* bas relief, Brooklyn Botanic Garden, New York. *Photo © Jackie Fortey.*

174 *Metasequoia* fossil and living leaf. *Courtesy Dr. Qin Leng and Wang Li, Nanjing Institute of Geology and Palaeontology, China.*

198 *A Fatal Case of Lampreycitis. From* Humors of History *by A. Moreland, reproduced from originals from* The Morning Leader, *1908.*

227 Tarsier. *Courtesy Ian Tattersall.*

238 Dodo souvenir ashtray from Mauritius. *Photo © Jackie Fortey.*

270 Mountain of buffalo skulls, Michigan Carbon Works, 1892. *Photo: Glenbow Archives NA-2242-2, courtesy Burton Historical Collection, Detroit Public Library.*

PLATES I

1. Horseshoe crabs, Delaware Bay. *Photo © Jackie Fortey.*

2. Trilobite fossils *Dikelokephalina brenchleyi,* Morocco. *Photo © Richard Fortey.*

3. Horseshoe crab moult. *Photo © Jackie Fortey.*

4. *Mesolimulus* fossil with trail, from the Solnhofen Limestone. *Photo © Jørn H. Hurum, Natural History Museum, Oslo, Norway. Specimen housed in Senckenberg Museum, Frankfurt, Germany.*

5. Fossil horseshoe crab from the Jurassic, Germany. *Photo © John Cancalcosi/Alamy.*

6. Giant trilobite *Ogyginus forteyi,* Arouca. *Photo © Jackie Fortey.*

7. Front portion of *Anomalocaris canadenis. With permission of Royal Ontario Museum and Parks Canada © ROM. Photo credit: J. B. Caron.*

8. *Lycopodium,* Karamea, South Island, New Zealand. *Photo © Jackie Fortey.*

9. Podocarps, Paparoa, South Island, New Zealand. *Photo © Rebecca Fortey.*

10. Looking for *Peripatus. Photo © Roger Cooper.*

11. *Peripatus,* Otaki Forks, Tararua Forest Park, Wellington, New Zealand. *Photo © Kellar Autumn.*

12. *Hallucigenia fortis* from the Chengjiang. *Photo © Derek J. Siveter, figured in* The Cambrian Fossils of Chengjiang, China: The Flowering of Early Animal Life *by Hou Xian-Guang, Richard J. Aldridge, Jan Bergström, David J. Siveter, Derek J. Siveter, Feng Xiang-Hong, Blackwell, 2004.*

13. Precambrian fossil, Mistaken Point, Newfoundland. *Photo © Jackie Fortey.*

14. Mistaken Point, Newfoundland. *Photo © Jackie Fortey.*

15. *Charniodiscus arboreus. Photo © Jim Gehling, South Australian Museum.*

16. *Spriggina floudersii. Photo © Jim Gehling, South Australian Museum.*

17. Stromatolites, Shark Bay, Western Australia. *Photo © Jackie Fortey.*

18. Fossil stromatolites, Nanjing Institute of Geology and Palaeontology, Chinese Academy of Sciences. *Photo © Jackie Fortey.*

19. Cyanobacteria *Cephalophytarion* fossil from the Bitter Springs Chert, central Australia and modern example. *Photo © Bill Schopf.*

20. Bitter Springs cyanobacteria chroococcacean colony. *Photo © Bill Schopf.*

21. *Bangiomorpha. Photo © Nick Butterfield.*

22. Red algae *Porphyra umbilicalis,* Sidmouth beach. *Photo © Jackie Fortey.*

23. Crested Pool, Upper Geyser Basin, Yellowstone National Park. *Photo © Jackie Fortey.*

24. Porcelain Basin, Norris Geyser Basin, thermophile runoff. *Photo © Jackie Fortey.*

25. Minerva Terrace, Yellowstone National Park. *Photo © Jackie Fortey.*

26. Iron Spring Creek, Porcelain Basin, Yellowstone National Park. *Photo © Jackie Fortey.*

27. *Zygogonium* detail. *Photo © David Patterson.*

28. Dragon's Mouth, Yellowstone National Park. *Photo © Jackie Fortey.*

29. *Sulfolobus acidicaldarius. Photo © Dr. Terry Beveridge/Visuals Unlimited/Corbis.*

30. Brothers black smoker. *Image courtesy of New Zealand American Submarine Ring of Fire 2007 Exploration, NOAA Vents Program, the Institute of Geological and Nuclear Sciences and NOAA-OE.*

31. Digging for *Lingula,* Hong Kong. *Photo © Jackie Fortey.*

32. *Lingula,* Hong Kong. *Photo © Jackie Fortey.*

33. Painting of *Lingula © Sally Bunker.*

34. *Solemya (Solemyarina) velesiana. Photo © Gonçalo Giribet.*

35. Nautilus living (ROV). *Photo © Justin Marshall, University of Queensland.*

36. Nautilus fossil. *Photo © Robert Francis.*

37. *Maotianoascus octonarius* (Chengjiang). *Photo © Derek J. Siveter, figured in* The Cambrian Fossils of Chengjiang, China: The Flowering of Early Animal Life *by Hou Xian-Guang et al., Blackwell, 2004.*

38. Sponge *Vaceletia,* Osprey Reef, Coral Sea. *Photo © Gert Wörheide, Ludwig-Maximillians-Universitaat, Munich.*

39. Jellyfish, Monterey Bay Aquarium, California. *Photo © Rebecca Fortey.*

40. *Caulophacus arcticus* (Hexactinellida) from the Arctic Sea off Svalbard, 2,400-metre depth. *Photo provided by Friederike Hoffmann, Uni Environment, Bergen, Norway. Picture taken by ROV Victor, R/V L'Atalante, Ifremer, France.*

41. *Huperzia,* Norway. *Photo © Anne Bruton.*

42. Tetrad. *Photo © Charles Wellman.*

43. *Baragwanathia* fossil. *Photo © Chris Berry.*

44. Liverwort, Oxfordshire. *Photo © Jackie Fortey.*

45. *Amborella. Photo courtesy of David H. Lorence, National Tropical Botanical Garden, Hawai'i.*

46. Horsetails. *Photo © Stuart Skeates.*

47. Ginkgo fossil leaf. *Courtesy of Smithsonian Institution. Photo by D. Hurlbert.*
48. Ginkgo, wild specimen, Tianmushan, China. *Photo © Jackie Fortey.*
49. *Araucaria,* Old Government House Garden, Brisbane. *Photo © Jackie Fortey.*
50. *Araucaria,* cones. *Photo © Jackie Fortey.*
51. Cycad, City Botanic Garden, Brisbane. *Photo © Jackie Fortey.*
52. *Welwitschia mirabilis. Photo © James O'Donoghue.*
53. *Branchiostoma belcheri. Photo © Paul Shin.*
54. *Myllokunmingia fengjiaoa. Photo © Derek J. Siveter, figured in* The Cambrian Fossils of Chengjiang, China: The Flowering of Early Animal Life *by Hou Xian-Guang et al., Blackwell, 2004.*
55. Richard with lungfish. *Photo © Leo Fortey.*
56. Coelacanth (*Latimeria*). *Painting © Gordon Howes.*
57. Fossil coelacanth, *Undina,* Messel. *Photo © Per Aas, Natural History Museum, Oslo, Norway. Specimen housed in Natural History Museum, Oslo.*

PLATES II

58. Brook lamprey. *Photo © Jackie Fortey.*
59. Fire salamander, Italy. *Photo © Jackie Fortey.*
60. Hellbender (giant salamander), Center for Hellbender Conservation, St. Louis Zoo. *Photo © Gonçalo M. Rosa.*
61. Tuatara, New Zealand. *Photo © Leo Fortey.*
62. Giant leatherback turtle, Trinidad. *Photo © Tony Rancombe.*
63. Fossil turtle *Archelon. Photo © Natural History Museum, Vienna.*
64. Echidna, Kangaroo Island, South Australia. *Photo © Leo Fortey.*
65. Platypus, Sydney Aquarium. *Photo © Jackie Fortey.*
66. Fossil primate, Ida (*Darwinius masillae*). *Photo © Per Aas, Natural History Museum, Oslo, Norway. Specimen housed in Natural History Museum, Oslo.*
67. Ida X-ray. *Photo © Jörg Habersetzer, Senckenberg Research Institute, Frankfurt, Germany.*

68. Dinobird reconstruction, based on *Deinonychus antirrhopus.* *Photo © Natural History Museum, Vienna.*
69. Moa skeleton, Bell Pettigrew Museum, St. Andrews University, Scotland. *Photo © Jackie Fortey.*
70. Tinamou. *Photo courtesy of Tom Friedel, © Copyright BirdPhotos .com. All rights reserved.*
71. Ferreret, Mallorca. *Photo © Jackie Fortey.*
72. Chillingham cattle. *Photo © Ann and Steve Toon/Alamy.*
73. Musk oxen, Greenland. *Photo © Paul Smith.*
74. Ibex, northern Italy. *Photo © Robi Janavel.*
75. Painting of ibex, Niaux Cave, France. *Photo © Bradshaw Foundation, Geneva.*
76. Bison outside Old Faithful Inn, Yellowstone National Park. *Photo © Jackie Fortey.*
77. Huangshan Mountain view. *Photo © Jackie Fortey.*
78. Chinese mountain painting. *Author's own collection.*
79. Magnolia. *Photo © Jackie Fortey.*
80. Ginkgo bonsai. *Photo © Jackie Fortey.*
81. Fossil cockroach. *Photo © Edward Baker.*
82. Living cockroach. *Photo © Edward Baker.*

QUOTATIONS

Page

169 From *Gingo biloba,* West-östlichen Divan, Johann Wolfgang von Goethe, 1819 (the "k" in *Gingko* was deliberately omitted from the title).
194 From *The Coelacanth,* Ogden Nash. Copyright © 1972 by Ogden Nash. Reproduced by permission of Curtis Brown Ltd.
199 From *To a Young Lady, With Some Lampreys, Poems on Several Occasions,* John Gay, 1720.
214 From *How Doth the Little Crocodile, Alice's Adventures in Wonderland,* Lewis Carroll (Macmillan, 1876).
299 From *There Are Too Many People, Last Poems,* D. H. Lawrence. Copyright © Angelo Ravagli and C. M. Weekley, Executors of the Estate of Friedo Lawrence Ravagli, 1964, 1971. Reproduced

by permission of Pollinger Limited and the Estate of Frieda Lawrence Ravagli.

While every effort has been made to trace the owners of copyright material reproduced herein, the publishers would like to apologise for any omissions and would be pleased to incorporate missing acknowledgements in any future editions.

FURTHER READING

Benton, Michael J. *When Life Nearly Died.* Thames & Hudson, 2008.

Brasier, Martin. *Darwin's Lost World.* Oxford University Press, 2009.

Braune, Wolfram, and M. D. Guiry. *Seaweeds.* Koeltz Books, 2011.

Briggs, Derek E. G., D. H. Erwin, and F. J. Collier. *The Fossils of the Burgess Shale.* Smithsonian Institution Press, 1994.

Carroll, Sean B. *Remarkable Creatures.* Quercus, Houghton Mifflin Harcourt (US), 2009.

Curtis, Gregory. *The Cave Painters: Probing the Mysteries of the World's First Artists.* Knopf, 2006.

Dawkins, Richard. *The Greatest Show on Earth: The Evidence for Evolution.* Bantam, 2009.

Erwin, Douglas H. *Extinction: How Life on Earth Nearly Ended 250 Million Years Ago.* Princeton University Press, 2006.

Fedonkin, M. A., J. G. Gehling, K. Grey, G. M. Narbonne, and P. Vickers-Rich. *The Rise of Animals: Evolution and Diversification in the Animal Kingdom.* Johns Hopkins University Press, 2008.

Flannery, Tim. *The Eternal Frontier.* Heinemann, 2001.

———. *The Future Eaters.* Grove Press, 1994.

Forey, P. L. *Coelacanth, Portrait of a Living Fossil.* Forrest Press, 2009.

Fortey, Richard. *The Earth: An Intimate History.* Harper Perennial, 2004.

———. *Trilobite! Eyewitness to Evolution.* Harper Perennial, 2000.

Gould, Stephen J. *Wonderful Life.* W. W. Norton, 1989.

Hou Xian-guang, R. J. Aldridge, J. Bergström, Derek J. Siveter, David J. Siveter, and Xiang-Hong Feng. *The Cambrian Fossils of Chengjiang, China.* Blackwell, 2004.

Jørgensen, Jordan, and Jean Joss, eds. *The Biology of Lungfishes.* CRC Press, 2011.

Knoll, Andrew H. *Life on a Young Planet: The First Three Billion Years of Evolution on Earth.* Princeton University Press, 2004.

Lengeler, Joseph W., G. Drews, and H. G. Schlegel, eds. *Biology of the Prokaryotes.* Oxford: Wiley-Blackwell, 1999.

Lott, Dale F. *The American Bison: A Natural History.* University of California Press, 2003.

Lutz, Dick. *Tuatara: A Living Fossil.* Salem, OR: DIMI Press, 2005.

Margulis, Lynn. *The Symbiotic Planet.* Basic Books, 1998.

Moran, Robbin C. *A Natural History of Ferns.* Timber Press, 2004.

Morton, Oliver. *Eating the Sun.* Fourth Estate, 2007.

Nield, Ted. *Incoming! Or, Why We Should Stop Worrying and Learn to Love the Meteorite.* Granta Books, 2011.

Parker, Andrew. *In the Blink of an Eye.* Perseus, 2003.

Raff, R. A. *The Shape of Life: Genes, Development, and the Evolution of Animal Form.* University of Chicago Press, 1996.

Rismiller, Peggy. *The Echidna, Australia's Enigma.* Hugh Lauter Levin Associates, 1999.

Saunders, W. Bruce, and N. H. Landman, "Nautilus: The Biology and Paleobiology of a Living Fossil." *Topics in Geobiology 6.* Springer, 2010.

Schopf, William J. *The Cradle of Life: The Discovery of Earth's Earliest Fossils.* Princeton University Press, 2001.

Sheehan, Kathy B., D. J. Patterson, B. L. Dicks, and J. M. Henson. *Seen and Unseen: Discovering the Microbes of Yellowstone.* Falcon Publishers, Globe Pequot Press, 2005.

Shipman, Pat. *Taking Wing.* Weidenfeld & Nicolson, Simon & Schuster (US), 1998.

Shubin, Neil. *Your Inner Fish.* Pantheon Books, 2008.

Shuster, Carl N., R. B. Barlow, and H. J. Brockmann, eds. *The American Horseshoe Crab.* Harvard University Press, 2003.

Taylor, Thomas N., E. L. Taylor, and M. Krings. *Paleobotany: The Biology and Evolution of Fossil Plants.* Prentice Hall, 2009.

Thompson, Keith. *The Legacy of the Mastodon.* Yale University Press, 2008.

Tudge, Colin. *Consider the Birds.* Allen Lane, 2008.

Tudge, Colin, and J. Young. *The Link: Uncovering Our Earliest Ancestor.* Little Brown, 2009.

Valentine, James W. *On the Origin of Phyla.* University of Chicago Press, 2004.

Vermeij, Geerat J. *A Natural History of Shells.* Princeton University Press, 1993.

Walker, Gabrielle. *Snowball Earth.* Bloomsbury, 2003.

Ward, Peter D. *The Call of Distant Mammoths.* Wiley-Blackwell, 1998.

———. *Rivers in Time: The Search for Clues to Earth's Mass Extinctions.* Columbia University Press, 2000.

Weinberg, Samantha. *A Fish Caught in Time: The Search for the Coelacanth.* Fourth Estate, 1999.

White, Mary E. *The Greening of Gondwana.* Reed Publishing, 1986.

Willis, K. J., and J. C. McElwain. *The Evolution of Plants.* Oxford University Press, 2002.

Woodford, James. *The Wollemi Pine.* Melbourne: Text Publishing, 2005.

INDEX

aboriginal people, 266–7
acanthodian "sharks," 203
Acanthostega, 206
Actinopterygii, 191
Adenosine triphosphate (ATP), 107
Adrover, R., 246
Africa, 32, 34, 58, 59, 78, 171, 185, 192, 214, 226,
    240, 255 and footnote, 278, 282, 288
agnathans *see* jawless fish
Ahlberg, Professor Per, 188
Alaska, 263
algae, 58, 83, 91, 112, 247, 301
    see also cyanobacteria (blue greens or
      blue-green algae)
alligators, 213–14
Allonautilus scrobiculatus, 142n, 293
Allosaurus, 240
Altamira caves, 259
alveolinid foraminifera (*Alveolinella*), 292
Alvin (deep-sea submersible), 116
amber, 42, 93
Amborella trichopoda, 182–3, 293
America, 206, 263–4, 266, 282
    Appalachian Mountains, 200, 290n
    Black Sand Basin, Yellowstone, 271
    Bronx Zoo, 270
    Cherry Creek Range, Nevada, 215
    Cincinnati, 130
    Delaware Bay, 3–7, 8–9, 14–15, 16–17,
      26, 30
    Grand Canyon, 89, 94, 98
    Grand Teton National Park, 96–7
    Great Lakes, 199
    Jackson Lake, 96–7
    John D. Rockefeller highway, 97
    Lamar Valley, Yellowstone, 271
    Maine, 15
    Mount St. Helens, 103, 161
    Nevada desert, 97
    New Jersey, 14
    New York, 168
    Ohio River, 256
    Oregon, 264
    Rocky Mountains, 96–7
    Washington State, 146
    Yellowstone National Park, 97–106, 109,
      110–18, 120–3, 270–1, 272
American bison (*Bison bison*), 268–72
American Bison Society, 270
American cave lion, 265
American Fur Company, 269
American mastodons, 265
ammonite, 141, 142–3, 281, 301
amoebae, 112
amoebocyte, 13, 301
amphibians, 204, 205–6, 254, 279
    see also frogs; midwife toad; toads
amphioxus *see* lancelet (*Branchiostoma*)
anaerobic, 106
anapsids, 209n
Anderson, Mike, 52
angel's hair, 113
Angiosperm phylogeny website, 182
angiosperms, 175, 179, 180–5
annelid worms, 57, 130
Anomalocaris (raptorial arthropod), 27
Anomia, 129
anoxic, 30, 136, 278, 301
ant, 272
Antarctic, Antarctica, 257n, 283
Anti-Atlas Mountains, Morocco, 278
antibiotic resistance, 119
apes, 226
Apollo Mission, 71
Araucaria, 174–6, 177, 182, 293
    A. araucana, 175
    A. cunninghamii, 175
    A. heterophylla, 175
archaea, 101, 102, 104, 107–9 and footnotes,
    111, 115–18, 122, 272

Archaean Era, 71, 75–6, 77, 79, 81, 94, 109
   and footnote, 118, 284
Archaefructus, 181
Archaeogolfingia, 128
Archaeopteryx, 239–40, 241
Archaeopteryx lithographica, 17
Archelon, 212, 213
Arctic, 8, 72–3, 166, 230, 291
Arctic Canada, 84, 195, 258
Arctic Circle, 255–6
Arctic fox, 257
Arctic Ocean, 149
Arctic poppy, 256
Arctic Russia, 49
   Kola Peninsula, 263
Arctic terns, 74, 256
Arkarua, 57
arsenic, 103
arthropods, 7, 9, 24–5, 26, 45–6, 48, 55, 57,
   101, 277, 294, 301
arum lilies, 232
arum (*Philodendron*), 232
ash (*Fraxinus*), 166
Ashmole, Elias, 238
Asia, 15, 227, 251, 253, 257, 260, 268, 283,
   292
asp, 215
Attenborough, Sir David, 228
auroch (*Bos primigenius*), 237, 259, 260–2,
   265
Australia, 61–5, 137, 149, 161, 171, 175–6, 193,
   206, 215, 254, 282, 283, 305
   Apex Chert, 75 and footnote, 114
   Blackall Mountains, 186
   Brisbane, 186
   Burnett River, 187
   Daintree Forest Reserve, 292
   Dirk Hartog Island, 63, 66
   Ediacara Hills, Flinders Range, 48–9, 54,
      56–7
   Fauré Island, 66
   Flagpole Landing, 66
   Francois Peron National Park, 63
   Fremantle, 66
   Glasshouse Mountains, 186
   Great Barrier Reef, 146
   Hamelin Pool, 64, 66–9, 76
   Kangaroo Island, 216–22
   Maleny, 186
   Manly Beach, Sydney, 175
   Mary River, 186–9, 190
   Monkey Mia beach resort, 63
   New South Wales, 176

   North Stradbroke Island, Moreton Bay
      (Queensland), 134–5, 285, 292
   Northern Territories, 213
   Osprey Reef, 139, 146
   Pearler Inn, Denham, 66
   Perth, 293
   Queensland, 186–90
   Queensland Plateau, 139, 143, 292
   Shark Bay, 61–71, 74, 94, 100–1, 283
   Sydney Aquarium, 222
   Traveston Crossing Dam (Mary River,
      Queensland), 187–9
   Useless Loop, 66
   Warrawong Sanctuary, Adelaide, 221–2
   Yabba Creek (off the Mary River,
      Queensland), 188
Aysheaia pedunculata, 43–5

bacteria, 13, 101, 103, 105, 106, 107, 108n,
   110–11, 112, 113, 114, 116–18, 122–3, 136,
   140, 145, 148, 159, 172, 247, 272, 276, 284,
   285, 286, 300, 301, 303
bacteriophages (or *phages*), 118
Bahamas, Exuma Island, 77
Balearic Islands, 250, 253
Balearic lizards, 250
bamboo, 165, 232
Bang, Fred, 13
Bangia (red alga), 84, 91
Bangiomorpha, 91
Bangiomorpha pubescens, 84
Banksia, 217
Baragwanathia (fossil herb), 156
Barghoorn, Elso Sterrenberg Jr., 78, 79
Barlow, Robert, 26
basalt, 279, 281
bats, 184, 225, 228, 230, 232
Bauhinia blakeana, 125
beard worms (formerly Pogonophora),
   130n
bears, 121–3, 265
beaver, 257
beech tree (*Fagus engleriana*), 289
bees, 183, 184, 281
beetles, 183, 247, 251, 262–3, 291
bellflower, 232
bennettites, 172
Bering Strait, 263
Beringia, 263, 268
Berkeley, Miles Joseph, 34
Bilateria animals, 57, 144, 152, 152n
bilby, 62
biofilm, 68, 77, 78, 110, 301

biophilia, 252
birds, 34–6, 230–42, 251, 253, 256, 281, 288
  see also Archaeopteryx; dodo;
  enantiornithines; flightless birds;
  Hesperornis; passerines; ratites;
  songbirds; tinamou
Birds of Ecuador (Ridgley & Greenfield), 234
Birmingham University, 262
bison, 121, 123, 259, 260, 268–72
Black smokers, 115–17
black water beetles, 247
Blattaria, 297–9
blind snakes (Scolecophida), 215
"blue greens" *see* cyanobacteria
blue-footed boobies, 211
Blumhardt, Jonathan, 207
bodhi tree, 167
Boletus, 155
bone, 201 and footnote
bone wars, 205
bonsai, 166, 168
bony fish (teleost), 199
Borneo, 214, 226
box jellyfish (*Chironex*), 148–9
brachiopod, 55, 127, 128, 130–1, 144, 202, 277,
  279, 280, 284, 301
bracken, 161
Branchiostoma, 201–2
Brasier, Martin, 54, 75n, 114, 278
  Darwin's Lost World, 278
brine shrimp (*Artemia*), 286–7, 287n
bristle worms, 130
Bristol University, 239
British toad (*Bufo bufo*), 248
bromeliads, 232
brook lampreys (*Lampetra planeri*), 197
brown seaweed, 90
Bruton, Anne, 155
Budd, Dr. Graham, 45
Buen Formation, 45
buffalo, 99
Buffon, Comte de, 72
Burgess Shale, 43–5, 129n, 130
Burke, Robert O'Hara, 161
Burroughs, Edgar Rice, *The Land That
  Time Forgot*, 283
bush wren, 242
Butterfield, Nicholas, 84
butterflies, 183, 184, 281

cacti, 185, 288
calcium carbonate, 112
Calvin cycle, 82

Cambrian Period, 23–5, 27, 42–3, 44–6, 55,
  57, 59, 59n, 70–1, 73, 127, 129n, 130, 132,
  134, 141, 144, 146, 147, 149, 150, 151, 152,
  201, 277–8, 284–5
Cambrian trilobite (*Paradoxides*), 10
Cambridge University, 72, 129n, 181
Cambrosipuncula, 128
camel, 66, 265
Campanile (sea snail), 292
Canada, 72, 184
  Athabasca "tar sands," 50
  Bay of Fundy, Nova Scotia, 210
  Burgess Shale, British Columbia, 43, 44n,
    129n, 130
  Gunflint Chert, 78
  Lake Superior, 78
Canadia spinosa, 130
Canadian Shield, 75, 78
carbon, 115
carbon dioxide, 82, 104, 106, 136, 157, 279
carbon fixation, 82
Carboniferous Period, 22, 42, 93, 155–6, 170,
  210, 231, 279, 288, 298n
Carcharodon megalodon, 204
Carcinocorpius rotundicaudatus (Asian
  horseshoe crab), 16
Caribbean islands, 268
carnivores, 225
carob trees (or algorrobas), 244, 245
Carroll, Lewis, 214
cassowary, 235, 240
cats, 222, 253
cave art, 258–60, 261
cave bear (*Ursus spelaeus*), 257, 274
Cecropia tree, 231, 232
cellulose, 122
Cephalaspis lyellii, 201
Cephalochordata, 202n
cephalopods, 138–42
Ceratodus, 190
Cerithidea (towered snails), 129
chanterelles, 155
Charniodiscus, 70, 152, 278
Charniodiscus masoni, 54
Chase, Mark W., 182
Chatwin, Bruce, *The Songlines*, 267
chelicerates, 301
chemistry of life, 107–10
chemoautotrophic symbionts, 136
cherry trees (*Exocarpos cupressiformis*), 216
chert, 74, 77, 78
Chile, 175
Chillingham herd, 261–2

chimpanzee, 282

China, 24, 124–5, 164–6, 181, 183, 206, 214, 239, 279
  Beijing, 140, 168
  Chengjiang fauna, 45, 48, 128, 147, 149, 150, 202, 277, 284
  Four Sides Peak, 166
  Guizhou, 212
  Hangzhou, 164
  Huangshan (Yellow Mountain), Anhui Province, 288–91, 293
  Hubei Province, 172–3
  Hunan Province, 173
  Kowloon, 125
  Nanjing, 149
  Sai Kung East Country Park, 125–6, 128
  Shandong Province, 169
  Tianmushan Forest Reserve, Zhejiang Province, 165–8

chitons, 133–4, 144, 302

chlorophyll, 92

chloroplasts, 79, 80, 82, 87, 88, 90, 302

choanoflagellates, 145

Choia xiaolantianensis, 147

chordates, 201–2, 277, 285, 302

Christmas bush, 216

chromosome, 302

chrysanthemum, 185

chytrid, 93–4, 249, 254, 302

City University, Hong Kong, 124

cladogram, 229, 302

clams, 66, 95, 116, 125, 127, 128, 129, 132, 136–8, 285, 286, 292

classification, 29n, 41n, 46, 52–3, 71–2, 88, 107, 127, 130 and footnote, 146, 182, 190–1, 205, 227–9, 235–6

climate change, 18, 173, 243, 256, 257–8, 276–7, 278–9, 281–2, 287–8, 291

climbing fern (*Diplopterygium bancrofti*), 232

cloud forest, 230–4, 237

Cloudina, 49n

Clovis culture, 263–6

Cnidaria, 149–50

coal, 17, 22, 42, 155–6, 210, 280

cockroaches, 297–9

coelacanth (*Latimeria chalumnae*), 19, 20, 191, 191–4, 276, 294

cold-blooded creatures (poikilothermic), 205

Colter, Mary, 98

comb jellies, 149, 151, 152n

Comoro Islands, 191, 193

computer methods, 25, 47–8, 229

conch (*Busycon*) (giant marine snail or whelk), 14

Conchocelis plants, 91

cones (*strobili*), 160, 171, 175, 176

conifers, 33, 101, 111, 112, 165, 166–7, 170, 172–5, 176, 179, 184, 272, 290

conjugation, 119–20

Conophyton ("cone plant"), 76–7

Cook, Captain James, 63–4

Coope, Russell, 262–3

Cooper, Alan, 226, 265n

Cope, John, 136–7

coprolites (human excrement), 264

Coprosma, 207

Coral Sea, 139

corals, 150–1, 279, 292

Cordaites, 170

Cortinarius, 155

Courtney-Latimer, Margaret, 192–3

crabs, 7–8, 10, 41n, 116, 125, 135

crested bulbuls, 125

Cretaceous Period, 142–3, 146, 156, 172, 173, 180, 181, 182–3, 192, 197, 211, 212, 214, 215, 226, 240, 281, 285, 290, 294, 304

crinoids *see* sea lilies

crocodiles, 209, 213–14, 276, 294

Crocodilia, 215

Croton, 232

Crustacea, 7, 10, 41n, 116, 149, 206, 256, 286, 302

Cryptomeria, 165, 167

Cryptozoon ("hidden life"), 72, 74

Cunninghamia, 167, 173, 290

Cupressus, 112

Cyanidium, 111

cyanobacteria (blue greens or blue-green algae), 66–70, 75, 77, 78, 82, 87, 112, 113, 114, 276, 284, 302

cycads, 170–2

cypress, 290

Cyttaria darwinii (fungus), 34

Dampier, William, 64

Dancing with Wolves (film), 266, 267, 269

Darwin, Charles, 34, 72, 77, 87, 130, 147, 159, 184, 201, 235, 251, 283, 293
  Origin of Species, 185, 274

Darwinius masillae (aka "Ida"), 227–9

dating techniques, 71, 302

Dawkins, Richard, 184, 275
  The Greatest Show on Earth, 83, 252

death cap (*Amanita phalloides*), 155

degrader communities, 159–60
Delaware Bay, 3–4, 8, 15, 16–17, 26, 30
    Kitt's Hummock, 4, 9
    Leipsic harbour, 5–6
    Little Creek Inn, 3
    Pickering Beach, 9
    Port Mahon, 6
    Sambo's restaurant, 5–6
Devonian Period, 22, 93, 131, 142, 161, 183,
    195, 200, 205, 279, 285
Diamond, Jared, 265
    Collapse, 267
diapsids, 209n
diatoms (algae), 70, 112
Dickinsonia, 57
dicots, 181
Dilwynites, 176–7
dinobirds, 239, 275
dinoflagellates, 58n, 150
dinosaurs, 15, 18, 42, 97, 142, 158, 161, 162,
    169, 170, 172, 173, 177, 179, 181, 191, 204,
    209, 211, 214, 223, 225, 239–40, 242, 254,
    280, 294
Dipteris, 162
dire wolf, 257, 265
Discoglossidae (aka Alytidae), 253–4
DNA, 17, 47, 80, 87, 91, 92, 94, 110n, 118–19,
    146, 179n, 180, 182, 202, 209, 238, 264
    and footnote, 272
dodo, 237–9, 261, 266
dogwood (*Cornus*), 166
dolphins, 211
dormice, 250
dowitchers, 8
dragonflies, 210, 298 and footnote
Dryas octopetala (mountain avens), 265n
duck-billed platypus (*Ornithorhynchus*),
    219, 221–3
dugongs, 63
Dunstan, Andy, 139–40
dwarf palms, 247

earthworm, 130
East London Museum, South Africa, 192
Easter Island, 266
Ecdysozoa, 129n
echidna (spiny anteater), 218–21, 222–3, 224,
    294, 295
echinoderms, 55, 57, 86, 303
ecosystems, 101, 112, 251
Ecuador, 160, 283
    Mindo cloud forest, 231–4
    Tandayapa Valley, 231

Ediacaran Period, 48–9, 49n, 52, 54–7, 70,
    77, 147, 152, 277, 278
eels, 199
elephant birds (*Aepyornis*), 237, 240
elephants, 230, 295
Elizabeth II, 199
elk, 97, 122
Elongation Factor gene, 107
emu, 235, 236, 240
enantiornithines, 241
Encephalartos woodii, 171
endosymbiont theory, 80
endosymbiosis, 88, 119–20, 122n, 303
endotoxins, 13
Engels, Friedrich, 280
England
    Boxford, Berkshire, 198
    Charnwood Forest, Leicestershire, 54
    Cornwall, 289
    Gloucester, 199
    Northumberland, 261
    Portland, 137
    River Lambourn, Berkshire, 197, 200
    Sidmouth, Devon, 85
    Suffolk, 134
    Worcestershire, 263
Enkianthus chinensis, 289
Entophysalis (stromatolite), 69
Environment Agency (UK), 198
Eoactinistia foreyi (fossil coelacanth), 193
Eocene Period, 292
Eosphaeria, 78
Ephedra (Mormon tea), 179n
ephydrid fly, 101
epiphytes, 160, 161, 231, 232
Erica carnea, 245
eucalypts, 27–8, 61, 216–17, 222
Eucalyptus, 135, 218
Euglena, 111–12
eukaryotes, 68, 80, 84 and footnote, 87–8,
    91, 108–9, 112, 303
Euphorbia, 185
European bison (aka wisent), 260, 265
extremophile, 102, 104, 110, 115, 276, 303

Family Kalotermitidae, 41
feather stars, 55
feathered dinosaurs, 239, 241–2
Fedonkin, Mikhail, 152
Feduccia, Professor Alan, 236
Felis domestica, 222, 253
ferns, 160–3, 231, 232, 294
ferreret *see* midwife toad

fig tree (*Ficus*), 167
Fijian longhorn beetle, 251
filmy ferns, 162–3, 233
finches, 251
fish, 186–204, 286
    see also coelacanth; lamprey; lungfish;
    sharks
Flannery, Tim, 265
flatworms, 57, 130
flightless birds, 235–8, 251, 253
Flinders, Matthew, 217
Flores (Indonesian island), 255, 257
flowering plants, 179–85, 232, 233, 256, 272,
    281, 283, 290, 292, 293
flying reptiles (pterosaurs), 209
Forey, Peter, 193
Fortey, Richard
    Dry Store Room No.1, 35n
    Life: An Unauthorised Biography, 83, 281n
    Trilobite! Eyewitness to Evolution, 25n
Fractofusus, 70
Fractofusus misrai, 53
Fragum hamelini (clam), 66, 95
freshwater snails, 188
Friesians, 261
frogs, 205, 249, 254
fruit fly, 251
Fuchsia, 232
fungi, 34–5, 58, 91–4, 122, 155, 159, 272

galahs (parrots), 61
Galapagos Islands, 203, 250, 293
Galionella, 111
Ganges gharial (*Gavialis gangeticus*), 214
gannets, 240
Gauvry, Glenn, 3, 10, 12
Gay, John, *To a Young Lady, with some
    Lampreys*, 199
gene therapy, 118n
genetic engineering, 118n
Germany, 228, 239, 261
    Messel fossil site, 227–8
geysers, 99–108
gharials, 214
giant antpitta, 234
giant capybara, 265
giant salamander (*Andrias*), 205–6, 294
Gibbs, George, 36–8
Ginkgo, 164–71, 171, 183, 290, 294
Ginkgo biloba, 169–70
gnathostomes (jaws), 195
Gnetophyta, 179 and footnote, 180
Gnetum, 170, 179, 216

gobies, 129
Goethe, J. W. von, 170
Golfingia, 128
Gondwana, 32, 33, 34, 35, 41, 58, 175, 176, 177,
    209, 225n, 240, 242, 279, 288, 290n, 303
Gonzalgunia, 233
gram-negative bacteria, 13
granite, 289, 291
Grant, Peter, 251
Grant, Rosemary, 251
graptolites, 127, 143
grass trees (*Xanthorrhoea*) (aka black boys
    or yaccas), 217
grasses, 181, 184–5, 271
Great Oxygenation Event (GOE), 83, 106,
    276–7, 303
greater black-backed gulls, 6
green alga (Chlorophyta), 88, 90, 91n
Greenfield, Paul J., 234
Greenland, 24, 45, 46, 115, 173, 195, 205, 258
    Thule Airport, 258
Greville, Robert Kaye, 85–6
ground sloths, 265
Grypania, 84
guinea fowl, 235
gums, 61
Gunflintia, 78
gymnosperms, 170, 175, 179, 180, 184
gypsum, 64

hagfish (*Myxine*), 200, 287
Hall, James, 72
Hallucigenia, 44
Hamamelis, 290 and footnote
Harris, W. F. (palynologist), 176
Hartline, Haldan K., 26
Hartog, Dirk, 63–4
hartstongue fern, 162
Hawai'i, 161, 182–3, 211, 251
Hayden survey (1871), 114
hazel (*Corylus*), 289
heath, 232, 245
heathers, 232
Hebe, 207
Heck, Heinz, 261
Heck, Lutz, 261
hedgehog fungi, 155
hedgehogs, 219
Hedley, Ron, 89–90
Helenodora, 42
hellbinder (*Cryptobranchus*), 206
hemichordates, 277
Henry I, 198–9

hepatics, 158
herbivores, 67, 122, 172, 223, 226, 264, 281
herbs, 154–9, 160, 164
Herefords, 261
hermit crabs, 135
herons, 135
Hesperornis, 240
heterotroph, 88–90
Himalayas, 58, 252–3, 283, 292
Hirase, Sakugoro, 170
holly oaks, 245
Homer, 29n
hominin, 254, 299, 303
Homo erectus, 255 and footnote
Homo floresiensis, 254–5
Homo sapiens, 225, 226, 260, 263–8, 282
honeycreepers, 251
honeyeaters, 217–18
Hong Kong, 124, 284, 286, 292
Hooper, John, 134, 135, 144–8
Hopper, Stephen, 293
hornbeam (*Carpinus*), 166
horse, 259, 260, 265
horsehair worms, 130
horseshoe crab (*Limulus polyphemus*), 40,
    41n, 42, 55, 64, 111, 121, 123, 210, 234,
    275–6, 279
  ancient relatives of, 9
  Asian species, 15–16
  blue blood, 12–13
  description of, 6–8
  eating, 15–16
  emergence of, 3–5
  evolutionary history, 16–20, 22–7
  life cycle, 11–12
  link with trilobites, 25–7, 29–31
  population of, 8
  resilience and toughness, 7, 11–12
  threats to, 14–15
horsetails, 163–4, 210
hot spots, 293
hot-blooded creatures, 205
Hughes, Norman, 181
Human Genome Project, 109
hummingbirds, 184, 232, 233, 237, 283
Huperzia, 155–8, 160, 210
Hurum, Jørn, 228, 229
Huxley, Thomas Henry, 235
hydrogen, 110, 116
hydrogen sulphide, 104, 122, 135
Hydrogenobaculum, 111
Hylonomus, 210
hyperthermophile, 107, 109, 111, 303

Iberian Peninsula, 253, 260
ibex, 260–1, 265, 267
ichthyosaurs, 209, 228
Ichthyostega, 205
Ida *see Darwinius masillae*
Illustrated London News, 192
India, 32, 84, 87, 196, 253
  Assam, 41
  Deccan traps, 281
Indochina, 214
Indonesian archipelago, 254, 257
infaunal animals, 56
insects, 7, 161, 179–80, 184, 208, 224, 228,
    242, 253, 283, 294, 297
invertebrates, 127–8, 131–2, 152, 277, 284
iodine, 87
iron, 83, 87, 110, 116
Italy, Solfatara, near Naples, 109
Ivar the Boneless, 201n

Jablonski, David, 146
Japan, 15–16, 169, 173, 173, 185
  Hiroshima, 169
  Tokyo Botanical Garden, 170
Java, 255
jawed fish, 202–4
jawless fish (Agnatha), 197–203, 279, 287–8
jellies, 148–53, 277
Jersey Zoo, 249
Johnson, Samuel, *Lives of the Poets*, 199
Joss, Jean, 189, 190, 203
Jurassic Period, 17, 137, 140, 143, 146, 172,
    239, 254, 281

kakapo (ground parrot), 35
kangaroos, 62, 214, 217, 225, 283
kauri pine (*Agathis*), 176, 177, 294
kea (mountain parrot), 35
Kelwasser Event, 279
Kerygmachela, 45, 46
Kew Gardens, 162, 169, 178, 182, 293
killdeer, 101
Kimberella, 57
Kind, Peter, 186, 187, 189
king brown snake, 215
kiwi, 35, 235, 236
Knoll, Andrew, 74, 79, 80, 83, 84
  Life on a Young Planet, 74
koala, 225
Korarchaeota, 111
Korea, 169
Koyaanisqatsi (film), 266
Kyoto University, 173

La grotte de Niaux, 259
Labandeira, Conrad, 179
labyrinthodonts, 206
Lactarius, 155
Lamarck, Jean-Baptiste, 136
lamprey, 195, 197–200, 203, 287
lancelet (*Branchiostoma*) (aka amphioxus), 201–2
language, 87, 195–9
Lascaux, 259
Last Glacial Maximum (LGM), 256, 309
Late Heavy Bombardment, 115, 303
Latimeria, 192
laver, 86–7, 90
Lawrence, D. H., 299
Lazarus taxa, 146
leatherback turtle, 212
lemurs, 226, 227, 228, 283
Lepidodendron (fossil tree), 156
Levin, Jack, 13
LGM *see* Last Glacial Maximum
lianas, 232
lichens, 58, 150, 255, 271
lilies, 181, 183, 185
Limulus amoebocyte lysate (LAL), 13, 294, 295
Lingula anatina, 126–9, 131 and footnote, 201, 284
Lingulella, 131n
Linnaeus, Carl, 182
Lissamphibia, 205
Lithocarpos, 166
Lithuania, Vilnius, 196–7
liverwort, 158 and footnote, 231, 233, 294
living fossils, 18–20, 97, 127, 146, 172–3, 191–4, 244, 245–50, 254, 275–6, 290
lizards, 205, 208, 210, 215, 250
lobopod, 45–9, 59 and footnote, 303
lobsters, 7, 10
loggerhead turtles, 11
long-beaked echidna (*Zaglossus*), 219, 221
long-horned bison, 265
lorises, 226
lungfish, 186–91, 194, 201, 203, 204, 206, 276, 288, 294
lycopod, 155, 160
lycopodiophyte, 155
Lycopodium (herb), 36, 156
Lyell, Charles, 201

Ma, Jinshuang, 173n
McGill, David, *Island of Secrets*, 206
Macquarie University, Sydney, 190

Madagascar, 226, 237, 240, 283, 288
Magnolia, 183, 290 and footnote
Magnolia sieboldiana, 183
Malay Peninsula, 132, 162, 214
Malaysia, 132, 171, 298
mallee trees, 64, 217
Mallorca, 251–2
    Las Palmas, 243–4
    Lluch Monastery, 245
    Muleta cave, 246
    Sierra de Tramontana, 244–8, 250
Maloof, Adam, 147
Mammalia, 220
mammals, 222–30, 257, 262, 263, 282, 283, 288, 291
mammoths, 257, 264–5, 288, 295
Mandelbrot, Dr. Benoit, 53
Maori, 266
Maotianoascus octonarius, 150
Maotianshan Formation, 45
Marchantiophyta, 158
Margulis, Lynn, 79, 81, 88
Marine Biological Laboratory (Woods Hole), 13
Marinoan glaciation, 276
marsupials, 62, 225 and footnote
mass extinction events, 9, 17, 18, 22, 42, 83, 131, 161, 169, 175, 181, 205, 211, 212, 240, 276–83, 287, 295, 298, 304
Massachusetts, 14
mastodons, 257, 265
Matheson, Major Dugald, 207
Matonia, 162
medusae, 150
Medusinites, 152
Megazostrodon, 223
Melastoma, 232
Memorial University, St. John's (Newfoundland), 50, 52
Mesolimulus walchi (ancient horseshoe crab), 17
Mesoproterozoic Period, 71, 83
Mesozoic Period, 131, 137, 142, 159, 171, 172, 175, 179, 225, 280, 290, 292
Metasequoia glyptostroboides (dawn redwood), 173–4, 290
methane, 104, 121
methane hydrates (methane "ice"), 282
microbes, 113, 120, 122, 272, 280, 284
microbialites, 74
Microcoleus, 70
Mid-Atlantic Ridge, 116
Middle East, 261, 292

midwife toad (*Alytes muletensis*) (aka ferreret), 245, 247–50, 253, 255, 276
Miki, Professor Shigero, 173
millipedes, 7
Miocene Period, 184, 292
Misra, S. B., 51–2
mites, 159, 288
mitochondria, 80, 87
mitochondrion, 304
moa, 35, 237, 240, 266
mollusc, 55, 57, 131–44, 202, 277, 280, 281, 286, 292, 294, 304
molybdenum, 84
Mongolia, 263
mongoose, 251
monkey puzzle tree (*A. araucana*), 174–6, 288
monkeys, 226
monocots, 181, 183
Monoplacophora, 132
monotremes, 219, 223
moon rocks, 71
moose, 96
Moran, Robbin, *A Natural History of Ferns*, 162
Moreton Bay pine (*A. cunninghamii*), 175
Morris, Simon Conway, 44, 129n
Morton, Oliver, *Eating the Sun*, 82
moss, 101, 159, 162, 255, 272
moth, 184, 232
moth (*Agathiphaga*), 294
Mount Toba volcano (Sumatra), 268
mountain ibex (*Capra ibex*) (aka *le bouquetin*), 260
mudskippers, 194
mushrooms, 92–3, 246
musk ox (*Ovibos moschatus*), 258, 262, 265, 268
mycelium, 92, 93
mycorrhiza, 159
Myllokunmingia fengjiaoa, 202
Myotragus, 246, 250, 252

Namibia, 78, 178
Narbonne, Guy, 52
nardoo (*Marsilea drummondii*), 161
Nash, Ogden, 191, 194
Native Americans, 267, 268–9
Natural History Museum, 9, 71, 89, 132, 192, 193, 225, 237, 241
Nature, 52, 195, 229
nautiloid, 138–42
Nautilus, 139–44, 151, 292, 294, 299

Nautilus pompilius, 139
Nazis, 207
nematodes, 159
Neoceratodus forsteri (lungfish), 187–91, 193
Neopilina, 132
Neoproterozoic Period, 71, 88–9
Neotrigonia margaritacea, 137, 292
New Caledonia, 139, 142n, 175, 182
New Guinea, 162, 219, 221
New Zealand, 32–7, 131, 160, 162, 177, 206, 237, 240, 242, 252, 253, 266
   Akatara Ridge, 36–7
   Day's Bay, Wellington, 206
   Honeycomb Hill, 35
   Karamea, 34, 35–6
   Somes Island (*Matiu* in Maori), 206–10, 254
   Wellington, 36
Newfoundland, 50–60
   Avalon Peninsula, 50
   Churchill Falls hydroelectric plant, 50
   Come-by-Chance oil refinery, 50
   flora and fauna, 63
   Mistaken Point, 51–8, 60, 69, 73, 90, 144, 277
   St. John's, 51
newts, 194, 205
Nile crocodile, 214
nitrate, 84
nitrogen, 119, 159
Noble, David, 176
noble savage, 265–7
Norfolk Island pine (*A. heterophylla*), 174–5 and footnote
Norris, Philetus, 102
Norway, 154–6
Nothofagus fusca (New Zealand red beech), 33, 34
notochord, 304

oak (*Quercus*), 166
oak (*Quercus stewardii*), 289
octopus, 131–2, 140
Odontochelys semitestacea, 212–13
Ogygia, island of, 29n
Ogyginus forteyi (giant trilobite), 29n
Olenellus (trilobite), 23
Oligocene Period, 230
olives, 244–5
Oman, 21, 132, 215
onychophora, 41–50, 304
orchids, 92, 181, 184, 232, 283
Order Sphenodontia, 209

Ordovician Period, 22, 28, 30, 73, 86, 127, 130, 133–4, 136, 141, 143, 146, 151, 152, 158, 278–9, 284, 292
organelles, 79–80, 122, 304
ostrich, 235, 239–40
Owen, Richard, 35n
Oxford University, 54, 75n, 264 and footnote
Oxford University Museum, 238–9
oxygen, 82–3, 90–1 and footnote, 106, 135, 189, 278, 284, 285, 303
oysters, 137
ozone layer, 94–5

Palaearctic flora, 166
Palaeognatha, 235
Palaeolimulus, 22
Palaeontological Museum, Oslo, 228
Palaeoproterozoic Period, 71, 78
Palaeozoic Era, 10, 22, 71, 129, 130, 155, 159, 163, 172, 204
Palicouria, 233
palms, 232, 247
Pangaea, 58, 208, 215, 226, 279, 282, 288, 292, 298, 305
"Paomu tree," 173
Paramecium, 112
parasites, 200
parrots, 35, 253
passenger pigeon, 15
passerines (perching birds), 240, 241–2
Paua (abalone) shells, 207
Paucipodia, 45
peanut worm *see* Phylum Sipunculida
pearly nautilus, 140, 293
Pei, I. M., 124
penis worm *see* Phylum Priapulida
Peperomia, 232
Permian Period, 42, 131, 142, 151, 156, 169, 205, 279–81, 285, 298
Perry, Jonathan, 229
phages, 118, 305
Philip, James, 269
Philippines, 226
Phormidium, 70
phosphate, 159
photosynthesis, 65, 81–4, 88, 94, 111, 115, 136, 149, 151, 156, 157, 178, 216, 284
Phragmites, 5
phycocyanin, 111
Phylum Annelida, 130
Phylum Arthropoda, 41n, 301
Phylum Brachiopoda, 127

Phylum Chordata, 127, 200
Phylum Cnidaria, 150, 151, 152
Phylum Ctenophora, 149
Phylum Echinodermata, 143
Phylum Mollusca, 127, 132
Phylum Nematoda, 130n
Phylum Nematomorpha, 130n
Phylum Onychophora, 41 and footnote
Phylum Platyhelmithes, 130n
Phylum Porifera, 144–9, 277, 292
Phylum Priapulida (penis worm), 59, 129n, 130 and footnote, 277, 284
Phylum Sipunculida (peanut worm), 59n, 128, 129, 130, 277, 284
Pilbaria perplexa, 76
pine trees (*Pinus huangshanensis*), 289
pink gums (*Eucalyptus fasculosa*), 216
Pinya, Samuel, 246–9
Pittosporum, 207
plankton, 256, 281
plasmids, 118–20
plate tectonics, 75, 97, 282, 284, 288, 305
plattenkalk, 17
platypus, 219, 221–3, 224–5
Playfair, John, 75
Pleistocene Period, 177, 250, 256, 257, 258, 262, 263, 265, 266, 274–5, 281–2, 288, 291
Pleurotomaria (marine snail), 137
plover, 101
podocarp trees, 34, 35, 58, 160, 175
Poland, 195, 260, 261
polar bear, 256, 257
pollen, 176–7, 262
polychaetes, 56, 130
Pope, Alexander, 199
Porphyra, 86–7, 90
　　P. dioica, 86
　　P. linearis (red alga [seaweed]), 86
　　P. purpurea, 86
　　P. umbilicalis, 86
porphyry, 90
Porter, Susannah, 88
Portugal, 180
　　Arouca, 27–30
　　Canelas, 28
possum, 36
potoo, 234
Prance, Sir Ghillean, 178
Precambrian Period, 52, 54, 55, 62, 69, 71–2, 75, 78, 87, 88, 97, 113, 114, 122, 159, 276, 277
priapulids, 129n, 285
primates, 214, 226–7

Princeton University, 147, 251
prokaryote–eukaryote division, 68, 108
prokaryotic organisms, 68, 69, 78, 79, 109, 121, 148, 272, 276, 284, 305
Protea, 36
Proterozoic Era, 71, 72, 77, 81, 84, 87 and footnote, 94, 114, 147
Protodonata (Carboniferous dragonflies), 298n
Prototaxites, 93
protozoa, 88
pseudoscorpions, 159
Psilophyton, 158
Psychotria, 233
pterodactyls, 254
Punch magazine, 130, 147
purple bacteria, 83, 122n
purple sulphur bacteria, 122 and footnote
Pyrobolus fumarii, 116–17
Pyrococcus furiosus, 109

Queen's University, Ontario, 52
Queensland Museum (Queensland, Australia), 134, 139, 144
quilted organisms, 57
Quinsywort, 158n

radiometric age *see* dating techniques
rails, 253
ratites, 235–6, 242, 288, 295, 305
rats, 251
rattlesnake, 215
Reamer, Robert, 98
red alga (Rhodophyta), 84–5, 88, 90, 91
red knot, 8, 14
red-capped parrot, 207
reptiles, 204–5, 207–15, 224, 225–6, 240, 253, 257, 279
Rhabdopleura, 143
rheas, 240
rhinoceros, 257
Rhipsalis (cactus), 288
rhizomes, 163
rhyolite, 103
ribosome, 108–9
ricinuleid arachnids, 288
Ridgley, Robert, 234
rifleman bird, 242
Rijksmuseum, Amsterdam, 64
rimu tree, 34, 36
Rismiller, Peggy, 219, 221n
river lampreys "lamperns" (*Lampetra*), 199–200

RNA (ribonucleic acid), 47, 108–9, 118
rock quartz, 145
Rogers, Richard, 124
Roosevelt, Theodore, 270
Rosen, Brian, 294
roses (*Rosa henryi*), 166
roundworms, 130
Rousseau, Jean-Jacques, 265
Royal Society, London, 260
Royal Society for the Protection of Birds, 92, 253
Rubisco, 82–3, 115
Rudkin, David, 22
rupturewort, 158n
Russula, 155

sabre-tooth cats, 257, 265
Sacks, Oliver, 20
sagebrush, 97
salamanders, 205, 294
Sambourne, Linley, 130, 147
Sanchíz, J., 246
sanderlings, 8
sandpipers, 8
Sangre de Grado (blood of the dragon) tree, 232
saprobes, 92–4, 305
Sarcopterygia, 191
Savery, Jan, 237
saxifrage, 256
scallops, 137
Schizothrix, 69
Schmidt, Herman, 207
Schopf, Bill, 75n, 79, 80, 114
scorpion flies, 179
scorpions, 20–2, 41n, 43, 279, 286–7
Scotland, 200, 201
    Rhynie Chert, 157
sea cucumbers, 143
sea gooseberries, 149
sea grass (*Zostera*), 135
sea lampreys (*Petromyzon*), 199
sea lilies (crinoids), 143, 280, 292
sea lily (*Saccocoma*), 17
sea lizards, 209
sea monkeys, 286–7
sea slugs, 148
sea snail, 67, 148, 292
sea urchins, 55, 143
seals, 256
seaweed, 68, 86–7, 90–1, 111
sedges, 181, 288
seed ferns, 163, 210

Selaginella (spikemoss), 160–1, 233
Seto Inland Sea, 15
sharks, 134, 140, 203–4, 211, 269
she-oaks, 61, 188
sheep, 32, 36, 217, 246, 258
shellfish, 127
Shin, Dr. Paul, 124–6
short-beaked echidna (*Tachyglossus*), 219, 221
short-faced bear, 265
shrews, 223–4, 250
shrimps, 7, 10, 41n, 86, 128, 129, 188, 286, 287
Shuster, Carl, 10–11, 12
  The American Horseshoe Crab, 10
Siberia, 79, 257, 263, 282
  Lena River, 278
Siberian "traps," 279
Siguang, Professor Li, 291
silica, 74, 78, 100, 114, 145, 147
siliceous rocks, 77
siliceous sinter, 100, 112
silicon dioxide, 145
Silurian Period, 131, 156, 159, 200, 280
silverfish (Thysanura), 297
simian line, 226, 227
Singapore, 194
sinter, 99, 100, 112
Sipuncula, 128
sipunculid worm (peanut worm), 59n, 128
skink, 207, 208, 210
sloths, 257, 265, 268
slugs, 131, 208
Smith, Dr. J. L. B., *Old Fourlegs*, 191
Smithsonian Institution, Washington, D.C., 43, 179
snails, 132, 202, 286
snakes, 204–5, 209, 214–15, 249, 250
"Snowball Earth," 276, 299, 305
soldier crabs (*Mictyris*), 135
Solemya (clam), 136–9, 285, 286, 292, 294
Solnhofen Limestone quarries, (Bavaria), 17
songbirds, 253
South Africa, 171
  Barberton Group (Hoogenoeg Formation), 114
  Cape Province, 293
South America, 8, 32, 34, 41, 163, 175, 212, 215, 225n, 234, 263, 266, 288
  Yucatán Peninsula, 9, 281
South Brothers Island, Cook Strait, 208

South Pole, 278
Spanish "moss," 231
Spencer, Herbert, 274
Sphenodon, 208–10
Sphinctozoa, 146
spiders, 7, 101, 288
Spillman's tapaculo, 233
spiny anteater (echidna), 218–21
Spirulina, 113
Spitsbergen, 72–4, 77, 255–6, 257
  Hinlopen Strait, 72
  Ny Friesland, 72
sponges (Phylum Porifera), 144–9, 277, 292
spoon worms (Echiura), 130n
spoonbills, 135
spores, 158–9, 160, 161, 162
Spriggina, 57
springtails, 159
Sri Lanka, 288
Staphylococcus aureus, 232
starfish, 143
Stensio, Erik, 203
stomata, 157
Stone, Rob, 207, 208
strata, 51, 71, 72, 90, 97, 160, 161, 163, 181, 190, 254, 305
  Cambrian, 23, 44, 48, 55, 73, 127, 149
  Cretaceous, 192, 214
  Devonian, 131, 195
  Ediacaran, 49 and footnote
  Jurassic, 137
  limestone, 112
  Ordovician, 136, 158
  Precambrian, 56, 72
  Silurian, 131, 156, 159
  Triassic, 22, 212
strawberry tree (*Arbutus*), 289
stromatolites, 91, 97, 112, 114, 160
  description of, 65–70, 74–6
  evolutionary history of, 70–3, 75–7
  search and discovery of, 64–5, 70, 78–81
  survival of, 65–6
  in Yellowstone Park, 101
sturgeon (*Acipenser* and relatives), 193 and footnote
Suborder Serpentes, 215
sugar gums (*Eucalyptus cladocalyx*), 216
Sulfolobus, 104
Sulfolobus solfataricus, 109
sulphur, 84–5, 97, 102, 103, 111, 115, 272, 279, 284, 285, 286
sulphuric acid, 104

Sumatra, 214, 226
supervolcano, 305
Svalbard Archipelago, 255, 258
swamp cypress, 276
Swaziland, 75, 78
Swedish Natural History Museum, Stockholm, 203
symbiosis, 88, 123, 149, 305–6

Tachypleus gigas (Asian horseshoe crab), 16
Tachypleus tridentatus (Asian horseshoe crab), 15
tadpole shrimp (*Triops*), 287n
tadpoles, 247–8, 254
Taimyr Peninsula, 258
tanager, 233
tapirs, 257
Taq polymerase, 110n
Tardigrada (water bears), 48
tarsier, 227
Tasmanian devil, 264
Tasmanian wolf (thylacine), 264
Teinolophos, 223
termites, 37, 41
Tertiary Period, 211, 214, 292
testate amoebae, 90
Tethys, 292
tetrapods, 191, 194, 195, 205, 212, 306
Thailand, 16, 132, 160
thermal fields, 99–107
Thermodesulfobacterium hydrogeniphilum, 109
thermophiles, 107, 110, 112, 306
Thermoplasma volcanium, 109
Thermus aquaticus, 110 and footnote
theropod dinosaurs, 239
Thiomicrospira, 136
Thomas, Bob and Carol, 3
Tierra del Fuego, 34
tiger shark, 63
time havens, 292
tinamou, 233–6, 239
titanium, 135
toads, 205, 246, 247–50, 254
Tolkien, J. R. R., *The Lord of the Rings*, 33
Tomistoma (false gharial), 214
tortoise, 209n, 212, 213
totara tree, 34
toucan barbet, 234
Tradescant, John, 238
travertine terraces, 112–14
Treatise on Invertebrate Paleontology, 131 and footnote

tree fern, 161, 171
trees, 166, 288–90, 293
    see also conifers; *Ginkgo*; monkey puzzle tree
Triassic Period, 22, 162, 175, 206, 211, 213, 223, 276, 280
tribal peoples, 265–7
trigoniids, 137
trilobites, 9–10, 41n, 70, 90, 127, 134, 138, 143, 151, 202, 255, 278, 279, 280, 306
    in Arouca, Portugal, 27–30
    description of, 25–6
    evolutionary history, 23–5
    link with horseshoe crabs, 25–7, 29–31
Trimnell, R. L., *Trim's Majorca Guide* (1954), 244
trout, 197
tuatara (*Sphenodon*), 206, 207–10, 215, 252, 276, 280, 294
Tudge, Colin, *The Link*, 228
tui (bird), 36
tulip tree (*Liriodendron*), 166
Turatao Island, 132–3
turtles, 11–12, 209n, 211–13, 294
Tyler, Stanley, 78
Tyrannosaurus rex, 202
tyrant flycatcher, 233

University of California, 69, 75n, 79
University of Chicago, 146
University of Copenhagen, 264
University of Illinois, 107
University of Pennsylvania, 26
University of Uppsala, 188
University of Wellington, 36
U.S. Fish and Fisheries Commission, 14
Ussher, Bishop, 72

Vaceletia (coralline sponge), 146, 292
Valhallfonna, 255
Vallisneria (waterweed), 187
Valongo Formation, 28
Vampyroteuthis infernalis, 138
vascular plants, 156–7
velvet worm (*Peripatus novae-zelandiae*), 55, 57, 64, 111, 121, 123, 127, 176, 228, 252, 277, 287, 288, 295
    description of, 37–41
    evolutionary history of, 41–9, 58–9
    search for, 36–7
Vendobionts, 57, 58, 306
Verne, Jules, *20,000 Leagues Under the Sea*, 139

vertebrates, 127, 186, 190–1, 195, 200, 202, 277
vestimentiferan worms, 116
Viburnum, 166
Victoria University, New Zealand, 210
Vienna Natural History Museum, 212
viperine snake, 248, 250
viruses, 118, 148
volcanoes, 103–4, 277, 279

wading birds, 8, 11, 56
Walcott, Charles Dolittle, 43
Wales, Carmarthen, 136
Walkabout (film), 267
Walker, Cyril, 241
wallabies, 217, 225, 283
Walmsley, John, 222
walnut (*Juglans*), 166
Washington University, 224
water lily (Nymphaeales), 180–1
waterweed, 187, 188, 197
Weinberg, Samantha, *A Fish Caught in Time*, 191
Wellman, Charles, 158
Welwitsch, Friedrich, 178
Welwitschia mirabilis, 178–80, 182, 287, 294
wetas, 208
Wettinia, 232
whale shark, 204, 211
whales, 225, 230
White, Gilbert, *The Natural History of Selborne*, 182
wild horse (*Equus przewalskii*), 260
wildebeest, 214
Willerslev, Eske, 264
Wills, William John, 161
Wilson, E. O., 252
Wilson, Richard, 224
winkles, 129
wisent *see* European bison
Woese, Carl, 107
Wollemi pine, 176–7, 276
Wollemia, 176–8, 294
wolves, 121
wombats, 225
woodlice (or pillbug), 7
Woods Hole Oceanographic Institute, 116
woolly mammoth, 257, 265, 295

worms, 129–30, 159, 208, 277
woundwort, 158n
wrens, 242

Xianguangia, 150

Yamada, Takeshi, 15
Yellowstone National Park, 109, 157
    Angel Terrace, 112
    bears and wolves in, 121–3
    "Black Saturday," (1988), 103
    Canary Springs, 113
    Castle Geyser, 99
    Chinaman Spring, 100
    colours of, 106, 110–11, 112–13
    description of, 97–107
    Dragon's Mouth cave, 105–6
    Ear Geyser, 99
    Echinus Geyser, 102, 106
    Firehole River, 100
    geological and evolutionary roots, 110–18
    Grand Prismatic Spring, 120–1
    journey to, 96–8
    Liberty Cap, 114
    Lion Geyser, 99
    Mammoth Springs, 112
    Minerva Terraces, 112
    Norris Geyser Basin, 102, 106, 111
    Nymph Creek, 111
    Obsidian Pool, 111
    Old Faithful Geyser, 99–102
    Old Faithful Inn, 98–9
    Palette Terraces, 112
    Porcelain Basin, 102
    Punch Bowl Spring, 100
    Steamboat Geyser, 102
    Sulphur Cauldron, 104
    West Thumb thermal field, 102
    Whirligig geyser, 102
    Yellowstone Lake, 103
Younger Dryas, 265 and footnote

Zaglossus attenboroughi (New Guinea anteater), 219
Zhou, Zhiyi, 164
Zhou, Zhiyan, 169, 170
zooxanthellae, 150

TRILOBITE
Eyewitness to Evolution

They lived in the oceans over five hundred million years ago, and they were as myriad then as the beetle is now. Their eyes were made of calcite lenses that could number from one to a thousand. Some were spiky while others were smooth; some were larger than lobsters while others were smaller than fleas. They were trilobites, crustacean-like animals whose span on Earth was so long—three hundred million years—that to study them is to glimpse a different world: ancient continents now positioned elsewhere and mighty oceans long since vanished. *Trilobite* is that rare volume as compelling for the layman as the specialist, and an elegant testimony to the joys of scientific discovery.

Science/Evolution

VINTAGE BOOKS
Available wherever books are sold.
www.vintagebooks.com